# Farming Factors' Influence on Animal Productions

# Farming Factors' Influence on Animal Productions

Guest Editor

**Daniel Simeanu**

MDPI

Basel • Beijing • Wuhan • Barcelona • Belgrade • Novi Sad • Cluj • Manchester

*Guest Editor*
Daniel Simeanu
Department of Food,
Expertise and Services
Ion Ionescu de la Brad
University of Life Sciences Iasi
Iasi
Romania

*Editorial Office*
MDPI AG
Grosspeteranlage 5
4052 Basel, Switzerland

This is a reprint of the Special Issue, published open access by the journal *Agriculture* (ISSN 2077-0472), freely accessible at: https://www.mdpi.com/journal/agriculture/special_issues/XBWM43JM3U.

For citation purposes, cite each article independently as indicated on the article page online and as indicated below:

| Lastname, A.A.; Lastname, B.B. Article Title. *Journal Name* **Year**, *Volume Number*, Page Range. |
|---|

**ISBN 978-3-7258-5643-5 (Hbk)**
**ISBN 978-3-7258-5644-2 (PDF)**
**https://doi.org/10.3390/books978-3-7258-5644-2**

# Contents

**About the Editor** . . . . . . . . . . . . . . . . . . . . . . . . . . . . . . . . . . . . . . . . . . . . . . . . **vii**

**Preface** . . . . . . . . . . . . . . . . . . . . . . . . . . . . . . . . . . . . . . . . . . . . . . . . . . . . . . **ix**

**Daniel Simeanu and Mădălina Matei**
Farming Factors' Influence on Animal Production
Reprinted from: *Agriculture* **2025**, *15*, 1346, https://doi.org/10.3390/agriculture15131346 . . . . **1**

**Alexandru Usturoi, Marius-Gheorghe Doliș, Roxana-Nicoleta Rațu, Cristina Simeanu, Florin Baltag, Cătălin-Emilian Nistor, et al.**
Effects of Ecological Sea Buckthorn Powder Supplementation on Egg Production and Quality in Free-Range Moravia Black Hens
Reprinted from: *Agriculture* **2025**, *15*, 104, https://doi.org/10.3390/agriculture15010104 . . . . . **8**

**Dragoș Mihai Lăpușneanu, Silvia-Ioana Petrescu, Cristina-Gabriela Radu-Rusu, Mădălina Matei and Ioan Mircea Pop**
Mycotoxicological Assessment of Broiler Compound Feed: A Multi-Year Analysis of Five Mycotoxins in a Romanian Feed Mill
Reprinted from: *Agriculture* **2025**, *15*, 84, https://doi.org/10.3390/agriculture15010084 . . . . . . **30**

**Mădălina-Alexandra Davidescu, Claudia Pânzaru, Bianca-Maria Mădescu, Răzvan-Mihail Radu-Rusu, Marius Gheorghe Doliș, Cristina Simeanu, et al.**
Genetic Diversity and Phylogenetic Analysis of the Endangered Transylvanian Pinzgau Cattle: A Key Resource for Biodiversity Conservation and the Sustainability of Livestock Production
Reprinted from: *Agriculture* **2024**, *14*, 2234, https://doi.org/10.3390/agriculture14122234 . . . . **47**

**Constantin Pascal, Claudia Pânzaru, Răzvan-Mihail Radu-Rusu, Vasile Maciuc, Alexandru Marian Florea and Ionică Nechifor**
The Influence of Season and Age on Specific Semen Traits and Reproductive Behavior in Carpatina Breed Bucks
Reprinted from: *Agriculture* **2024**, *14*, 2092, https://doi.org/10.3390/agriculture14112092 . . . . **64**

**Mădălina Matei, Silvia Ioana Petrescu, Bianca Maria Mădescu, Dragoș Mihai Lăpușneanu, Daniel Simeanu, Paul Corneliu Boișteanu and Ioan Mircea Pop**
The Impact of Feed Management Technologies on Mineral Oil Hydrocarbons (MOH) Contamination: A Comparative Farm Level Approach
Reprinted from: *Agriculture* **2024**, *14*, 2008, https://doi.org/10.3390/agriculture14112008 . . . . **83**

**Marius Gheorghe Doliș, Claudia Pânzaru, Marius Giorgi Usturoi, Alexandru Usturoi, Cristina-Gabriela Radu-Rusu and Mădălina Alexandra Davidescu**
On the Digestibility of Mulberry Leaf Fed to *Bombyx mori* Larvae
Reprinted from: *Agriculture* **2024**, *14*, 1394, https://doi.org/10.3390/agriculture14081394 . . . . **102**

**Vasile-Cosmin Andronachi, Cristina Simeanu, Mădălina Matei, Răzvan-Mihail Radu-Rusu and Daniel Simeanu**
Melatonin: An Overview on the Synthesis Processes and on Its Multiple Bioactive Roles Played in Animals and Humans
Reprinted from: *Agriculture* **2025**, *15*, 273, https://doi.org/10.3390/agriculture15030273 . . . . . **120**

**Marius Giorgi Usturoi, Roxana Nicoleta Rațu, Ioana Cristina Crivei, Ionuț Dumitru Veleșcu, Alexandru Usturoi, Florina Stoica and Răzvan-Mihail Radu Rusu**
Unlocking the Power of Eggs: Nutritional Insights, Bioactive Compounds, and the Advantages of Omega-3 and Omega-6 Enriched Varieties
Reprinted from: *Agriculture* **2025**, *15*, 242, https://doi.org/10.3390/agriculture15030242 . . . . . **149**

**Claudia Pânzaru, Marius Gheorghe Doliș, Răzvan-Mihail Radu-Rusu, Constantin Pascal, Vasile Maciuc and Mădălina-Alexandra Davidescu**
Equine Milk and Meat: Nutritious and Sustainable Alternatives for Global Food Security and Environmental Sustainability—A Review
Reprinted from: *Agriculture* **2024**, *14*, 2290, https://doi.org/10.3390/agriculture14122290 . . . . **178**

**Andrei Claudiu Proca, Loredana Horodincu, Carmen Solcan and Gheorghe Solcan**
The Potential of Grape Polyphenols Additive in Pig Nutrition: Chemical Structure, Bioavailability and Their Effect on Intestinal Health of Pigs
Reprinted from: *Agriculture* **2024**, *14*, 1142, https://doi.org/10.3390/agriculture14071142 . . . . **208**

# About the Editor

**Daniel Simeanu**

Daniel Simeanu, PhD, University hab. professor, Iași University of Life Sciences, Romania, acquired his higher education degrees in Romania, from his preparatory degree in 1999 to his installment as a tenured professor in 2021. He defended his doctoral thesis in 2023 and his habilitation thesis in 2021, to become a PhD supervisor. His research areas include nutrition and feeding and the influence of feeding on animal production quality. He has authored over 200+ papers published in national and international journals, including 64 in Web of Science journals. He has managed 5 research projects and participated as member in 25 grant teams. He has published 5 books as the sole or first author and co-authored 15 university books and handbooks. He was awarded two prizes, one from the Romanian Academy (2004) and another form The EURO INVENT Exhibition of Creativity and Innovation (2019), for two referential books in the field of animal sciences engineering ("Treaty of Aviculture"—coord. by Vacaru-Opriș I. and "Animal productions"—Simeanu D., Doliș M.G.).

# Preface

Our goal in developing this Special Issue was to gather as many scientific articles as possible to highlight the multitude of factors that can interfere with livestock production both quantitatively and qualitatively. This aim was guided by the idea that the influence of farming factors on animal production is a complex topic involving a multitude of variables that can affect animals' health, productivity and welfare status.

As the Guest Editor, I contacted many fellow researchers from all over the world, receiving the submission of 20 articles on topics relevant to this exploration, of which 10 have ultimately been published after a thorough evaluation by reviewers, chosen according to their expertise, especially in the field of animal sciences. All papers were improved as a result of this review process, and the final publishing decisions were based on their invaluable comments and recommendations.

Given this, this Special Issue "Farming Factors' Influence on Animal Productions" comprises 10 articles (6 original research studies and 4 literature reviews). The original research articles included present the latest scientific concerns regarding the use of sea buckthorn in the diet of laying hens, the digestibility of mulberry leaves in Bombyx mori larvae, the contamination of dairy cow feedstuffs with hydrocarbons from mineral oils, the mytoxicological evaluation of compound feeds for broiler chickens, the phylogenetic analysis of the Pinzgau cattle breed and a study on the influence of both goats' age and breeding season on the quality of the semen and reproduction performance. As for the reviews, these addressed topics related to the impact of grape polyphenols on intestinal health in pigs, the nutritional value of equine meat and milk, nutritional perspectives of hen eggs and the role of melatonin on the animal and human body. Therefore, the addressed topics encompass the range of issues related to this broad topic, touching on several of the factors that can influence livestock production, from both quantitative and qualitative perspectives.

It is our hope that this publication will be of use to researchers, postgraduates and students studying animal sciences, veterinary science, zoology and biochemistry.

**Daniel Simeanu**
*Guest Editor*

agriculture

MDPI

*Editorial*

# Farming Factors' Influence on Animal Production

**Daniel Simeanu and Mădălina Matei ***

"Ion Ionescu de la Brad" Iasi University of Life Sciences, 3 Mihail Sadoveanu Alley, 700489 Iasi, Romania; daniel.simeanu@iuls.ro
* Correspondence: madalina.matei@iuls.ro

The influence of farm factors on animal production is a complex topic, involving a multitude of elements that can impact the health, productivity, and welfare of animals. These factors can be categorized as genetic, environmental, technological, economic, and even socio-political. The interaction among these factors determines the efficiency and sustainability of animal production systems. To optimize productivity, farmers must consider all these influences and adopt integrated approaches that enhance animal health and welfare while maximizing output. Ongoing research and innovation are essential to address emerging challenges and to improve the sustainability of animal production under changing environmental conditions.

Agriculture and animal husbandry are two closely interconnected fields, with agricultural factors directly influencing livestock production. The agricultural sector serves as the primary source of feed for animal husbandry, and both the quantity and quality of agricultural production directly influence the quantity and quality of animal products. The cultivation of plants with low nutritional value, along with the poor management of harvested crops (e.g., inadequate preservation or storage), can result in low-quality feed that fails to meet the nutritional requirements of the animals. Consequently, this leads to reduced yields and a lower quality of animal products (milk, meat, and eggs) [1].

Modern agriculture faces numerous challenges, such as soil degradation caused by intensive farming practices and water scarcity for irrigation. However, climate change represents the most important threat to agricultural output, especially because it cannot be controlled. Climate change can also negatively impact the productivity of farm animals, both directly by altering weather patterns that affect the animals' ability to adapt to new environmental conditions, and indirectly by reducing the quality of field crops and, consequently, influencing the nutritional value of animal feed. Therefore, it is important for farmers in the agricultural and animal husbandry fields to constantly find new methods to mitigate these adverse effects. Potential solutions could be the development of new crop hybrids that are more resilient to climate change and the implementation of artificial climate control systems for raising and managing farm animals [1–3].

The Special Issue "*Farming Factors' Influence on Animal Production*" includes 10 articles (six original research papers and four literature reviews). The original research articles explore recent scientific concerns such as the use of sea buckthorn in the diet of laying hens, the digestibility of mulberry leaves in *Bombyx mori* larvae, the contamination of dairy cow feed with mineral oil hydrocarbons, the mycotoxicological evaluation of compound feeds for broiler chickens, phylogenetic analysis of the Pinzgau bull breed, and a study on how goats' age and breeding season affect semen quality. The literature reviews address topics such as the impact of grape polyphenols on intestinal health in pigs, the nutritional value of equine meat and milk, nutritional perspectives on hen eggs, and the role of melatonin in both animal and human physiology. As evidenced by the diversity of these topics, the

Special Issue covers a wide range of the factors that can influence livestock production, both in terms of quantity and quality. The 10 scientific contributions were developed by a total of 65 researchers from Romania.

Despite the thematic diversity of the papers included in this Special Issue, they are all interconnected through their focus on the broader objective of enhancing animal production by addressing key agricultural factors. Whether investigating nutritional strategies, reproductive efficiency, genetic resources, or functional bioactive compounds, each contribution explores innovative approaches to improving animal performance, health, or sustainability within agricultural systems. Together, these studies provide a comprehensive overview of how diverse yet complementary research areas converge to address the central challenge of optimizing animal production under various environmental and economic pressures.

A study by Usturoi A. et al. [4] investigated the effects of organic sea buckthorn powder supplementation on egg production and quality in Moravian Black hens. The research assessed the impact of the dietary inclusion of organic sea buckthorn powder (*Hippophae rhamnoides*) on the performance and egg quality of 600 Moravian Black laying hens raised in an open-air system. Three groups were included: a control group fed a standard compound feed specific to this species and category, and two experimental groups whose diets were supplemented with 1% and 2% sea buckthorn powder, respectively. Over an 11- week period, parameters such as egg production, feed intake, and various egg quality traits—including weight, volume, shell thickness, and yolk color—were monitored. Supplementation with 2% sea buckthorn powder significantly improved egg production, egg weight, shell strength, and yolk carotenoid content, while maintaining stable feed intake and negligible mortality. These findings confirm the beneficial effects of sea buckthorn as a natural dietary additive, capable of enhancing productive performance and health status in laying hens. These results also highlight the major potential of sea buckthorn powder in supporting sustainable and higher-quality egg production.

Two of the research articles focused on mycotoxicological and hydrocarbon contamination in farm animal feed. The article "Mycotoxicological evaluation of broiler compound feeds: a multiannual analysis of five mycotoxins in a Romanian compound feed factory" by the research team led by Lăpușneanu D. [5] presents a five-year study on the contamination of compound feeds intended for broiler chickens at all production stages (starter, grower, and finisher). The study targeted five major mycotoxins: total aflatoxins (AFT), deoxynivalenol (DON), fumonisins (FUMs), ochratoxin A (OTA), and zearalenone (ZEN). AFT was detected in 49.3–72.2% of samples, with concentrations ranging from 0.01 to 5.2 μg/kg. DON showed the highest prevalence (77.6–98.9%), with maximum concentrations between 330 and 1740 μg/kg. FUMs were presented in 42.7–87.2% of samples, reaching levels of 460 to 1400 μg/kg. OTA was found in 44.2–87.9% of the samples, with peak concentrations of 21.4 μg/kg. ZEN exhibited consistently high incidence (86.5–97.4%) across all feed stages, with maximum values of up to 89.4 μg/kg. The co-occurrence of mycotoxins was frequently observed, with the most common combination of four mycotoxins identified in 38.51% of the samples. Samples were collected from storage silos, homogenized, and analyzed in certified laboratories. Sampling procedures varied depending on batch size to ensure representativeness. This study highlights the importance of ongoing mycotoxin monitoring to protect animal health and food safety. It also highlights the need to investigate mycotoxin transfer into animal products and the potential combined effects of multiple mycotoxins on animal health, including synergistic or antagonistic interactions.

The second research article addressing the quality of farm animal feed focused on the impact of feed management technologies on mineral oil hydrocarbon (MOH) contamination. This comparative approach at the farm level was conducted by a research team coordinated by Matei M. [6]. The objective was to evaluate the extent of mineral oil hydrocarbon (MOH) contamination in feed and to identify the technological factors contributing to this problem, especially focused on mechanized harvesting and processing. Three dairy farms were selected and classified according to their estimated contamination risk (low, medium, and high). A total of fifteen feed samples were analyzed using coupled liquid chromatography–gas chromatography with the flame ionization detection (LC-GC-FID) method, incorporating a microwave-assisted saponification (MAS) step to quantify the levels of mineral oil saturated (MOSHs) and aromatic hydrocarbons (MOAHs). The study revealed important differences in contamination levels based on the technological development of each farm. The MOSH levels ranged from 11.4 mg/kg to 81.40 mg/kg, while the MOAH levels varied between 0.5 mg/kg and 4.6 mg/kg. MOAHs accounted for 4.74% of the total MOH content. The results demonstrated a clear association between feed production technologies and MOH contamination levels. Factors such as the degree of mechanization, the type and condition of agricultural machinery, and storage conditions appeared to contribute to contamination. Chemical treatments did not show a direct impact, although potential risks were acknowledged. Contamination levels varied among farms, suggesting that some sources may be external to technological factors. The study recommends the implementation of advanced technological solutions and proper maintenance of equipment as key strategies to reduce the risk of MOH contamination in animal feed.

Another noteworthy article presents research on the digestibility of mulberry leaves administered to *Bombyx mori* larvae. Considering that sericulture is an important branch of animal husbandry—not only for silk production, but also as a valuable source of high-value protein—the research team led by Doliș M.G. [7] considered this investigation to be highly relevant. The study, conducted in the summer of 2021, aimed to assess the nutritional value and digestibility of mulberry leaves from two varieties: the Japanese Kokuso 21 and the Romanian Eforie. These varieties were used as feed for the Romanian Triumf hybrid of *Bombyx mori* larvae. A total of 600 larvae were divided into two main groups (300 per variety), each subdivided into six replicates of 50 larvae. The larvae were reared in paper trays and grouped according to age and size. The results showed that mulberry leaves had an average digestibility of 54.46%. The aging of the leaves altered their chemical composition, generally reducing nutrient digestibility as the larvae progressed over the growth period, with the exception of crude fiber [8]. Fiber digestibility remained stable in the early larval stages and increased to 26.78% toward the end of the experiment. Overall, the Kokuso 21 variety demonstrated better digestibility parameters than Eforie, suggesting its superior nutritional profile for silkworm feeding. A conclusion of the study was the need for further investigations to evaluate the extent of nutrient metabolism and their conversion efficiency into silk production.

A further research article explored the influence of age and season on specific sperm parameters and reproductive behavior in Carpathian goats. Coordinated by Pascal C. [9], the study investigated how these two factors affect reproductive potential in male Carpathian goats. The animals were divided into three age categories: young (14–23 months; L14), adult (3–4 years; L34), and older bucks (5–6 years; L56). Scrotal biometrics were assessed using a measuring tape, while testicular volume was determined by completely immersing the testicles in a container filled with water and measuring the displaced water. Semen samples were collected and analyzed across all four seasons, with evaluations including ejaculate volume, color, pH, motility, sperm concentration, and morphology. A computer-assisted sperm analysis (CASA) system was employed for precise measurements of motility and morphology, and testosterone levels were determined from seasonal blood samples. Sexual behavior was monitored based on mating desire and the male's response to female presence. Key findings showed that testicular volume was significantly influenced by both age and season, with the most substantial differences observed between the youngest and oldest groups, especially during autumn. Sperm quality—parameters such as ejaculate volume, sperm concentration, and motility—varied seasonally, being lowest in younger goats. Testosterone levels increased with age and peaked in the autumn season. Behavioral observations showed that young males exhibited lower sexual activity, although this improved in autumn as well. Additionally, a strong correlation was found between body weight and testicular volume in adult bucks ($R = 0.942$, $p$-value = 0.016 for L34; $R = 0.797$, $p$-value = 0.022 for L56), suggesting age-related reproductive development. The study confirms that although Carpathian goats can reproduce year-round, autumn provides optimal conditions for sperm quality and breeding performance. These findings support improvements in breeding strategies tailored to seasonal and age-specific reproductive patterns in temperate continental climates.

The final research paper included in this Special Issue was conducted by a team led by Davidescu M.A. [10], who investigated the genetic diversity and phylogenetic background of the Pinzgau cattle of Transylvania, a local breed currently facing the threat of extinction. This breed represents a valuable genetic resource for biodiversity conservation and for promoting the sustainability of livestock systems, particularly under the increasing pressures of climate change and disease. Animal genetic biodiversity is essential for maintaining the functionality of local food systems and for ensuring sustainable livelihoods. Since 2000, the Food and Agriculture Organization of the United Nations (FAO) has highlighted the decline of cattle populations, including the Pinzgau breed of Transylvania, Romania. Known for its resistance, adaptability, disease resistance, and adaptability to environmental stressors, the Transylvanian Pinzgau is considered an important genetic asset for improving livestock productivity. The study focused on the genetic assessment of 24 animals from the Transylvanian region by analyzing two mtDNA markers, cytochrome b and D-loop sequences, both widely recognized for their relevance and importance in studies of genetic diversity and phylogenetic reconstruction in cattle. The findings, obtained through a statistical analysis of nucleotide sequences using specialized software, indicated that the cattle belonged to the ancestral haplogroup T, a lineage tracing back to *Bos taurus*. These findings underscore the importance of conserving genetic diversity within local breeds and support the development of targeted breeding and crossbreeding programs. Such efforts could improve the genetic base of commercial cattle breeds while contributing to the preservation of agri-biodiversity.

The first article from the literature review group addressed the potential of grape polyphenols as feed additives in pig nutrition, focusing on their chemical structure, bioavailability, and their effects on gut health. The group of authors led by Proca A.C. [11] states the growing interest in natural feed additives following the ban on antibiotic use in animal production. Grapes, as well as by-products derived from the wine industry, including grape marc and seed extracts, are rich in bioactive constituents such as flavonoids, stilbenes, and phenolic acids, which exhibit a wide range of health-promoting properties. The aim of this review was to synthesize the existing knowledge on the impact of grape polyphenols on intestinal health in pigs. The first section of the paper discusses the chemical structure of the major polyphenolic compounds found in grapes, along with their bioavailability and metabolic pathways in pigs. The core section summarizes findings from experimental investigations that highlight the antioxidant, antimicrobial, and prebiotic properties of these compounds, as well as their role in modulating intestinal barrier functions through cellular signaling mechanisms. Although fewer studies have been conducted specifically in pigs compared to other species, the available evidence supports the efficacy of incorporating up to 9% grape by-products in pig diets, leading to increased performance parameters. The authors conclude that supplementing grape polyphenols as natural feed additives improves antioxidant capacity, humoral and cellular immune responses, and the biodiversity of the intestinal ecosystem, ultimately contributing to better animal health and production outcomes.

Another bibliographic study published in this Special Issue focused on horse milk and meat as sustainable nutritional alternatives for global food security and environmental sustainability. The review, coordinated by Pânzaru C. [12], highlights the potential of equine-derived products that require further development in response to current global challenges, such as malnutrition, limited access to conventional animal products, and ecological concerns. Equine milk is distinguished by its bioavailable nutrients, essential fatty acids, and hypoallergenic properties, making it a viable substitute for individuals with allergies, lactose intolerance, or restricted access to traditional dairy sources. Likewise, equine meat, known for its high-quality protein content, low fat content, and essential micronutrients such as iron and zinc, provides an affordable and sustainable source of protein for food-insecure populations. The ability of equines to thrive on marginal lands, coupled with their lower environmental impact compared to traditional livestock (such as ruminants), underscores their potential within sustainable agricultural systems. This review concludes by underscoring the need for further research to address key challenges related to the integration of equine products in combating global hunger, highlighting their nutritional benefits, environmental advantages, and the need for further research to address the challenges of versatility, cultural acceptance, and policy integration.

Another article included in this Special Issue presents a comprehensive bibliographic study on the nutritional quality of eggs, with a focus on the development of omega-3- and omega-6-enriched varieties The paper, authored by Usturoi M.G. et al. [13], explores the role of eggs as functional foods, emphasizing their potential contribution to improving public health outcomes through targeted nutritional enhancement. The study outlines the biochemical and physiological importance of omega-3 and omega-6 polyunsaturated fatty acids, particularly in the context of cardiovascular protection, anti-inflammatory action, and cognitive support. Special attention is given to the techniques used to enrich eggs, specifically dietary modifications with flaxseed or marine algae for laying hens. These methods have been shown to improve the content of beneficial lipids and bioactive compounds in eggs, especially omega-3 content and in balancing the omega-6-omega-3 ratio. New research indicates that enriched eggs provide higher levels of essential fatty acids and bioactive compounds than conventional eggs, offering an accessible alternative to traditional omega-3 sources, especially for populations with limited access to fish, for example. In addition to highlighting the nutritional advantages of enriched eggs, the study further addresses the challenges of consumer perception, the sustainability of enrichment practices, and the regulatory constraints governing the commercialization of functional foods. The findings highlight that omega-enriched eggs represent a valuable nutritional and functional food that aligns with health-related dietary trends and encourages further research to refine enrichment methods and ensure broader accessibility in the market.

The final contribution to this Special Issue is a bibliographic study presented by Andronachi V.C. et al. [14], offering an overview of melatonin, its synthesis processes, and its multiple bioactive functions in both animals and humans. Melatonin, a natural hormone synthesized primarily by the pineal gland of vertebrates and, secondarily, by other tissues and organs, is increasingly recognized as a multifunctional bioactive molecule with physiological, metabolic, and regulatory roles across animals and humans. The review synthesizes scientific evidence up to the year 2024 regarding the endogenous and exogenous sources of melatonin, including its occurrence in various plant species and bacterial strains. Particular attention is given to the factors influencing melatonin biosynthesis and secretion in animals, emphasizing how nutritional and environmental factors, such as light exposure, feeding methods, and physiological status, can modulate circulating melatonin levels and even affect its transfer into animal food products, notably milk. Furthermore, the study explores the complex interactions between melatonin and other bioactive compounds within animals and humans, aiming to clarify its functions and roles in biological systems.

This Special Issue of *Agriculture*, dedicated to the influence of agricultural factors on animal production, brings together the diverse approaches of original research and bibliographic studies that reflect some of the most recent and relevant topics in the field. The contributions address both fundamental and applied aspects of animal science, highlighting the role of genetics, nutrition, environmental conditions, and sustainable practices in optimizing animal productivity and welfare. We are pleased to have successfully completed the editorial process for this issue, despite the challenges encountered along the way. We would like to acknowledge the valuable scientific contributions of all the authors whose work has enriched this volume. Notably, many of the contributors are young researchers whose involvement demonstrated the promising future of agricultural and animal science research.

**Author Contributions:** Conceptualization, D.S. and M.M.; investigation, D.S. and M.M.; writing—original draft preparation, D.S.; writing—review and editing, M.M.; visualization, D.S. and M.M.; All authors have read and agreed to the published version of the manuscript.

**Acknowledgments:** We would like to sincerely thank all authors who submitted papers to the Special Issue of *Agriculture* entitled "Farming Factors' Influence on Animal Productions", from the reviewers of these papers for their constructive comments and thoughtful suggestions to the editorial staff of *Agriculture*.

**Conflicts of Interest:** The authors declare no conflicts of interest.

## References

1. Rojas-Downing, M.M.; Nejadhashemi, A.P.; Harrigan, T.; Woznicki, S.A. Climate change and livestock: Impacts, adaptation, and mitigation. *Clim. Risk Manag.* **2017**, *16*, 145–163. [CrossRef]
2. Malhi, G.S.; Kaur, M.; Kaushik, P. Impact of Climate Change on Agriculture and Its Mitigation Strategies: A Review. *Sustainability* **2021**, *13*, 1318. [CrossRef]
3. Escarcha, J.F.; Lassa, J.A.; Zander, K.K. Livestock Under Climate Change: A Systematic Review of Impacts and Adaptation. *Climate* **2018**, *6*, 54. [CrossRef]
4. Usturoi, A.; Doliș, M.-G.; Rațu, R.-N.; Simeanu, C.; Baltag, F.; Nistor, C.-E.; Pânzaru, C.; Radu-Rusu, R.-M.; Davidescu, M.A.; Usturoi, M.-G. Effects of Ecological Sea Buckthorn Powder Supplementation on Egg Production and Quality in Free-Range Moravia Black Hens. *Agriculture* **2025**, *15*, 104. [CrossRef]
5. Lăpușneanu, D.M.; Petrescu, S.-I.; Radu-Rusu, C.-G.; Matei, M.; Pop, I.M. Mycotoxicological Assessment of Broiler Compound Feed: A Multi-Year Analysis of Five Mycotoxins in a Romanian Feed Mill. *Agriculture* **2025**, *15*, 84. [CrossRef]
6. Matei, M.; Petrescu, S.I.; Mădescu, B.M.; Lăpușneanu, D.M.; Simeanu, D.; Boișteanu, P.C.; Pop, I.M. The Impact of Feed Management Technologies on Mineral Oil Hydrocarbons (MOH) Contamination: A Comparative Farm Level Approach. *Agriculture* **2024**, *14*, 2008. [CrossRef]
7. Doliș, M.G.; Pânzaru, C.; Usturoi, M.G.; Usturoi, A.; Radu-Rusu, C.-G.; Davidescu, M.A. On the Digestibility of Mulberry Leaf Fed to *Bombyx mori* Larvae. *Agriculture* **2024**, *14*, 1394. [CrossRef]
8. Doliș, M.G.; Simeanu, C.; Usturoi, A.; Simeanu, D. Research Regarding Chemical Composition and the Digestibility of the Mulberry Leaves from Eforie Variety. *Rev. Chim.* **2017**, *68*, 151–156. [CrossRef]
9. Pascal, C.; Pânzaru, C.; Radu-Rusu, R.-M.; Maciuc, V.; Florea, A.M.; Nechifor, I. The Influence of Season and Age on Specific Semen Traits and Reproductive Behavior in Carpatina Breed Bucks. *Agriculture* **2024**, *14*, 2092. [CrossRef]
10. Davidescu, M.-A.; Pânzaru, C.; Mădescu, B.-M.; Radu-Rusu, R.-M.; Doliș, M.G.; Simeanu, C.; Usturoi, A.; Ciobanu, A.; Creangă, Ș. Genetic Diversity and Phylogenetic Analysis of the Endangered Transylvanian Pinzgau Cattle: A Key Resource for Biodiversity Conservation and the Sustainability of Livestock Production. *Agriculture* **2024**, *14*, 2234. [CrossRef]
11. Proca, A.C.; Horodincu, L.; Solcan, C.; Solcan, G. The Potential of Grape Polyphenols Additive in Pig Nutrition: Chemical Structure, Bioavailability and Their Effect on Intestinal Health of Pigs. *Agriculture* **2024**, *14*, 1142. [CrossRef]

12. Pânzaru, C.; Doliș, M.G.; Radu-Rusu, R.-M.; Pascal, C.; Maciuc, V.; Davidescu, M.-A. Equine Milk and Meat: Nutritious and Sustainable Alternatives for Global Food Security and Environmental Sustainability—A Review. *Agriculture* **2024**, *14*, 2290. [CrossRef]
13. Usturoi, M.G.; Rațu, R.N.; Crivei, I.C.; Veleșcu, I.D.; Usturoi, A.; Stoica, F.; Radu Rusu, R.-M. Unlocking the Power of Eggs: Nutritional Insights, Bioactive Compounds, and the Advantages of Omega-3 and Omega-6 Enriched Varieties. *Agriculture* **2025**, *15*, 242. [CrossRef]
14. Andronachi, V.-C.; Simeanu, C.; Matei, M.; Radu-Rusu, R.-M.; Simeanu, D. Melatonin: An Overview on the Synthesis Processes and on Its Multiple Bioactive Roles Played in Animals and Humans. *Agriculture* **2025**, *15*, 273. [CrossRef]

agriculture

MDPI

*Article*

# Effects of Ecological Sea Buckthorn Powder Supplementation on Egg Production and Quality in Free-Range Moravia Black Hens

**Alexandru Usturoi [1], Marius-Gheorghe Doliș [2], Roxana-Nicoleta Rațu [3], Cristina Simeanu [2], Florin Baltag [1], Cătălin-Emilian Nistor [2], Claudia Pânzaru [2], Răzvan-Mihail Radu-Rusu [2], Mădălina Alexandra Davidescu [1,*] and Marius-Giorgi Usturoi [2]**

[1] Department of Control, Expertise and Services, Faculty of Food and Animal Sciences, "Ion Ionescu de la Brad" Iasi University of Life Sciences, 700489 Iasi, Romania; alexandru.usturoi@iuls.ro (A.U.); b.florin7819@yahoo.com (F.B.)
[2] Department of Animal Resources and Technologies, Faculty of Food and Animal Sciences, "Ion Ionescu de la Brad" Iasi University of Life Sciences, 700489 Iasi, Romania; marius.dolis@iuls.ro (M.-G.D.); cristina.simeanu@iuls.ro (C.S.); catalin.nistor@iuls.ro (C.-E.N.); claudia.panzaru@iuls.ro (C.P.); razvan.radu@iuls.ro (R.-M.R.-R.); marius.usturoi@iuls.ro (M.-G.U.)
[3] Department of Food Technologies, Faculty of Agriculture, "Ion Ionescu de la Brad" Iasi University of Life Sciences, 700489 Iasi, Romania; roxana.ratu@iuls.ro
* Correspondence: madalina.davidescu@iuls.ro

**Abstract:** The growing demand for sustainable and healthier egg production systems, combined with the need to reduce the use of synthetic additives in poultry feed, has led to an increased interest in natural feed supplements. This study evaluated the effects of dietary supplementation with ecological sea buckthorn (*Hippophae rhamnoides*) powder on the performance and egg quality of 600 laying hens of the Moravia Black breed, raised in a free-range system. Three groups were included: one control group with standard feed and two experimental groups supplemented with 1% and 2% sea buckthorn powder. Over 11 weeks, parameters such as egg production, feed consumption, and egg quality, including egg weight, volume, shell thickness, and yolk color, were monitored. The 2% supplementation significantly improved egg production, egg weight, shell strength, and yolk carotenoid content, with stable feed consumption and negligible mortality, confirming the additive's safety. These results highlight the potential of sea buckthorn powder as a natural feed additive to enhance poultry productivity and product quality, supporting sustainable and healthier egg production.

**Keywords:** egg production; egg quality; laying hens; sea buckthorn powder; sustainable poultry farming

## 1. Introduction

Eggs represent an important source of nutrients, ranging from high-biological-value proteins to vitamins and minerals essential for biochemical processes [1]. Some studies have shown that most consumers' daily diets include low levels of essential fatty acids, with implications for cardiovascular health [2]. As a result, solutions have been sought to modify the lipid composition of eggs, either by reducing cholesterol or altering the fatty acid profile, as well as increasing their nutritional value, specifically producing functional eggs with a higher content of bioactive substances [3]. Generally, it can be stated that the quality of eggs for consumption is dictated by numerous factors, with breed (hybrid used), rearing system, and diet being essential [4,5].

A key factor for obtaining eggs with higher biological value, analyzed in this study, is the breed or hybrid. In the poultry industry, significant progress has been made in creating laying hybrids with outstanding productive performances, the result of research in genetics and improvement. Thus, studies in this field have highlighted multiple differences between purebred hens and commercial hybrids [6]. For instance, Sussex hens demonstrated the best welfare indicators among pure breeds, while ISA Brown recorded the lowest levels among commercial lines. Comparing these two groups of birds revealed higher mortality and aggressiveness in purebred hens, indicating lower welfare levels compared to commercial hybrids, which adapt better to confined spaces [7].

The second factor considered in this study is the poultry rearing system, regulated by Directive 74/1999/EC of the European Union Council, which aims to ensure minimal comfort conditions for laying hens [8,9]. Consequently, alternative rearing solutions have been adopted that allow the productive potential of birds to be expressed while ensuring welfare conditions [10]. One widely accepted and consumer-appreciated alternative is free-range rearing [11]. This method has proven highly beneficial for bird populations used for egg production, positively impacting both the general health of the flocks and the quality of the eggs produced. Advantages include reduced mortality rates [12,13], minimization of stress-inducing factors [11] (affecting quantitative egg production [14]), and improved egg quality [15,16].

Closely related to the rearing system is egg hygiene, a parameter of interest because eggs can carry microbiological or physical contaminants. Contamination pathways stem from farm biosecurity or the quality of raw materials used in feed production [17].

The third element investigated focuses on the nutritional characteristics of the combined feeds given to laying hens and their effects on welfare and egg quality. From this perspective, direct correlations were identified between the levels of essential amino acids (methionine, cystine) and the incidence of feather pecking or losses due to cannibalism [18].

Administering diets with low protein levels (14%) resulted in reduced productive parameters and significant decreases in certain blood parameters (uric acid, triglycerides, albumin) compared to birds receiving 16% crude protein [19]. In the same context, it was observed that low-protein combined feeds decreased serum proteins and cholesterol only in birds raised on permanent litter, with these parameters unaffected in battery-housed birds [20]. Using formulas with reduced metabolizable energy and crude protein lowered feeding costs and improved serum creatine kinase activity but also reduced serum triglyceride and cholesterol levels [21]. Another study showed that a wheat-rich diet (50% vs. 25% wheat) negatively affected plumage quality, with mortality and cannibalism rates influenced by the rearing system [22,23].

Using unconventional resources (dried olive pulp at doses of 2–6%) improved egg quality (increased polyunsaturated fatty acids and reduced saturated fatty acids) and lipid health indices (lower AI and TI and higher h/H ratio), in a dose-dependent manner [24]. Studies show that the inclusion of 2% of SBM in the experimental diet led to a more than 25% increase in vitamin E and an almost 50% increase in xanthophylls compared to the control. The markers specific to the coronary risk decreased significantly in the experimental group compared to the control, showing a beneficial effect of dietary SBM on the quality of yolk lipids [25].

In conclusion, the concept of quality, as reflected in the welfare of birds and the quality of their production, is directly influenced by the careful selection of the hybrid used, aligning the rearing system with the birds' ethological characteristics, and ensuring proper nutrition [26]. The thesis of this study is that dietary supplementation with ecological sea buckthorn powder can improve performance and egg quality in Moravia Black hens, the objective of this study being to evaluate the effects of this supplementation on egg production and quality. This case study represents research conducted under production conditions, focusing on the performance of Moravia Black hens reared in one of the most valued systems (free-range) and fed a diet supplemented with a natural biostimulator (organic sea buckthorn powder). The use of ecological sea buckthorn powder in the diet of laying hens has gained attention in poultry nutrition due to its rich bioactive composition. Sea buckthorn *(Hippophae rhamnoides)* is a natural source of vitamins (A, C, E), antioxidants, polyunsaturated fatty acids, and carotenoids, all of which are essential for enhancing the overall health of birds. Scientific studies have demonstrated that supplementing the diets of laying hens with natural biostimulants like sea buckthorn can strengthen their immune systems, reduce oxidative stress, and improve resistance to diseases [27]. This is particularly beneficial in sustainable poultry farming practices, where the focus is on reducing the reliance on synthetic additives and enhancing the natural health and productivity of birds [28,29].

Incorporating sea buckthorn powder into poultry feed has been shown to significantly enhance egg quality, particularly in terms of yolk color, shell strength, and nutrient profile. Research highlights that the carotenoids and polyunsaturated fatty acids present in sea buckthorn are directly deposited in the yolk, leading to richer pigmentation and improved nutritional value. Studies also report a reduction in saturated fatty acids and an increase in beneficial omega-3 and omega-6 fatty acids in the eggs of hens fed sea buckthorn, contributing to a healthier lipid profile. Such eggs are increasingly sought after by health-conscious consumers, aligning with the growing demand for functional foods that provide additional health benefits [30–32].

Regardless of the technology used for rearing laying hens, the fundamental goal remains the provision of eggs with superior quality parameters suitable for a healthy diet [33]. This goal can be achieved through various methods, including incorporating natural preparations derived from wild or cultivated flora, or by-products of their processing, into the birds' feed.

Moreover, ecological sea buckthorn powder aligns with sustainable farming practices by utilizing renewable and organic resources. Research has shown that using natural additives like sea buckthorn can enhance the welfare of laying hens, reducing stress and promoting better productivity, especially in systems such as free-range rearing. This approach not only improves the overall quality of eggs but also contributes to environmental sustainability and animal welfare [34]. As highlighted in various scientific works, integrating ecological ingredients like sea buckthorn into poultry diets represents a step forward in producing high-quality, nutritious eggs while adhering to eco-friendly and ethical farming practices.

## 2. Materials and Methods

### 2.1. Biological Material

The Negru de Moravia breed originates from South Moravia, Czechia, where it was selected for egg production. The first chickens were black, but later, other color varieties were developed.

These birds have an elegant body, not very large but strong and compact, with a small head and a narrow comb. Their neck is of medium length, with feathers evenly distributed, and their wings are short but well proportioned.

The Negru de Moravia has a body weight of 2.0–2.2 kg and an excellent egg production rate of approximately 300 eggs per year. The eggs are beige in color and weigh about 60 g.

They are active birds with a gentle temperament, efficiently utilizing food resources for egg production. They also exhibit a good reproduction rate and excellent adaptability to a wide range of environmental conditions.

The biological material studied consisted of 600 laying hens of the Moravia Black breed, raised in a free-range system. Their diet was supplemented with ecological sea buckthorn powder, and a series of parameters related to egg production and quality were subsequently monitored. The tested product was "Ecological Sea Buckthorn Powder", produced by S.C. Eco Catena S.R.L., Vulturești-Bacău, Romania, with the following characteristics: energy value = 374 kcal/100 g; protein = 18.21%; lipids = 3.38%; total carbohydrates = 67.76%.

The birds (600 individuals) were divided into three groups: one control group (200 birds) and two experimental groups (200 birds each). The control group (coded M-m) was fed a standard combined feed without sea buckthorn powder. In the experimental group M-1, 1% sea buckthorn powder was added to the standard feed, while in the experimental group M-2, the feed was supplemented with 2% sea buckthorn powder.

The experiment was conducted over a period of 11 weeks, spanning from early September to late November 2023, which corresponds to the peak laying period of the hens, raised in a free-range system.

The birds were monitored over 11 weeks (from the 30th week of life to the 40th week), during which the main production indicators were recorded, including body weight, flock losses, egg production, laying intensity, feed consumption, and the structure of egg production. The second objective involved determining specific egg quality indicators, assessed weekly (egg weight, egg volume, shape index, specific gravity, egg structure, shell thickness and strength, yolk color, and carotenoid content).

### 2.2. Description of the Experimental Unit

The poultry farm where the research was conducted was established in 2018, with its main activity being the raising of laying hens and the commercialization of table eggs (Figure 1). The farm is located on a plateau in the southwest of Târgu Frumos town, Iași County, and covers an area of 9000 m$^2$, enclosed by a steel wire mesh fence mounted on concrete poles. To the south, it borders the E583 road (Iași-Bacău), to the north, it is surrounded by a forest strip, and to the east and west by two agricultural fields.

**Figure 1.** Overview of the poultry farm where the research was carried out (original photo).

The area where the farm is located is characterized by a temperate climate, which in recent years has undergone a series of changes predominantly marked by temperature fluctuations and shifts in seasonal patterns. For instance, temperatures ranged between 15 and 26 °C in September (night and day), between 12 and 18 °C in October, and between 6 and 10 °C in November. However, as previously mentioned, there were fluctuations, with some days in September seeing temperatures rise to 31 °C, while in November, they dropped to 0–5 °C.

The rearing halls are constructed of wood, thermally insulated, and equipped with windows for natural lighting. They are also fitted with solar panels to provide artificial lighting. Regarding infrastructure, the shelters feature reinforced concrete block foundations beneath the structural pillars and interrupted reinforced concrete foundations for the non-structural pillars on the facades and gables. The flooring consists of lightly reinforced concrete, 15 cm thick, placed on a 15 cm compacted ballast fill.

The superstructure of the halls is made of wooden frames (pillars and beams) to which double panels of pressed wood are attached, with basalt wool inserted between them for insulation. Additionally, there are non-structural pillars for gables and window and door frames. To comply with welfare standards and ensure bird comfort, the halls are equipped with slatted floors, perches for resting, and an appropriate number of drinkers, feeders, and nesting boxes to accommodate the housed population.

### 2.3. *The Technological Flow Applied in the Experimental Unit*

The facility operates a complete production flow, which includes raising replacement pullets as well as adult hens. The halls are equipped with paddocks enclosed by metal mesh fences, where fruit trees have been planted to provide shade and reduce wind speed. Access to the paddocks is through sliding doors in the walls of the halls, allowing the hens to explore the natural environment and forage on grass and insects.

The farm carries out a range of activities common to both sectors, such as preparing the halls for population, populating the halls, supplying feed, and daily care (feeding, watering, cleaning the halls and paddocks, ensuring the microclimate, and monitoring general health).

Replacement pullets are purchased at one day old and introduced into the youth hall, which is cleaned and disinfected beforehand. Antibiotics are not used in the facility, and only the two mandatory PPA vaccines (administered on days 9 and 21) are applied. The farm is serviced by a local veterinarian who regularly monitors the health of the flock. During the research, no parasites were observed, as this parameter is controlled by maintaining proper hygiene.

Adult hen halls are populated when the pullets reach 14 weeks of age. At this point, the birds undergo a 2-week acclimation period to the new environment (Figure 2). The laying period begins when the hens are 19 weeks old and typically lasts until they are 80–85 weeks old. The average laying rate is 25% at 22 weeks of age, peaks at 93% during weeks 28–33, and then gradually declines to 65% by the time the hens are 80 weeks old.

At the end of the production period, the adult hen hall is depopulated (hens are sold as culls), followed by a 42-day cycle for cleaning and sanitary rest. During this time, the hall is washed, disinfected, and sealed.

The eggs are collected manually and transferred to a dedicated room where they are sorted and packed into cartons of 10 or 30 eggs (Figure 3). These cartons are wrapped and delivered according to orders. Non-conforming eggs (cracked or deformed) are considered waste and are temporarily stored in a refrigerated container.

**Figure 2.** House for laying hens.

**Figure 3.** Unsorted eggs produced by Moravia Black hens.

### *2.4. The Structure of the Fodder Recipe*

The birds forming the three groups were fed a combined feed with similar nutritional characteristics, with the main differences being the inclusion of sea buckthorn powder in the diets of groups M-1 and M-2 (Table 1).

**Table 1.** The structure of the combined feed administered to the studied birds.

| Specification | Experimental Batches | | |
|---|---|---|---|
| | M-m | M-1 (1%) | M-2 (2%) |
| Corn kernels | 23.5 | 23.5 | 23.5 |
| Soybean meal (46% protein) | 5 | 5 | 5 |
| Granulated alfalfa | 3 | 3 | 3 |
| Peas | 3 | 3 | 3 |
| Wheat bran | 1 | 1 | 1 |
| Shell grit | 1 | 1 | 1 |
| Calcium | 2.2 | 2.2 | 2.2 |
| Premix | 0.8 | 0.8 | 0.8 |
| Unfiltered sunflower oil | 0.5 | 0.5 | 0.5 |
| Organic sea buckthorn powder | - | 0.4 | 0.8 |

**Table 1.** *Cont.*

| Specification | Experimental Batches | | |
|---|---|---|---|
| | M-m | M-1 (1%) | M-2 (2%) |
| | Characteristics | | |
| Protein (%) | 12.10 | 12.54 | 13.16 |
| Fat (%) | 3.71 | 3.77 | 3.84 |
| Fiber (%) | 4.26 | 4.38 | 4.58 |
| Ash (%) | 11.51 | 12.13 | 13.89 |
| Starch (%) | 39.45 | 37.88 | 36.02 |

The primary ingredient was corn grain (23.5%), supplemented with soybean meal (46% crude protein—5%), pelleted alfalfa (3%), peas (3%), wheat bran (1%), shell grit (1%), calcium (2.2%), a vitamin–mineral premix (0.8%), unfiltered sunflower oil (0.5%), and sea buckthorn powder (1–0.4% for M-1 and 2–0.8% for M-2).

As a result, for the control group (M-m), the protein level was 12.1%, the fat content was 3.71%, the fiber content was 4.26%, the ash level was 11.51%, and the starch content was 39.54%.

The control batches recorded higher values of protein, specifically 12.54% for M-1 and 13.16% for M-2, compared to 12.1% observed in the control sample M-m. Regarding the fat content, slight differences were identified, with an increase of 0.07% (in M-1) and 0.13% (in M-2) compared to the value identified for the control batch (M-m).

The fiber percentage ranged from a minimum value of 4.26% in M-m to a maximum of 4.58% in M-2, with the M-1 batch being intermediate at 4.38%.

For the ash content, the control batch (M-m) showed a percentage of 11.51%, while increases of 0.62% (for M-1) and 2.38% (for M-2) were observed in the other batches.

Starch was present at a percentage of 39.45% in the control batch (M-m), which was lower by 1.57 percentage points in M-1 and by 3.43 points in M-2.

### 2.5. Methods

The indicators tracked were determined using the following methods:

- Weight growth dynamics: In each group, 20 individual birds were selected and weighed at the start of each control week.
- Flock losses: Mortality cases were calculated relative to the flock size at the beginning of the respective week.
- Laying intensity: Calculated as the ratio between the weekly egg production and the average flock size of the respective group.
- Egg production structure: Each control week, the eggs obtained were categorized into four size classes: XL (over 73 g), L (63–72.9 g), M (53–62.9 g), and S (under 53 g). The distribution was then reported relative to the weekly egg production.
- Feed consumption: Recorded as total weekly feed consumption per group (kg feed/week) and average daily consumption (g feed/bird/day).
- Egg weight: Determined as the average weekly weight of the eggs.
- Shape index: Calculated as the percentage ratio of the large diameter to the small diameter of the eggs.
- Egg volume: Computed using the formula $V(cm^3) = 0.519 \times D \times d^2$, where D is the large diameter (cm) and d is the small diameter (cm).
- Egg structure: The three egg components (albumen, yolk, and shell) were weighed and expressed as percentages of the total egg weight.
- Shell thickness: Measured as the average of three readings (sharp end, rounded end, and equatorial zone) using a gauge with a comparator dial.

- Shell strength: Calculated using the formula R(gf/cm$^2$) = shell thickness × 230.
- Yolk color: Assessed by comparison with the La Roche color scale.
- Carotenoid content: Expressed as double the La Roche score plus one.

Data Processing

The experimental data were processed using calculation algorithms in Microsoft Excel. The comparison of variability between groups and within groups was performed using the One-Way ANOVA (Analysis of Variance) test.

## 3. Results

### *3.1. Productive Performances of the Herds Studied*

#### 3.1.1. Body Weight of Birds

The first productive parameter observed was the body weight of the studied birds. At the beginning of the research, during the 30th week of life, the body weight recorded values that were nearly equal across the three groups (M-m = 2412.4 g, M-1 = 2412.0 g, M-2 = 2412.2 g). This similarity was due to the rigorous selection process conducted when forming the study groups. The average body weights were within the limits defined by the breed standard, specifically ranging from 2346 g to 2478 g.

Analyzing the dynamics of this parameter, it was observed that for all groups, there was an increase in the average weight achieved. Thus, at the midpoint of the period (week 35), maximum values of 2526.8 g were recorded for M-m, while for M-1 and M-2, these values were 0.13% and 0.25% lower, respectively. Although there were some differences, they did not exceed the breed standard (2471–2531 g) (Table 2).

**Table 2.** The evolution of weekly and total body weight of birds based on the percentage of sea buckthorn powder administered.

| Age (Weeks) | Standard Weight (g) | Realized Weight (g) | | |
|---|---|---|---|---|
| | | **M-m** | **M-1** | **M-2** |
| 30 | 2346–2478 | 2412.4 | 2412.0 | 2412.2 |
| 31 | 2360–2483 | 2439.2 | 2438.1 | 2437.4 |
| 32 | 2372–2492 | 2470.7 | 2469.5 | 2467.0 |
| 33 | 2402–2510 | 2494.3 | 2492.1 | 2490.4 |
| 34 | 2450–2518 | 2509.8 | 2507.3 | 2503.8 |
| 35 | 2471–2531 | 2526.8 | 2523.4 | 2520.5 |
| 36 | 2478–2539 | 2540.5 | 2535.8 | 2532.2 |
| 37 | 2482–2544 | 2565.4 | 2560.9 | 2557.1 |
| 38 | 2486–2565 | 2588.4 | 2583.7 | 2579.5 |
| 39 | 2491–2573 | 2611.9 | 2606.6 | 2602.1 |
| 40 | 2496–2584 | 2640.4 | 2635.8 | 2631.3 |
| The evolution of body weight during the 30–40-week period | | | | |
| | M-m | | M-1 | M-2 |
| $\overline{X} \pm s_{\overline{X}}$ | 2527.25 ± 71.32 | | 2524.11 ± 69.69 | 2521.23 ± 68.24 |
| V% | 2.50 | | 2.45 | 2.41 |
| L1 vs. L2 = n.s. [F(1, 20) = 0.010, $p$ = 0.842]. | | | | |
| L1 vs. L3 = n.s. [F(1, 20) = 0.041, $p$ = 0.842]. | | | | |
| L2 vs. L3 = n.s. [F(1, 20) = 0.009, $p$ = 0.842]. | | | | |

M-m—control group (no organic sea buckthorn powder was used); M-1—group where 1% organic sea buckthorn powder was used; M-2—group where 2% organic sea buckthorn powder was used; $\overline{X} \pm s_{\overline{X}}$—mean ± standard deviation; V%—coefficient of variation; n.s.—not significant.

By the end of the study, in week 40, the body weights continued to rise. Over the five weeks, the body weight surpassed the standard for the Negru de Moravia breed (2496–2584 g). As a result, in the final week of the study, the average weight of the birds in

M-m was 2640.4 g, in M-1 it was 2635.8 g, and in M-2 it was 2631.3 g. Specifically, compared to the average weight suggested by the standard (2540 g), the values obtained were higher by 3.95% in M-m, 3.77% in M-1, and 3.6% in M-2.

#### 3.1.2. The Situation of Exits from the Workforce

Another major aspect of poultry growth is the situation of exits from the flock, which impacts quantitative production on one hand, but also aligns with the applied raising system. In this case, the age of the birds at the start of the study (30 weeks) and the health benefits provided by the free-range system were two key elements that led to the identification of a minimal loss percentage. Over the course of the 11 weeks of the study, the losses represented 1% for the M-m group and 0.5% for the M-1 and M-2 groups (Table 3). Within the three groups, the exits from the flock were accidental, caused by mechanical injuries.

**Table 3.** The situation of exits from the workforce.

| Age (Weeks) | M-m | | | M-1 | | | M-2 | | |
|---|---|---|---|---|---|---|---|---|---|
| | Effective Weekly | | Death Rate (%) | Effective Weekly | | Death Rate (%) | Effective Weekly | | Death Rate (%) |
| | Beginning (Heads) | End (Heads) | | Beginning (Heads) | End (Heads) | | Beginning (Heads) | End (Heads) | |
| 30 | 200 | 200 | - | 200 | 200 | - | 200 | 200 | |
| 31 | 200 | 200 | - | 200 | 199 | 0.50 | 200 | 200 | - |
| 32 | 200 | 199 | 0.50 | 199 | 199 | - | 200 | 200 | - |
| 33 | 199 | 199 | - | 199 | 199 | - | 200 | 200 | - |
| 34 | 199 | 199 | - | 199 | 199 | - | 200 | 200 | - |
| 35 | 199 | 199 | - | 199 | 199 | - | 200 | 200 | - |
| 36 | 199 | 198 | 0.50 | 199 | 199 | | 200 | 200 | - |
| 37 | 198 | 198 | - | 199 | 199 | - | 200 | 199 | 0.5 |
| 38 | 198 | 198 | - | 199 | 199 | - | 199 | 199 | - |
| 39 | 198 | 198 | - | 199 | 199 | - | 199 | 199 | - |
| 40 | 198 | 198 | - | 199 | 199 | - | 199 | 199 | - |
| 30–40 | - | - | 1.0 | - | - | 0.50 | - | - | 0.50 |

#### 3.1.3. Numerical Egg Production and Laying Intensity

The numerical egg production for the M-m group amounted to 13,719 eggs for the period from 30 to 40 weeks. In the first week of control, the laying intensity was calculated at 86.29%, while the peak laying intensity occurred in week 33, reaching 93.18%. A similar situation was observed for the control groups M-1 and M-2, where the peak laying occurred in week 33 as well, but with higher values: 93.32% (M-1) and 93.79% (M-2). In the first week of control, the laying intensity for M-1 was calculated at 89.07% and for M-2 at 90.00% (Table 4).

**Table 4.** The effect of administering organic sea buckthorn powder on the numerical egg production and laying intensity in the studied birds.

| Experimental Batch | M-m | | M-1 | | M-2 | |
|---|---|---|---|---|---|---|
| Total numerical egg production (11 weeks) | 13,719 | | 13,793 | | 13,907 | |
| Statistical indicator/studied parameter | $\overline{X} \pm s_{\overline{x}}$ | V% | $\overline{X} \pm s_{\overline{x}}$ | V% | $\overline{X} \pm s_{\overline{x}}$ | V% |
| Effective average—reported over the 11 weeks of research (head) | 198.82 ± 0.78 | 0.36 | 199.68 ± 0.46 | 0.28 | 199.68 ± 0.46 | 0.28 |
| Average egg production per week—reported over the 11 weeks of research (eggs/week) | 1247.18 ± 42.36 | 3.11 | 1253.91 ± 37.26 | 3.12 | 1264.27 ± 37.79 | 3.14 |

**Table 4.** *Cont.*

| Experimental Batch | M-m | | M-1 | | M-2 | |
|---|---|---|---|---|---|---|
| Average laying intensity—reported over the 11 weeks of research (%) | 89.61 ± 2.87 | 3.39 | 89.95 ± 2.67 | 3.10 | 90.44 ± 2.52 | 2.93 |
| Numerical egg production | | | | | | |
| M-m vs. M-1 = n.s. (F(1, 20) = 0.1564, $p$ = 0.6967)<br>M-m vs. M-2 = n.s. (F(1, 20) = 0.9971, $p$ = 0.3300)<br>M-1 vs. M-2 = n.s. (F(1, 20) = 0.4194, $p$ = 0.5246) | | | | | | |
| Laying intensity | | | | | | |
| M-m vs. M-1 = n.s. (F(1, 20) = 0.0839, $p$ = 0.7750)<br>M-m vs. M-2 = n.s. (F(1, 20) = 0.5244, $p$ = 0.4774)<br>M-1 vs. M-2 = n.s. (F(1, 20) = 0.1972, $p$ = 0.6617) | | | | | | |

M-m—control group (no organic sea buckthorn powder was used); M-1—group where 1% organic sea buckthorn powder was used; M-2—group where 2% organic sea buckthorn powder was used; $\overline{X} \pm s_{\overline{X}}$—mean ± standard deviation; V%—coefficient of variation; n.s.—not significant.

The higher values of laying intensity were based on higher total egg production, with 13,793 eggs for M-1 and 13,907 eggs for M-2.

The overall situation regarding laying intensity shows that the best average results were identified in the M-2 group (90.45%), followed by M-1 (89.95%) in second place, and M-m in last place (89.61%).

#### 3.1.4. The Structure of Egg Production

An interesting element is the structure of egg production by weight categories, as this parameter is directly related to the price of the eggs. In the case of the control group M-m, the largest share was held by eggs classified in the L and M categories, accounting for 51.77% and 45.32%, respectively. Very large eggs (XL) were found at a percentage of 2.24%, while eggs in the S category made up only 0.67% (Table 5).

**Table 5.** The influence of organic sea buckthorn powder on the structure of egg production according to European commercial standards.

| Eggs Category | M-n | | M-1 | | M-2 | |
|---|---|---|---|---|---|---|
| | $\overline{X} \pm s_{\overline{X}}$ | V% | $\overline{X} \pm s_{\overline{X}}$ | V% | $\overline{X} \pm s_{\overline{X}}$ | V% |
| XL | 2.24 ± 0.41 | 16.39 | 2.32 ± 0.42 | 15.20 | 2.38 ± 0.43 | 14.86 |
| L | 51.77 ± 5.15 | 6.10 | 51.96 ± 5.34 | 6.53 | 52.29 ± 5.53 | 6.83 |
| M | 45.32 ± 4.36 | 7.33 | 45.06 ± 4.59 | 7.97 | 44.70 ± 4.81 | 8.49 |
| S | 0.67 ± 1.39 | 207.18 | 0.66 ± 1.38 | 203.28 | 0.63 ± 1.37 | 214.44 |
| Category XL | | | | | | |
| M-m vs. M-1 = n.s. (F(1, 20) = 0.2115, $p$ = 0.6506)<br>M-m vs. M-2 = n.s. (F(1, 20) = 0.6461, $p$ = 0.4310)<br>M-1 vs. M-2 = n.s. (F(1, 20) = 0.1226, $p$ = 0.7299) | | | | | | |
| Category L | | | | | | |
| M-m vs. M-1 = n.s. (F(1, 20) = 0.0073, $p$ = 0.9329)<br>M-m vs. M-2 = n.s. (F(1, 20) = 0.0517, $p$ = 0.8225)<br>M-1 vs. M-2 = n.s. (F(1, 20) = 0.0199, $p$ = 0.8892) | | | | | | |
| Category M | | | | | | |
| M-m vs. M-1 = n.s. (F(1, 20) = 0.0190, $p$ = 0.8917)<br>M-m vs. M-2 = n.s. (F(1, 20) = 0.0312, $p$ = 0.8616)<br>M-1 vs. M-2 = n.s. (F(1, 20) = 0.0996, $p$ = 0.7556) | | | | | | |

**Table 5.** *Cont.*

| Eggs Category | M-n | M-1 | M-2 |
|---|---|---|---|
| | | Category S | |
| | M-m vs. M-1 = n.s. (F(1, 20) = 0.0002, $p$ = 0.9879)<br>M-m vs. M-2 = n.s. (F(1, 20) = 0.0059, $p$ = 0.393)<br>M-1 vs. M-2 = n.s. (F(1, 20) = 0.0038, $p$ = 0.9513) | | |

M-m—control group (no organic sea buckthorn powder was used); M-1—group where 1% organic sea buckthorn powder was used; M-2—group where 2% organic sea buckthorn powder was used; European standards (XL—extra large; L—large; M—medium; S—small); $\overline{X} \pm s_{\overline{X}}$—mean ± standard deviation; V%—coefficient of variation; n.s.—not significant.

For the M-1 group, the results were similar, with the highest proportion of eggs being in the L (51.96%) and M (45.06%) categories. Only 2.32% were XL eggs, while the S category accounted for 0.66%.

The M-2 group stood out by producing more eggs in the L and M categories (specifically 52.29% for L and 44.70% for M). In this group, a higher percentage of XL eggs was observed (2.38%), while the S category had a lower proportion (0.63%).

#### 3.1.5. Consumption of Combined Feeds

The feed consumption was expressed weekly (kg per group per week), and based on this, the average daily consumption (grams per bird per day) was determined. Finally, the average values for both parameters were calculated for all 11 weeks of the study.

Thus, the control group M-m recorded a total consumption of 2145.6 kg over the period, with very similar values identified for M-1 (2145.2 kg) and M-2 (2146.1 kg).

The average daily consumption per group varied from 139.6 g per bird per day for M-2 to 140.2 g per bird per day for M-m, with M-1 being intermediate at 139.9 g per bird per day (Table 6).

**Table 6.** The comparative evolution of compound feed consumption relative to the percentage of organic sea buckthorn powder administered.

| Experimental Batch | M-m | | M-1 | | M-2 | |
|---|---|---|---|---|---|---|
| **Studied Parameter/Statistical Indicator** | $\overline{X}\pm s_{\overline{X}}$ | V% | $\overline{X}\pm s_{\overline{X}}$ | V% | $\overline{X}\pm s_{\overline{X}}$ | V% |
| Total (kg/batch) | 195.05 ± 0.26 | 0.13 | 195.02 ± 0.397034 | 0.21354 | 195.10 ± 0.28 | 0.12 |
| Daily average (g/head) | 140.16 ± 0.43 | 0.27 | 139.90 ± 0.40 | 0.249883 | 139.58 ± 0.32 | 0.19 |
| | | | Total combined feed consumption (kg/batch) | | | |
| | | | M-m vs. M-1 = n.s. (F(1, 20) = 0.0648, $p$ = 0.8016)<br>M-m vs. M-2 = n.s. (F(1, 20) = 0.1508, $p$ = 0.7019)<br>M-1 vs. M-2 = n.s. (F(1, 20) = 0.3047, $p$ = 0.5870) | | | |
| | | | Daily average combined feed consumption (g/head) | | | |
| | | | M-m vs. M-1 = n.s. (F(1, 20) = 2.1935, $p$ = 0.1542)<br>M-m vs. M-2 = * $p$ = 0.0548 (F(1, 20) = 4.1610)<br>M-1 vs. M-2 = ** $p < 0.05$ (F(1, 20) = 12.9211, $p$ = 0.0018) | | | |

M-m—control group (no organic sea buckthorn powder was used); M-1—group where 1% organic sea buckthorn powder was used; M-2—group where 2% organic sea buckthorn powder was used; $\overline{X} \pm s_{\overline{X}}$—mean ± standard deviation; V%—coefficient of variation; n.s.—not significant; *—significant differences; **—distinct significant differences.

### 3.2. *Quality Indicators of Laid Eggs*

#### 3.2.1. Egg Weight

The weight of the eggs did not experience major fluctuations between the groups under study, with the general trend being an increase in the value of the indicator from one week of control to another. Analyzing the results as a whole, it can be observed that the largest egg weights were recorded in the M-2 group, at 63.80 g, while the smallest were in the M-m group, at 63.46 g. The average value determined for the M-1 group was 63.53 g (Table 7).

**Table 7.** The evolution of egg weight based on the percentage of organic sea buckthorn powder administered.

| Age (Weeks) | M-m (N = 30) | | M-1 (N = 30) | | M-2 (N = 30) | |
|---|---|---|---|---|---|---|
| | $\overline{X} \pm s_{\overline{X}}$ (g) | V% | $\overline{X} \pm s_{\overline{X}}$ (g) | V% | $\overline{X} \pm s_{\overline{X}}$ (g) | V% |
| 30 | 62.89 ± 0.55 | 4.82 | 62.90 ± 0.47 | 4.11 | 62.89 ± 0.45 | 3.92 |
| 31 | 63.00 ± 0.93 | 8.05 | 63.10 ± 0.51 | 4.44 | 63.15 ± 0.46 | 4.01 |
| 32 | 63.11 ± 0.92 | 7.98 | 63.20 ± 0.54 | 4.72 | 63.32 ± 0.50 | 4.29 |
| 33 | 63.24 ± 0.89 | 7.68 | 63.31 ± 0.57 | 4.98 | 63.50 ± 0.53 | 4.58 |
| 34 | 63.37 ± 0.86 | 7.48 | 63.40 ± 0.58 | 5.03 | 63.68 ± 0.53 | 4.58 |
| 35 | 63.46 ± 0.91 | 7.86 | 63.58 ± 0.61 | 5.29 | 63.84 ± 0.55 | 4.69 |
| 36 | 63.57 ± 0.94 | 8.13 | 63.62 ± 0.67 | 5.80 | 63.91 ± 0.58 | 5.01 |
| 37 | 63.66 ± 0.99 | 8.55 | 63.74 ± 0.68 | 5.87 | 64.08 ± 0.61 | 5.23 |
| 38 | 63.78 ± 1.01 | 8.69 | 63.85 ± 0.70 | 6.02 | 64.11 ± 0.63 | 5.55 |
| 39 | 63.89 ± 1.04 | 8.95 | 63.94 ± 0.71 | 6.10 | 64.54 ± 0.66 | 5.63 |
| 40 | 64.04 ± 1.06 | 9.07 | 64.20 ± 0.72 | 6.13 | 64.77 ± 0.67 | 5.68 |
| | The average egg weight during the 30–40-week period | | | | | |
| 30–40 | 63.46 ± 0.37 | 8.53 | 63.53 ± 0.39 | 5.55 | 63.80 ± 0.57 | 4.80 |
| | M-m vs. M-1 = n.s. (F(1, 20) = 0.2144, $p$ = 0.6483)<br>M-m vs. M-2 = n.s. (F(1, 20) = 2.7915, $p$ = 0.1103)<br>M-1 vs. M-2 = n.s. (F(1, 20) = 1.6509, $p$ = 0.2135) | | | | | |

M-m—control group (no organic sea buckthorn powder was used); M-1—group where 1% organic sea buckthorn powder was used; M-2—group where 2% organic sea buckthorn powder was used; N—number of eggs for which the determination was made; $\overline{X} \pm s_{\overline{X}}$—mean ± standard deviation; V%—coefficient of variation; n.s.—not significant.

### 3.2.2. Egg Format Index

The determination of the format index revealed normal values for this indicator across all three groups. Specifically, for M-m, a value of 73.25% was determined, for M-1 it was 73.30%, and for M-2 it was 73.38% (Table 8).

**Table 8.** The influence of the addition of organic sea buckthorn powder in the birds' diet on the size index.

| Age (Weeks) | M-m (N = 30) | | M-1 (N = 30) | | M-2 (N = 30) | |
|---|---|---|---|---|---|---|
| | $\overline{X} \pm s_{\overline{X}}$ (%) | V% | $\overline{X} \pm s_{\overline{X}}$ (%) | V% | $\overline{X} \pm s_{\overline{X}}$ (%) | V% |
| 30 | 71.85 ± 1.00 | 7.59 | 71.85 ± 0.86 | 6.55 | 71.86 ± 0.85 | 6.48 |
| 31 | 72.15 ± 1.40 | 10.62 | 72.19 ± 0.99 | 7.52 | 72.23 ± 0.87 | 6.58 |
| 32 | 72.44 ± 1.43 | 10.78 | 72.48 ± 1.02 | 7.68 | 72.52 ± 0.92 | 6.96 |
| 33 | 72.46 ± 1.43 | 11.11 | 72.50 ± 1.02 | 7.70 | 72.58 ± 0.93 | 6.99 |
| 34 | 72.50 ± 1.53 | 11.57 | 72.56 ± 1.06 | 7.98 | 72.62 ± 0.93 | 7.04 |
| 35 | 73.04 ± 1.59 | 11.95 | 73.10 ± 1.07 | 8.02 | 73.18 ± 0.96 | 7.16 |
| 36 | 73.56 ± 1.64 | 12.23 | 73.61 ± 1.09 | 8.13 | 73.70 ± 0.98 | 7.28 |
| 37 | 73.97 ± 1.69 | 12.56 | 74.02 ± 1.28 | 9.45 | 74.11 ± 1.07 | 7.91 |
| 38 | 74.11 ± 1.74 | 12.87 | 74.20 ± 1.29 | 9.56 | 74.31 ± 1.08 | 7.97 |
| 39 | 74.69 ± 1.78 | 13.04 | 74.73 ± 1.31 | 9.60 | 74.84 ± 1.11 | 8.11 |
| 40 | 75.03 ± 1.99 | 14.56 | 75.11 ± 1.37 | 9.98 | 75.25 ± 1.12 | 8.13 |
| | The size index analyzed for the 30–40-week period. | | | | | |

**Table 8.** *Cont.*

| Age (Weeks) | M-m (N = 30) | | M-1 (N = 30) | | M-2 (N = 30) | |
|---|---|---|---|---|---|---|
| | $\overline{X} \pm s_{\overline{X}}$ (%) | V% | $\overline{X} \pm s_{\overline{X}}$ (%) | V% | $\overline{X} \pm s_{\overline{X}}$ (%) | V% |
| 30–40 | 73.25 ± 1.07 | 1.39 | 73.30 ± 1.09 | 1.41 | 73.38 ± 1.13 | 1.44 |

M-m vs. M-1 = n.s. ($F(1, 20) = 0.0116$, $p = 0.9152$)
M-m vs. M-2 = n.s. ($F(1, 20) = 0.073$, $p = 0.7898$)
M-1 vs. M-2 = n.s. ($F(1, 20) = 0.027$, $p = 0.8722$)

M-m—control group (no organic sea buckthorn powder was used); M-1—group where 1% organic sea buckthorn powder was used; M-2—group where 2% organic sea buckthorn powder was used; N—number of eggs for which the determination was made; $\overline{X} \pm s_{\overline{X}}$—mean ± standard deviation; V%—coefficient of variation; n.s.—not significant.

### 3.2.3. Specific Gravity of Eggs

The determination of the specific weight was carried out on a sample of 30 eggs from each group, with the analyses repeated weekly. The highest value of the indicator was found in the M-2 group (1.100), while the lowest was in the M-m group, at 1.087. The M-1 group had an intermediate value, with a determined specific weight of 1.093 (Table 9).

**Table 9.** Changes in the specific weight of eggs relative to the addition of organic sea buckthorn powder used.

| Age (Weeks) | M-m (N = 30) | | M-1 (N = 30) | | M-2 (N = 30) | |
|---|---|---|---|---|---|---|
| | $\overline{X} \pm s_{\overline{X}}$ (g/cm$^3$) | V% | $\overline{X} \pm s_{\overline{X}}$ (g/cm$^3$) | V% | $\overline{X} \pm s_{\overline{X}}$ (g/cm$^3$) | V% |
| 30 | 1.061 ± 0.014 | 7.16 | 1.062 ± 0.014 | 7.03 | 1.062 ± 0.009 | 6.22 |
| 31 | 1.068 ± 0.017 | 8.64 | 1.070 ± 0.011 | 5.86 | 1.073 ± 0.010 | 4.99 |
| 32 | 1.074 ± 0.017 | 8.88 | 1.078 ± 0.012 | 5.93 | 1.082 ± 0.010 | 5.03 |
| 33 | 1.079 ± 0.018 | 8.92 | 1.083 ± 0.012 | 6.12 | 1.088 ± 0.010 | 5.28 |
| 34 | 1.082 ± 0.018 | 8.97 | 1.089 ± 0.012 | 6.18 | 1.094 ± 0.011 | 5.47 |
| 35 | 1.088 ± 0.019 | 9.14 | 1.096 ± 0.014 | 6.87 | 1.100 ± 0.012 | 5.66 |
| 36 | 1.091 ± 0.019 | 9.58 | 1.099 ± 0.014 | 6.93 | 1.107 ± 0.012 | 5.81 |
| 37 | 1.095 ± 0.019 | 9.26 | 1.104 ± 0.013 | 6.45 | 1.115 ± 0.013 | 6.11 |
| 38 | 1.098 ± 0.020 | 9.94 | 1.109 ± 0.014 | 7.04 | 1.120 ± 0.014 (10-R1) | 6.23 |
| 39 | 1.104 ± 0.020 | 10.11 | 1.115 ± 0.014 | 7.18 | 1.126 ± 0.014 | 6.78 |
| 40 | 1.118 ± 0.021 | 11.20 | 1.122 ± 0.015 | 7.97 | 1.130 ± 0.014 | 6.80 |
| The average specific weight during the 30–40-week period | | | | | | |
| 30–40 | 1.087 ± 0.016 | 8.36 | 1.093 ± 0.018 | 6.51 | 1.100 ± 0.022 | 5.75 |

M-m vs. M-1 = n.s. ($F(1, 20) = 0.684$, $p = 0.418$)
M-m vs. M-2 = n.s. ($F(1, 20) = 2.289$, $p = 0.146$)
M-1 vs. M-2 = n.s. ($F(1, 20) = 0.524$, $p = 0.478$)

M-m—control group (no organic sea buckthorn powder was used); M-1—group where 1% organic sea buckthorn powder was used; M-2—group where 2% organic sea buckthorn powder was used; N—number of eggs for which the determination was made; $\overline{X} \pm s_{\overline{X}}$—mean ± standard deviation; V%—coefficient of variation; n.s.—not significant.

### 3.2.4. Egg Volume

Another quality indicator was the egg volume, where the highest values were recorded in the M-2 group at 56.41 cm$^3$, and the lowest in the M-m group at 56.02 cm$^3$. The values for the M-1 group were intermediate, but without significant differences, with a value of 56.25 cm$^3$ (Table 10).

**Table 10.** The influence of the addition of organic sea buckthorn powder on the volume of eggs laid.

| Age (Weeks) | M-m (N = 30) | | M-1 (N = 30) | | M-2 (N = 30) | |
|---|---|---|---|---|---|---|
| | $\overline{X} \pm s_{\overline{x}}$ (cm³) | V% | $\overline{X} \pm s_{\overline{x}}$ (cm³) | V% | $\overline{X} \pm s_{\overline{x}}$ (cm³) | V% |
| 30 | 54.74 ± 0.99 | 5.71 | 54.78 ± 0.57 | 3.28 | 54.76 ± 0.62 | 3.58 |
| 31 | 54.89 ± 1.07 | 6.15 | 55.07 ± 0.64 | 3.69 | 55.12 ± 0.46 | 2.64 |
| 32 | 55.08 ± 1.04 | 5.98 | 55.21 ± 0.68 | 3.89 | 55.44 ± 0.53 | 3.02 |
| 33 | 55.29 ± 1.10 | 6.27 | 55.65 ± 0.78 | 4.45 | 55.82 ± 0.61 | 3.45 |
| 34 | 55.71 ± 1.11 | 6.97 | 55.92 ± 0.86 | 4.86 | 56.03 ± 0.59 | 3.33 |
| 35 | 56.12 ± 1.27 | 7.14 | 56.38 ± 0.81 | 4.52 | 56.61 ± 0.64 | 3.58 |
| 36 | 56.47 ± 1.36 | 7.59 | 56.74 ± 0.74 | 4.13 | 56.92 ± 0.74 | 4.11 |
| 37 | 56.69 ± 1.47 | 8.17 | 56.93 ± 0.89 | 4.97 | 57.04 ± 0.77 | 4.25 |
| 38 | 56.90 ± 1.62 | 8.99 | 57.09 ± 0.92 | 5.11 | 57.21 ± 0.85 | 4.69 |
| 39 | 57.11 ± 1.71 | 9.45 | 57.28 ± 0.95 | 5.23 | 57.54 ± 0.86 | 4.73 |
| 40 | 57.27 ± 1.80 | 9.93 | 57.65 ± 1.04 | 5.69 | 57.99 ± 0.89 | 4.88 |
| The volume of eggs determined for the 30–40-week period | | | | | | |
| 30–40 | 56.02 ± 0.92 | 7.54 | 56.25 ± 0.97 | 4.58 | 56.41 ± 1.04 | 3.66 |
| M-m vs. M-1 = n.s. (F(1, 20) = 0.296, $p$ = 0.593)<br>M-m vs. M-2 = n.s. (F(1, 20) = 0.824, $p$ = 0.375)<br>M-1 vs. M-2 = n.s. (F(1, 20) = 0.141, $p$ = 0.712) | | | | | | |

M-m—control group (no organic sea buckthorn powder was used); M-1—group where 1% organic sea buckthorn powder was used; M-2—group where 2% organic sea buckthorn powder was used; N—number of eggs for which the determination was made; $\overline{X} \pm s_{\overline{X}}$—mean ± standard deviation; V%—coefficient of variation; n.s.—not significant.

### 3.2.5. Weight of Egg Components

Since egg consumption in Romania primarily focuses on shell eggs, it was deemed appropriate to determine the proportion of their components. Although the results obtained were similar, the overall analysis allowed us to rank the M-m and M-2 groups first in terms of yolk proportion, with a value of 32.11% for both. The M-1 group was ranked second, with a value of 32.10%.

The proportion of egg white was 59.25% for M-2, 59.37% for M-1, and 59.43% for M-m. Regarding the total egg structure, the mineral content of the shell ranged from 8.46% for M-m to 8.64% for M-2 (Table 11).

**Table 11.** The proportion of the main components of eggs based on the percentage of organic sea buckthorn powder used.

| Experimental Batch | M-m (N = 30) | | M-1 (N = 30) | | M-2 (N = 30) | |
|---|---|---|---|---|---|---|
| Indicator | $\overline{X} \pm s_{\overline{x}}$ | V% | $\overline{X} \pm s_{\overline{x}}$ | V% | $\overline{X} \pm s_{\overline{x}}$ | V% |
| Yolk (%) | 32.18 ± 0.13 | 0.39 | 32.11 ± 0.10 | 0.31 | 32.12 ± 0.17 | 0.53 |
| Egg white (%) | 59.38 ± 0.03 | 0.05 | 59.37 ± 0.03 | 0.04 | 59.24 ± 0.09 | 0.15 |
| Shell (%) | 8.44 ± 0.15 | 1.68 | 8.52 ± 0.13 | 1.49 | 8.64 ± 0.08 | 0.95 |
| Yolk (%) | | | | | | |
| effectM-m vs. M-1 = n.s. (F(1, 20) = 1.60, $p$ = 0.221)<br>M-m vs. M-2 = n.s. (F(1, 20) = 0.0051, $p$ = 0.944)<br>M-1 vs. M-2 = n.s. (F(1, 20) = 0.0051, $p$ = 0.944) | | | | | | |
| Egg white (%) | | | | | | |
| M-m vs. M-1 = n.s. (F(1, 20) = 0.4697, $p$ = 0.5010)<br>M-m vs. M-2 = ** $p < 0.05$ (F(1, 20) = 15.6774, $p$ = 0.0008)<br>M-1 vs. M-2 = ** $p < 0.05$ (F(1, 20) = 13.1192, $p$ = 0.0017) | | | | | | |

**Table 11.** *Cont.*

| Experimental Batch | M-m (N = 30) | M-1 (N = 30) | M-2 (N = 30) |
|---|---|---|---|
| | Shell (%) | | |
| | M-m vs. M-1 = n.s. (F(1, 20) = 1.2145, $p$ = 0.2835)<br>M-m vs. M-2 = ** $p < 0.05$ (F(1, 20) = 10.3560, $p$ = 0.0043)<br>M-1 vs. M-2 = * $p < 0.05$ (F(1, 20) = 4.6823, $p$ = 0.0428) | | |

M-m—control group (no organic sea buckthorn powder was used); M-1—group where 1% organic sea buckthorn powder was used; M-2—group where 2% organic sea buckthorn powder was used; $\overline{X} \pm s_{\overline{X}}$—mean ± standard deviation; V%—coefficient of variation; n.s.—not significant; *—significant differences; **—distinct significant differences.

### 3.2.6. The Thickness of the Mineral Shell of Eggs

The integrity of eggs during handling is directly influenced by the thickness of the mineral shell. Analyzing this indicator, a reduction in shell thickness was observed from one week of control to the next, a trend that was consistent across all three groups.

The results showed that the thickest mineral shell was found in the M-2 group (with a value of 0.787 mm), while the thinnest was in the M-m group (with a value of 0.728 mm). The M-1 group had an intermediate value, with a determined thickness of 0.776 mm (Table 12).

**Table 12.** The impact of using different percentages of organic sea buckthorn powder on the mineral shell thickness of the eggs studied.

| Age (Weeks) | M-m (N = 30) | | M-1 (N = 30) | | M-2 (N = 30) | |
|---|---|---|---|---|---|---|
| | $\overline{X} \pm s_{\overline{X}}$ (mm) | V% | $\overline{X} \pm s_{\overline{X}}$ (mm) | V% | $\overline{X} \pm s_{\overline{X}}$ (mm) | V% |
| 30 | 0.792 ± 0.021 | 14.25 | 0.793 ± 0.015 | 10.41 | 0.792 ± 0.014 | 10.02 |
| 31 | 0.790 ± 0.021 | 14.87 | 0.793 ± 0.015 | 10.59 | 0.794 ± 0.014 | 10.23 |
| 32 | 0.787 ± 0.021 | 14.28 | 0.791 ± 0.016 | 10.97 | 0.795 ± 0.014 | 10.45 |
| 33 | 0.783 ± 0.021 | 15.01 | 0.788 ± 0.016 | 11.11 | 0.795 ± 0.015 | 10.56 |
| 34 | 0.780 ± 0.022 | 15.69 | 0.786 ± 0.016 | 11.25 | 0.793 ± 0.015 | 10.69 |
| 35 | 0.775 ± 0.023 | 15.97 | 0.780 ± 0.016 | 11.69 | 0.790 ± 0.016 | 10.71 |
| 36 | 0.774 ± 0.023 | 16.28 | 0.778 ± 0.017 | 12.13 | 0.787 ± 0.016 | 11.38 |
| 37 | 0.768 ± 0.024 | 16.83 | 0.773 ± 0.017 | 12.58 | 0.784 ± 0.017 | 11.69 |
| 38 | 0.763 ± 0.024 | 17.11 | 0.770 ± 0.018 | 12.61 | 0.781 ± 0.017 | 11.87 |
| 39 | 0.744 ± 0.023 | 17.09 | 0.766 ± 0.018 | 12.87 | 0.777 ± 0.017 | 12.12 |
| 40 | 0.702 ± 0.023 | 17.86 | 0.751 ± 0.018 | 13.45 | 0.774 ± 0.017 | 12.38 |
| | The mineral shell thickness determined for the 30–40-week period | | | | | |
| Average | 0.769 ± 0.026 | 15.43 | 0.779 ± 0.013 | 11.67 | 0.787 ± 0.007 | 10.98 |
| | M-m vs. M-1 = n.s. (F(1, 20) = 1.3084, $p$ = 0.2662)<br>M-m vs. M-2 = * $p < 0.05$ (F(1, 20) = 5.1252, $p$ = 0.0349)<br>M-1 vs. M-2 = n.s. (F(1, 20) = 3.4234, $p$ = 0.0791) | | | | | |

M-m—control group (no organic sea buckthorn powder was used); M-1—group where 1% organic sea buckthorn powder was used; M-2—group where 2% organic sea buckthorn powder was used; $\overline{X} \pm s_{\overline{X}}$—mean ± standard deviation; V%—coefficient of variation; n.s.—not significant; *—significant differences.

### 3.2.7. Crack Resistance of the Mineral Shell

For the reasons mentioned earlier, we were also interested in determining the breakage resistance of the mineral shell. In terms of this parameter, greater differences were observed between the experimental groups. Thus, the lowest resistance was observed in the M-m group, with a value of 176.90 g f/cm$^2$, followed by the M-1 group with a value of 179.17 g f/cm$^2$, and the M-2 group, where a value of 181.11 g f/cm$^2$ was determined (Table 13)

**Table 13.** The changes in breakage resistance of the mineral shell relative to the percentage of organic sea buckthorn powder used in the diet of Moravia Black hens.

| Age (Weeks) | M-m (n = 30) | | M-1 (n = 30) | | M-2 (n = 30) | |
|---|---|---|---|---|---|---|
| | $\overline{X}\pm s_{\overline{X}}$ (g f/cm$^2$) | V% | $\overline{X}\pm s_{\overline{X}}$ (g f/cm$^2$) | V% | $\overline{X}\pm s_{\overline{X}}$ (g f/cm$^2$) | V% |
| 30 | 182.16 ± 5.50 | 16.55 | 182.39 ± 4.39 | 13.18 | 182.16 ± 3.87 | 11.63 |
| 31 | 181.70 ± 5.58 | 16.83 | 182.39 ± 4.66 | 14.01 | 182.62 ± 4.00 | 12.01 |
| 32 | 181.61 ± 5.57 | 16.80 | 181.93 ± 4.46 | 13.44 | 182.85 ± 4.14 | 12.42 |
| 33 | 180.09 ± 5.52 | 16.79 | 181.24 ± 4.72 | 14.28 | 182.85 ± 4.20 | 12.59 |
| 34 | 179.40 ± 5.56 | 16.98 | 180.78 ± 4.82 | 14.61 | 182.39 ± 4.32 | 12.98 |
| 35 | 178.25 ± 5.71 | 17.56 | 179.40 ± 4.87 | 14.89 | 181.70 ± 4.20 | 12.67 |
| 36 | 178.02 ± 5.73 | 17.63 | 178.94 ± 4.87 | 14.91 | 181.01 ± 4.33 | 13.12 |
| 37 | 176.64 ± 5.96 | 18.49 | 177.79 ± 4.89 | 15.07 | 180.32 ± 4.28 | 13.01 |
| 38 | 175.49 ± 6.14 | 19.19 | 177.10 ± 4.88 | 15.11 | 179.63 ± 4.23 | 12.89 |
| 39 | 171.12 ± 6.30 | 20.16 | 176.18 ± 4.87 | 15.15 | 178.71 ± 4.30 | 13.20 |
| 40 | 161.46 ± 5.97 | 20.27 | 172.73 ± 4.79 | 15.20 | 178.02 ± 4.34 | 13.35 |
| The breakage resistance of the mineral shell for the 30–40-week period | | | | | | |
| 30–40 | 176.90 ± 6.05 | 16.46 | 179.17 ± 3.03 | 14.67 | 181.11 ± 1.71 | 12.98 |
| M-m vs. M-1 = n.s. ($F(1, 20) = 1.2331, p = 0.2800$)<br>M-m vs. M-2 = * $p < 0.05$ ($F(1, 20) = 4.9279, p = 0.0381$)<br>M-1 vs. M-2 = n.s. ($F(1, 20) = 3.4234, p = 0.0791$) | | | | | | |

M-m—control group (no organic sea buckthorn powder was used); M-1—group where 1% organic sea buckthorn powder was used; M-2—group where 2% organic sea buckthorn powder was used; $\overline{X} \pm s_{\overline{X}}$—mean ± standard deviation; V%—coefficient of variation; n.s.—not significant; *—significant differences.

3.2.8. The Color of the Yolk

A parameter that influences the purchasing decision of consumers, the color of the yolk was determined on a sample of 30 eggs from each group, each week of the experiment, using the La Roche scale.

At the end of the study, the average values obtained were calculated, highlighting a more intense orange color in the M-2 group (with a determined value of 12.12 points), a less intense color in M-1 (with a value of 10.22 points), and a paler shade in the M-m group (with a value of 8.33 points) (Table 14).

**Table 14.** The impact of adding organic sea buckthorn powder on the modification of yolk color in eggs produced by Moravia Black hens.

| Age (Weeks) | M-m (N = 30) | | M-1 (N = 30) | | M-2 (N = 30) | |
|---|---|---|---|---|---|---|
| | $\overline{X}\pm s_{\overline{X}}$ (Points) | V% | $\overline{X}\pm s_{\overline{X}}$ (Points) | V% | $\overline{X}\pm s_{\overline{X}}$ (Points) | V% |
| 30 | 8.37 ± 0.08 | 5.62 | 8.38 ± 0.06 | 4.21 | 8.37 ± 0.06 | 3.78 |
| 31 | 8.35 ± 0.09 | 5.89 | 9.72 ± 0.08 | 4.57 | 11.15 ± 0.07 | 3.56 |
| 32 | 8.35 ± 0.08 | 5.44 | 10.25 ± 0.09 | 4.63 | 12.04 ± 0.07 | 3.12 |
| 33 | 8.34 ± 0.09 | 6.19 | 10.44 ± 0.09 | 4.89 | 12.59 ± 0.07 | 3.09 |
| 34 | 8.34 ± 0.09 | 5.97 | 10.44 ± 0.07 | 3.75 | 12.62 ± 0.09 | 4.11 |
| 35 | 8.33 ± 0.09 | 6.28 | 10.48 ± 0.08 | 3.98 | 12.65 ± 0.09 | 4.07 |
| 36 | 8.32 ± 0.09 | 6.01 | 10.50 ± 0.08 | 4.10 | 12.70 ± 0.10 | 4.27 |
| 37 | 8.31 ± 0.08 | 5.55 | 10.50 ± 0.09 | 4.63 | 12.78 ± 0.09 | 3.92 |
| 38 | 8.31 ± 0.08 | 5.67 | 10.52 ± 0.09 | 4.57 | 12.78 ± 0.09 | 3.85 |
| 39 | 8.30 ± 0.10 | 6.45 | 10.55 ± 0.08 | 4.22 | 12.81 ± 0.10 | 4.37 |
| 40 | 8.29 ± 0.11 | 7.03 | 10.61 ± 0.10 | 4.99 | 12.86 ± 0.10 | 4.14 |
| The yolk color determined during the 30–40-week period | | | | | | |

**Table 14.** *Cont.*

| Age (Weeks) | M-m (N = 30) | | M-1 (N = 30) | | M-2 (N = 30) | |
|---|---|---|---|---|---|---|
| | $\overline{X} \pm s_{\overline{X}}$ (Points) | V% | $\overline{X} \pm s_{\overline{X}}$ (Points) | V% | $\overline{X} \pm s_{\overline{X}}$ (Points) | V% |
| 30–40 | 8.33 ± 0.02 | 5.25 | 10.22 ± 0.66 | 4.12 | 12.12 ± 1.34 | 4.22 |
| M-m vs. M-1 = *** $p < 0.001$ (F(1, 20) = 90.983, $p < 0.0001$)<br>M-m vs. M-2 = *** $p < 0.001$ (F(1, 20) = 87.997, $p < 0.0001$)<br>M-1 vs. M-2 = ** $p < 0.001$ (F(1, 20) = 17.908, $p = 0.0004$) | | | | | | |

M-m—control group (no organic sea buckthorn powder was used); M-1—group where 1% organic sea buckthorn powder was used; M-2—group where 2% organic sea buckthorn powder was used; $\overline{X} \pm s_{\overline{X}}$—mean ± standard deviation; V%—coefficient of variation; n.s.—not significant; **—very significant differences; ***—extremely significant differences.

### 3.2.9. The Carotenoid Content of the Yolk

The carotenoid quantities, which ultimately determine the color of the yolk, were also determined for 30 eggs from each group, each week of the experiment, and then the average values were calculated. Higher values were identified for the M-2 group, where the quantity was 25.25 mg/g, followed by the M-1 group (with a quantity of 19.43 mg/g) and the M-m group, with only 17.66 mg/g (Table 15).

**Table 15.** Determining the carotenoid content in the yolk based on the percentage of organic sea buckthorn powder administered to Moravia Black hens.

| Age (Weeks) | M-m (N = 30) | | M-1 (N = 30) | | M-2 (N = 30) | |
|---|---|---|---|---|---|---|
| | $\overline{X} \pm s_{\overline{X}}$ (mg/g) | V% | $\overline{X} \pm s_{\overline{X}}$ (mg/g) | V% | $\overline{X} \pm s_{\overline{X}}$ (mg/g) | V% |
| 30 | 17.74 ± 0.072 | 2.22 | 17.76 ± 0.038 | 1.17 | 17.74 ± 0.033 | 1.02 |
| 31 | 17.70 ± 0.092 | 2.86 | 20.44 ± 0.048 | 1.28 | 23.30 ± 0.042 | 0.98 |
| 32 | 17.70 ± 0.083 | 2.57 | 21.50 ± 0.053 | 1.36 | 25.08 ± 0.034 | 0.75 |
| 33 | 17.68 ± 0.087 | 2.69 | 21.88 ± 0.050 | 1.25 | 26.18 ± 0.053 | 1.11 |
| 34 | 17.68 ± 0.097 | 3.01 | 21.88 ± 0.043 | 1.08 | 26.24 ± 0.049 | 1.03 |
| 35 | 17.66 ± 0.088 | 2.74 | 21.96 ± 0.060 | 1.49 | 26.30 ± 0.047 | 0.97 |
| 36 | 17.64 ± 0.100 | 3.11 | 22.00 ± 0.063 | 1.58 | 26.40 ± 0.062 | 1.29 |
| 37 | 17.62 ± 0.095 | 2.96 | 22.00 ± 0.052 | 1.29 | 26.56 ± 0.095 | 1.97 |
| 38 | 17.62 ± 0.097 | 3.02 | 22.04 ± 0.065 | 1.61 | 26.56 ± 0.077 | 1.58 |
| 39 | 17.60 ± 0.100 | 3.11 | 22.10 ± 0.069 | 1.70 | 26.62 ± 0.055 | 1.13 |
| 40 | 17.58 ± 0.115 | 3.57 | 22.22 ± 0.052 | 1.29 | 26.72 ± 0.048 | 0.98 |
| The carotenoid content determined for the 30–40-week period | | | | | | |
| 30–40 | 17.66 ± 0.04 | 2.40 | 21.43 ± 1.31 | 1.35 | 25.24 ± 2.68 | 1.05 |
| M-m vs. M-1 = *** $p < 0.001$ (F(1, 20) = 90.983, $p < 0.0001$)<br>M-m vs. M-2 = *** $p < 0.001$ (F(1, 20) = 17.908, $p = 0.0004$)<br>M-1 vs. M-2 = *** $p < 0.001$ (F(1, 20) = 87.997, $p = 9.17 \times 10^{-9}$) | | | | | | |

M-m—control group (no organic sea buckthorn powder was used); M-1—group where 1% organic sea buckthorn powder was used; M-2—group where 2% organic sea buckthorn powder was used; $\overline{X} \pm s_{\overline{X}}$—mean ± standard deviation; V%—coefficient of variation; ***—extremely significant differences.

## 4. Discussion

The results of this study demonstrate the positive impact of sea buckthorn powder on the productive and qualitative characteristics of hens. At the outset of the experiment (30 weeks), the body weight of the birds was nearly identical across all three groups, with slight variations (M-m: 2412.4 g, M-1: 2412.0 g, M-2: 2412.2 g). By 40 weeks, the birds in all groups showed similar weight trends, consistent with the standard growth pattern for the breed. However, the slightly lower weights in the experimental groups, especially M-1 and M-2, could be attributed to the increased laying intensity observed in these groups. In comparison, a study on ISA Brown hens fed with black soldier fly larvae reported minor weight differences between control and experimental groups, reinforcing the notion that dietary supplements often have minimal effects on overall body weight [35]. Similarly, research on dietary blackberry supplementation found that egg weights steadily increase throughout the laying period, aligning with the positive effects of natural feed additives like sea buckthorn [36].

The mortality rate among the hens was very low in all groups, reflecting the breed's inherent resilience and the beneficial conditions provided by free-range housing. In the control group, two birds were lost, while only one bird was lost per experimental group, a finding that is consistent with the relatively low mortality rates observed in other studies on free-range poultry [35]. This suggests that both the breed's robustness and the environmental conditions contributed to the positive outcomes of the study.

Egg production was another area where significant results were observed. Hens in all three groups exhibited high laying intensity, peaking at 33 weeks. The experimental groups showed a slight increase in egg production, with group M-2 (2% sea buckthorn powder) yielding the highest total egg production (13,907 eggs) and individual production per hen (69.65 eggs). These results align with studies on the supplementation of feed with natural additives such as humic substances and essential oils, which have been shown to improve egg production parameters, including laying rate and feed conversion efficiency [37,38]. Meta-analyses on essential oils also report improvements in egg production, feed efficiency, and eggshell quality, attributed to the antioxidant properties of such supplements, which are likely present in sea buckthorn as well.

Egg production structure also improved with sea buckthorn supplementation, particularly in the M-2 group, where the proportion of larger eggs (L and XL) was higher compared to the control group. This result is in line with findings by Panaite et al. (2020), who reported that dietary additives like sea buckthorn could increase egg yolk weight and improve egg quality [39]. This suggests that sea buckthorn may positively influence the size and quality of eggs, potentially through its nutritional content and bioactive compounds.

Regarding feed consumption, all three groups showed similar patterns, with minimal differences in the average daily feed intake. This consistency suggests that sea buckthorn supplementation had a negligible effect on the overall feed consumption, which aligns with other studies that have found minimal impacts of natural additives on feed intake while still improving production efficiency [40–42]. The results further indicate that while feed consumption remained stable, supplementation with sea buckthorn led to better egg production and quality, thus enhancing overall productivity.

In terms of egg quality, the supplementation of sea buckthorn showed clear benefits. The average egg weight for the control group was 63.46 g, with the M-1 and M-2 groups showing slight increases, reaching 63.53 g and 63.80 g, respectively. These findings support other studies that suggest dietary supplementation, such as with sea buckthorn, can enhance egg weight and overall quality [43]. Furthermore, the shape index of the eggs remained within the optimal range, with only slight variations between groups. Notably, the eggs from the M-1 and M-2 groups showed better uniformity in shape, a characteristic often associated with the use of natural additives that help improve egg quality and consistency [44].

The specific gravity of eggs, which is an important indicator of egg quality, also improved with sea buckthorn supplementation. The M-2 group showed the highest specific gravity values, consistent with studies that have demonstrated the positive effects of sea buckthorn on egg characteristics, including shell quality and yolk weight [45,46]. This suggests that the bioactive compounds in sea buckthorn may enhance the nutritional quality of the eggs, particularly in terms of yolk and shell quality.

Egg volume, another quality indicator, was also positively influenced by sea buckthorn supplementation. The eggs from the M-1 and M-2 groups had higher average volumes compared to the control group, with the M-2 group showing the highest volume, indicating that the addition of sea buckthorn might contribute to larger eggs without significantly altering feed consumption. These findings are consistent with those of other studies that have observed improvements in egg size and volume with the use of dietary supplements [47,48].

When examining the composition of the eggs, particularly the proportion of yolk, albumen, and mineral shell, the results were consistent with known patterns of egg composition changes as hens age. The yolk proportion increased over time, while the mineral shell proportion decreased, which is typical for laying hens. Sea buckthorn supplementation did not drastically alter these trends but slightly improved the homogeneity of the egg components, particularly in the M-1 and M-2 groups. This could be a reflection of the bioactive compounds in sea buckthorn that may support overall egg quality and nutritional balance.

The thickness of the eggshells in all groups decreased over the experimental period, a typical result due to the increased size of the eggs. However, eggs from the experimental groups, particularly those receiving 2% sea buckthorn powder, had slightly thicker shells compared to the control group, aligning with the findings of studies that suggest dietary supplementation can have positive effects on calcium metabolism and eggshell quality [41,49,50]. The increased resistance to cracking in the eggs from the M-1 and M-2 groups further supports the hypothesis that sea buckthorn may improve eggshell strength and overall quality.

Finally, yolk color, an important factor for consumer preference, was more intense in the experimental groups, particularly in the M-2 group, which showed the highest La Roche points. This finding is in line with studies that have demonstrated the ability of dietary supplements like sea buckthorn to enhance yolk color consistency, making eggs more appealing to consumers [51–53].

The supplementation of sea buckthorn powder in the diet of hens led to improved productive and qualitative characteristics, including increased egg production, better egg quality, and enhanced uniformity in egg traits. These results align with previous studies on natural feed additives, supporting the potential of sea buckthorn as a valuable supplement for poultry production. Further studies could explore the underlying mechanisms of these effects and their long-term impact on poultry health and productivity.

Future research should address the limitations of this study by extending the trial duration, increasing the sample size, and controlling for external variables to better assess the long-term impact of sea buckthorn supplementation on egg production and quality.

## 5. Conclusions

This study evaluated the impact of sea buckthorn powder supplementation (0%, 1%, and 2%) on the performance, egg quality, and feed consumption of laying hens from 30 to 40 weeks of age. Results showed that hens receiving 2% sea buckthorn powder had the highest egg production (69.65 eggs/hen) and improved egg weight, specific gravity, and shell thickness compared to the control group. The shape index and egg volume were also enhanced, indicating better overall quality. Feed consumption varied slightly but was not significantly affected by the supplementation. Mortality rates were low, demonstrating the robustness of the hens and the suitability of the feeding conditions.

Based on the results, it is recommended that poultry producers incorporate 2% ecological sea buckthorn powder into the diets of laying hens. This supplementation can improve egg production, quality, and overall performance, while also supporting more sustainable and health-conscious poultry farming practices.

**Author Contributions:** Conceptualization, A.U., M.A.D., M.-G.U. and F.B.; methodology, C.S. and M.-G.D.; software, R.-M.R.-R. and R.-N.R.; validation, A.U., M.A.D., C.P. and C.-E.N.; formal analysis, A.U., R.-N.R., F.B. and M.-G.U.; investigation, A.U., M.A.D., and F.B.; resources, R.-M.R.-R., R.-N.R. and C.S.; data curation, C.-E.N. and M.-G.D.; writing—original draft preparation, A.U., C.P. and M.-G.U.; writing—review and editing, A.U., R.-M.R.-R. and C.S.; visualization, M.-G.U.; supervision, A.U. All authors have read and agreed to the published version of the manuscript.

**Funding:** This research received no external funding.

**Institutional Review Board Statement:** This study did not involve experiments that would harm the welfare of the birds in the experimental groups, and approval from an ethics committee was not required.

**Data Availability Statement:** The data present in this study are available on request from the first author and corresponding author.

**Conflicts of Interest:** The authors declare no conflicts of interest.

## References

1. Ozenturk, U.; Yildiz, A. Assessment of Egg Quality in Native and Foreign Laying Hybrids Reared in Different Cage Densities. *Braz. J. Poult. Sci.* **2021**, *22*, eRBCA-2020. [CrossRef]
2. Sparks, N.H.C. The hen's egg—Is its role in human nutrition changing. *World's Poult. Sci. J.* **2005**, *62*, 308–315. [CrossRef]
3. Lima, H.J.D.; Souza, L.A.Z. Vitamin A in the diet of laying hens: Enrichment of table eggs to prevent nutritional deficiencies in humans. *World's Poult. Sci. J.* **2018**, *74*, 619–626. [CrossRef]
4. da Silva, W.C.; Araújo, L.N.; da Silva, É.B.R.; de Sousa, E.D.V.; da Gato, A.P.C.; da Silva, J.A.R. Systematic review and scientometrics of commercial eggs production in Brazil. *Res. Soc. Dev.* **2020**, *9*, e1399108459. [CrossRef]
5. Ren, Y.; Wu, J.; Renema, R. Nutritional and Health Attributes of Eggs. In *Handbook of Poultry Science and Technology*; Wiley Online Library: New York, NY, USA, 2010; Volume 1, pp. 533–578. [CrossRef]
6. Usturoi, A.I.; Usturoi, M.G.; Avarvarei, B.V.; Pânzaru, C.; Simeanu, C.; Usturoi, M.I.; Spătaru, M.; Radu-Rusu, R.M.; Doliş, M.G.; Simeanu, D. Research Regarding Correlation between the Assured Health State for Laying Hens and Their Productivity. *Agriculture* **2022**, *3*, 86. [CrossRef]
7. Sosnowka-Czajka, E.; Herbut, E.; Skomorucha, I.; Muchacka, R. Welfare levels in heritage breed vs. Commercial laying hens in the litter system. *Ann. Anim. Sci.* **2011**, *11*, 585–595. [CrossRef]
8. Craig, J.V.; Swanson, J.C. Welfare perspectives on hens kept for egg-production. *Poult. Sci.* **1994**, *73*, 921–938. [CrossRef]
9. Savory, C.J.; Jack, M.C.; Sandilands, V. Behavioural responses to different floor space allowances in small groups of laying hens. *Br. Poult. Sci.* **2006**, *47*, 120–124. [CrossRef] [PubMed]
10. Rodriguez-Aurrekoetxea, A.; Estevez, I. Use of space and its impact on the welfare of laying hens in a commercial free-range system. *Poult Sci.* **2016**, *95*, 2503–2513. [CrossRef]
11. De Haas, E.N.; Kemp, B.; Bolhuis, J.E.; Groothuis, T.; Rodenburg, T.B. Fear, stress, and feather pecking in commercial white and brown laying hen parent-stock flocks and their relationships with production parameters. *Poult. Sci.* **2013**, *92*, 2259–2269. [CrossRef] [PubMed]
12. Dedousi, A.; Kritsa, M.Z.; Stojcic, M.D.; Sfetsas, T.; Sentas, A.; Sossidou, E. Production Performance, Egg Quality Characteristics, Fatty Acid Profile and Health Lipid Indices of Produced Eggs, Blood Biochemical Parameters and Welfare Indicators of Laying Hens Fed Dried Olive Pulp. *Sustainability* **2022**, *14*, 3157. [CrossRef]
13. Corona, J.; Trompiz, J.; Jerez, N.; Gomez, A.; Rincon, H. Effect of warehouse type on productive variables and egg quality in laying hens Isa Brown. *Rev. De La Fac. De Agron. De La Univ. Del Zulia* **2016**, *32*, 345–360.
14. De Oliveira Pereira, D.C.; Da Silva Miranda, K.O.; Dematte Filho, L.C.; Pereira, G.V.; De Stefano Piedade, S.M.; Berno, P.R. Presence of roosters in an alternative egg production system aiming at animal welfare. *Rev. Bras. De Zootec. -Braz. J. Anim. Sci.* **2017**, *46*, 175–184. [CrossRef]
15. Guinebretiere, M.; Mika, A.; Michel, V.; Balaine, L.; Thomas, R.; Keita, A.; Pol, F. Effects of Management Strategies on Non-Beak-Trimmed Laying Hens in Furnished Cages that Were Reared in a Non-Cage System. *Animals* **2020**, *10*, 399. [CrossRef] [PubMed]
16. Zhang, C.Y.; Kang, X.T.; Zhang, T.Y.; Huang, J. Positive Effects of Resveratrol on Egg-Laying Ability, Egg Quality, and Antioxidant Activity in Hens. *J. Appl. Poult. Res.* **2019**, *28*, 1099–1105. [CrossRef]
17. Roberts, J. Welfare standards for laying hens Achieving sustainable production of eggs Volume 2. *Anim. Welf. Sustain.* **2017**, *17*, 85–97.

18. Kjaer, J.B.; Sorensen, P. Feather pecking and cannibalism in free-range laying hens as affected by genotype, dietary level of methionine plus cystine, light intensity during rearing and age at first access to the range area. *Appl. Anim. Behav. Sci.* **2002**, *76*, 21–39. [CrossRef]
19. Wan, Y.I.; Ma, R.; Li, Y.; Liu, W.; Li, J.Y.; Zhan, K. Effect of a Large-sized Cage with a Low Metabolizable Energy and Low Crude Protein Diet on Growth Performance, Feed Cost, and Blood Parameters of Growing Layers. *J. Poult. Sci.* **2021**, *58*, 70–77. [CrossRef] [PubMed]
20. Torki, M.; Nasiroleslami, M.; Ghasemi, H.A. The effects of different protein levels in laying hens under hot summer conditions. *Anim. Prod. Sci.* **2017**, *57*, 927–934. [CrossRef]
21. Viana, E.F.; De Carvalho Mello, H.H.; Carvalho, F.B.; Cafe, M.B.; Mogyca Leandro, N.S.; Arnhold, E.; Stringhini, J.H. Blood biochemical parameters and organ development of brown layers fed reduced dietary protein levels in two rearing systems. *Anim. Biosci.* **2022**, *35*, 444–452. [CrossRef]
22. Mens, A.J.W.; van Krimpen, M.M.; Kwakkel, R.P. Nutritional approaches to reduce or prevent feather pecking in laying hens: Any potential to intervene during rearing. *World's Poult. Sci. J.* **2020**, *76*, 591–610. [CrossRef]
23. Rozempolska-Rucinska, I.; Czech, A.; Kasperek, K.; Zieba, G.; Ziemianska, A. Behaviour and stress in three breeds of laying hens kept in the same environment. *S. Afr. J. Anim. Sci.* **2020**, *50*, 272–280. [CrossRef]
24. Marcq, C.; Marlier, D.; Beckers, Y. Improving adjuvant systems for polyclonal egg yolk antibody (IgY) production in laying hens in terms of productivity and animal welfare. *Vet. Immunol. Immunopathol.* **2015**, *165*, 54–63. [CrossRef]
25. Ma, J.S.; Chang, W.H.; Liu, G.H.; Zhang, S.; Zheng, A.J.; Li, Y.; Xie, Q.; Liu, Z.Y.; Cai, H.Y. Effects of flavones of sea buckthorn fruits on growth performance, carcass quality, fat deposition and lipometabolism for broilers. *Poult. Sci.* **2015**, *94*, 2641–2649. [CrossRef]
26. Usturoi, M.G.; Radu-Rusu, R.M.; Usturoi, A.l.; Simeanu, C.; Dolis, M.G.; Rațu, R.N.; Simeanu, D. Impact of Different Levels of Crude Protein on Production Performance and Meat Quality in Broiler Selected for Slow Growth. *Agriculture* **2023**, *13*, 427. [CrossRef]
27. Chen, Y.; Cai, Y.F.; Wang, K.; Wang, Y.S. Bioactive Compounds in Sea Buckthorn and their Efficacy in Preventing and Treating Metabolic Syndrome. *Foods* **2023**, *12*, 1985. [CrossRef] [PubMed]
28. Abdel-Wareth, A.A.A.; Lohakare, J. Moringa oleifera Leaves as Eco-Friendly Feed Additive in Diets of Hy-Line Brown Hens during the Late Laying Period. *Animals* **2021**, *11*, 1116. [CrossRef]
29. Song, Q.L.; Zou, Z.H.; Chen, X.L.; Ai, G.X.; Xiong, P.W.; Song, W.J.; Liu, G.H.; Zheng, A.J.; Chen, J. Effect of Moringa oleifera Leaf Powder Supplementation on Growth Performance, Digestive Enzyme Activity, Meat Quality, and Cecum Microbiota of Ningdu Yellow Chickens. *Agric. Basel* **2024**, *14*, 1523. [CrossRef]
30. Vlaicu, P.A.; Untea, A.E.; Oancea, A.G. Sustainable Poultry Feeding Strategies for Achieving Zero Hunger and Enhancing Food Quality. *Agriculture* **2024**, *14*, 1811. [CrossRef]
31. Ma, Y.; Yao, J.X.; Zhou, L.; Zhao, M.J.; Wang, W.; Liu, J.K.; Marchioni, E. Comprehensive untargeted lipidomic analysis of sea buckthorn using UHPLC-HR-AM/MS/MS combined with principal component analysis. *Food Chem.* **2023**, *430*, 136964. [CrossRef]
32. Rațu, R.N.; Usturoi, M.G.; Simeanu, D.; Simeanu, C.; Usturoi, A.l.; Doliş, M.G. Research regarding dynamics of chemical content from pasteurized egg melange stored in polyethylene type packing. *Rev. Mater. Plast.* **2017**, *54*, 368–374. [CrossRef]
33. Usturoi, A.l.; Simeanu, C.; Usturoi, M.G.; Doliş, M.G.; Rațu, R.N.; Simeanu, D. Influence of packaging type on the dynamics of powdered eggs chemical composition. *Rev. Mater. Plast.* **2017**, *54*, 380–385. [CrossRef]
34. Davidescu, M.A.; Panzaru, C.; Usturoi, A.l.; Radu-Rusu, R.M.; Creanga, S.t. An Appropriate Genetic Approach to Endangered Podolian Grey Cattle in the Context of Preserving Biodiversity and Sustainable Conservation of Genetic Resources. *Agriculture* **2023**, *13*, 2255. [CrossRef]
35. Montalbán, A.; Madrid, J.; Hernández, F.; Schiavone, A.; Ruiz, E.; Sánchez, C.J.; Ayala, L.; Fiorilla, E.; Martínez-Miró, S. The Influence of Alternative Diets and Whole Dry Black Soldier Fly Larvae (Hermetia illucens) on the Production Performance, Blood Status, and Egg Quality of Laying Hens. *Animals* **2024**, *14*, 2550. [CrossRef]
36. Anene, D.O.; Akter, Y.; Thomson, P.C.; Groves, P.; O'Shea, C.J. Variation and Association of Hen Performance and Egg Quality Traits in Individual Early-Laying ISA Brown Hens. *Animals* **2020**, *10*, 1601. [CrossRef]
37. Salahuddin, M.; Abdel-Wareth, A.A.A.; Stamps, K.G.; Gray, C.D.; Aviña, A.M.W.; Fulzele, S.; Lohakare, J. Enhancing Laying Hens' Performance, Egg Quality, Shelf Life during Storage, and Blood Biochemistry with Spirulina platensis Supplementation. *Vet. Sci.* **2024**, *11*, 383. [CrossRef]
38. Orzuna-Orzuna, J.F.; Lara-Bueno, A. Essential Oils as a Dietary Additive for Laying Hens: Performance, Egg Quality, Antioxidant Status, and Intestinal Morphology: A Meta-Analysis. *Agriculture* **2023**, *13*, 1294. [CrossRef]
39. Panaite, T.D.; Vlaicu, P.A.; Saracila, M.; Cismileanu, A.; Varzaru, I.; Voicu, S.N.; Hermenean, A. Impact of Watermelon Rind and Sea Buckthorn Meal on Performance, Blood Parameters, and Gut Microbiota and Morphology in Laying Hens. *Agriculture* **2022**, *12*, 177. [CrossRef]

40. Shaker, M.M.; Al-Beitawi, N.A.; Bláha, J.; Mahmoud, Z. The effect of sea buckthorn (*Hippophae rhamnoides* L.) fruit residues on performance and egg quality of laying hens. *J. Appl. Anim. Res.* **2018**, *46*, 422–426. [CrossRef]
41. Chand, N.; Naz, S.; Irfan, M.; Khan, R.U.; Rehman, Z.U. Effect of Sea Buckthorn (*Hippophae rhamnoides* L.) Seed Supplementation on Egg Quality and Cholesterol of Rhode Island RedxFayoumi Laying Hens. *Korean J. Food Sci. Anim. Resour.* **2018**, *38*, 468–475. [PubMed]
42. Untea, A.E.; Panaite, T.D.; Varzaru, I.; Turcu, R.P.; Gavris, T.; Lupu, A. Study on the influence of dietary sea buckthorn meal on nutritional properties of laying hen eggs. *Czech J. Anim. Sci.* **2021**, *66*, 225–234. [CrossRef]
43. Nour, V.; Panaite, T.D.; Corbu, A.R.; Ropota, M.; Turcu, R.P. Nutritional and bioactive compounds in dried sea-buckthorn pomace. *Erwerbs-Obstbau* **2021**, *63*, 91–98. [CrossRef]
44. Yao, B.N.; Liao, F.Y.; Yang, J.Y.; Liu, A.; Wang, J.; Zhu, B.G.; Yang, S.L. Effect of sea buckthorn extract on production performance, serum biochemical indexes, egg quality, and cholesterol deposition of laying ducks. *Front. Vet. Sci.* **2023**, *10*, 112–117. [CrossRef]
45. Saracila, M.; Panaite, T.D.; Untea, A.; Varzaru, I.; Dragotoiu, D.; Criste, R.D. Use of the dietary sea buckthorn meal as phytoaditive in heat-stressed broiler. *Sci. Pap. Ser. D Anim. Sci.* **2020**, *63*, 83–91.
46. Kang, H.K.; Kim, J.H.; Kim, C.H. Effect of dietary supplementation of Lactobacillus-fermented and non-fermented sea buckthorn on the laying performance and meat lipid peroxidation of Hy-line Brown laying hens. *Eur. Poul. Sci.* **2015**, *79*, 1–10.
47. BenMahmoud, Z.T.; Sherif, B.M.; Elfituri, A.M. Effect of partial replacing of wheat by sea buckthorn (Hippophae rhamnoides L.) fruit residues in broiler diets on performance and skin pigmentation. *Open Vet. J.* **2022**, *11*, 780–788.
48. He, Q.; Yang, K.L.; Wu, X.Y.; Zhang, C.H.; He, C.N.; Xiao, P.G. Phenolic compounds, antioxidant activity and sensory evaluation of sea buckthorn (*Hippophae rhamnoides* L.) leaf tea. *Food Sci. Nutr.* **2023**, *11*, 1212–1222. [CrossRef] [PubMed]
49. Biswas, A.; Bhart, V.K.; Acharya, S.; Pawar, D.D.; Singh, S.B. Sea buckthorn: New feed opportunity for poultry in cold arid Ladakh region of India. *World's Poult Sci J.* **2010**, *66*, 707–714. [CrossRef]
50. Lokaewmanee, K.; Yamauchi, K.; Okuda, N. Effects of dietary red pepper on egg yolk colour and histological intestinal morphology in laying hens. *J. Anim. Physiol. Anim. Nutr.* **2013**, *97*, 986–995. [CrossRef] [PubMed]
51. Dvorak, P.; Suchy, P.; Straková, E.; Dolezalová, J. The effect of a diet supplemented with sea-buckthorn pomace on the colour and viscosity of the egg yolk. *Acta Vet. BRNO* **2017**, *86*, 303–308. [CrossRef]
52. Vilas-Franquesa, A.; Saldo, J.; Juan, B. Potential of sea buckthorn-based ingredients for the food and feed industry—A review. *Food Prod. Process. Nutr.* **2020**, *2*, 17. [CrossRef]
53. Beardsworth, P.M.; Hernandez, J.M. Yolk colour—An important egg quality attribute. *Int. Poult. Prod.* **2004**, *12*, 17–18.

*agriculture*

MDPI

*Article*

# Mycotoxicological Assessment of Broiler Compound Feed: A Multi-Year Analysis of Five Mycotoxins in a Romanian Feed Mill

**Dragoș Mihai Lăpușneanu, Silvia-Ioana Petrescu *, Cristina-Gabriela Radu-Rusu, Mădălina Matei and Ioan Mircea Pop**

Department of Control, Expertise and Services, Faculty of Food and Animal Sciences, "Ion Ionescu de la Brad" Iasi University of Life Sciences, 8 Mihail Sadoveanu Alley, 700489 Iasi, Romania; dragos.lapusneanu@iuls.ro (D.M.L.); cristina.radurusu@iuls.ro (C.-G.R.-R.); madalina.matei@iuls.ro (M.M.); mircea.pop@iuls.ro (I.M.P.)

* Correspondence: silvia.petrescu@iuls.ro

**Abstract:** Mycotoxins are secondary metabolites of filamentous fungi that cause massive agricultural losses worldwide and constitute a significant health problem for humans and animals. The aim of this five-year study was to investigate the contamination of compound feed for broiler chickens at all stages (starter, grower and finisher) from a feed mill in Romania with mycotoxins such as total aflatoxins (AFT), deoxynivalenol (DON), fumonisins (FUMs), ochratoxin A (OTA) and zearalenone (ZEN). AFT was detected in 49.3–72.2% of the samples with concentrations ranging from 0.01 to 5.2 μg/kg. DON was detected in 77.6–98.9% of the samples, with maximum concentrations ranging from 330 to 1740 μg/kg. FUM contamination ranged from 42.7% to 87.2%, with maximum levels between 460 and 1400 μg/kg. OTA was present in 44.2–87.9% of the samples, with maximum concentrations reaching 21.4 μg/kg. ZEN was consistently elevated at all feeding stages, being detected in 86.5–97.4% of the samples, with maximum levels of 89.4 μg/kg. Mycotoxin co-occurrence was common in the samples, with the most common combination of four mycotoxins occurring in 38.51% of the samples. Samples were collected from storage bunkers, homogenised and analysed in certified laboratories, with sampling procedures varying according to batch size to ensure representativeness. Investigation of the transfer of mycotoxins into animal products and the combined effects of mycotoxins on animal health, including potential synergistic or antagonistic interactions, is particularly relevant. This study emphasises the essential role of comprehensive and continuous monitoring of mycotoxins in protecting animal health and food safety.

**Keywords:** mycotoxins; compound feed; broiler; co-occurrence; food safety

## 1. Introduction

Food and feed safety is a major concern for both animal and human health due to the frequent contamination of food and feed with various contaminants [1,2].

The word "mycotoxin" is derived from the Greek words "myco" and "toxin", meaning "mould" and "poison" produced by a living organism [3]. Mycotoxins are a group of toxic chemical compounds produced as secondary metabolites by certain mould species, primarily within the genera *Aspergillus, Fusarium* and *Penicillium* [4–6].

Based on the literature data, the overall prevalence of mycotoxins in food crops varies widely depending on many factors such as the mycotoxin in question, the analytical


https://doi.org/10.3390/agriculture15010084

methods used and the reporting of results, but the prevalence for detected mycotoxins is reported to be up to 60–80% [7], and this is considered an unavoidable and unpredictable problem that poses a challenge to food safety [8].

Contaminated feed and food products pose high risks to animal health and human metabolic conditions, which can range from acute symptoms of severe disease to long-term effects [9–11]. Because of incidents of mycotoxin poisoning [12,13], most countries or regions have regulatory levels for the presence of mycotoxins in certain food staples or feeds; therefore, testing for those specific regulated mycotoxins is required [14], using specific and selective analytical techniques adapted to verify food safety and protect public health [3]. The maximum levels or guideline levels for mycotoxins in products intended for animal feed in the European Union are highlighted in Table 1, as they have been found in the legislative support mentioned [15,16].

**Table 1.** European Union mycotoxin limits or guidance levels in animal feed.

| Mycotoxin | Products Intended for Animal Feed | Maximum Content/Guidance Value Relative to a Feed with a Moisture Content of 12% (μg/kg) | Legislative Support |
|---|---|---|---|
| Aflatoxin B1 | Feed materials | 20 | Reg. (EU) No 574/2011 [15] |
| | Compound feed for young poultry | 5 | |
| | Compound feed for poultry | 20 | |
| Deoxynivalenol | Cereals and cereal products | 8000 | Reg. (EU) 576/2006 [16] |
| | Maize by-products | 12,000 | |
| | Compound feed for poultry | NR | |
| Fumonisin B1 + B2 | Maize and maize products | 60,000 | |
| | Compound feed for poultry | 20,000 | |
| Ochratoxin A | Cereals and cereal products | 2500 | |
| | Compound feed for poultry | 100 | |
| Zearalenone | Cereals and cereal products | 2000 | |
| | Maize by-products | 3000 | |
| | Compound feed for poultry | NR | |

NR = no/without regulation.

Aflatoxins (AFTs) are a class of carcinogenic mycotoxins produced by Aspergillus species, especially *Aspergillus flavus* and *A. parasiticus* [17–19]. When grains such as maize are grown in an environment with high ambient temperatures (day > 32 °C; night > 24 °C), the grains become more susceptible to aflatoxin formation. Maize grains can contain up to 400,000 μg/kg of aflatoxin, so sampling is very important when analysing contamination levels [5]. All primary transformations of aflatoxin $B_1$ involve conversion to hydroxyl metabolites, the most important resulting toxin in terms of toxicity being aflatoxin $M_1$. Aflatoxin $B_1$ is immunosuppressive in animals, with particularly strong effects on cell-mediated immunity. According to expert studies, aflatoxin $B_1$ is genotoxic, inducing genetic mutations and chromosomal changes [19].

Based on what we know so far, the presence of AFTs in feed leads to suppression of the immune response in birds, onset of oxidative stress and disruption of liver enzyme activity [20]. Furthermore, a recent study highlights the effects of long-term exposure to $AFB_1$, which may lead to decreased bone density in broiler chickens not only as a result of impaired vitamin D or calcium and phosphorus absorption, but the mycotoxin itself at levels of 230 μg/kg causes decreased bone mass in poultry [21].

Deoxynivalenol (DON), also known as vomitoxin, is produced by *Fusarium geaminearum* and, in certain geographical areas, by *F. culmorum* [22]. The main crops affected are maize and small grains such as wheat, oats and barley. In maize, "stem and ear rot" caused by *F. graminearum* may appear as purple or pink kernels with visible pink mould growths on affected areas of the cob. Storage under optimum conditions (<14% humidity) will minimise further toxin production by pathogenic fungi [5]. Contamination of feed with DON even in low concentrations, below 1900 μg/kg, can lead to severe intestinal pathologies according to current studies, affecting not only the morphostructural activity of the intestinal villi of broilers but also possibly leading to decreased response of digestive enzymes [23].

Fumonisins (FUMs) include a group of relatively recently discovered mycotoxins (mainly fumonisins $B_1$, $B_2$ and $B_3$), primarily produced by *F. verticillioides* and *F. proliferatum*, with maize being the main commodity affected [24]. Grains damaged by insects, birds or cracked kernels will often contain the highest levels of toxin and cause serious disease in animals [25]. Worldwide reports have documented ppm levels of fumonisin $B_1$ contamination. Human exposure occurs at levels ranging from micrograms to milligrams per day and is highest in regions where maize products are a staple. Based on toxicological evidence, the IARC (International Agency for Research on Cancer) has classified fumonisin $B_1$ as possibly carcinogenic to humans (group 2B) [26].

Ochratoxin A (OTA) is a naturally occurring fluorescent compound, and its detection during analysis typically relies on this property [5]. Following aflatoxins, ochratoxin A represents the most significant mycotoxin in terms of its impact. OTA is produced by members of the genera *Aspergillus* (*A. ochraceus, A. carbonarius*) and *Penicillium* (*P. verrucosum*). It has been observed that contamination with OTA is a global phenomenon, as evidenced by studies [27,28]. The initial fungal growth in cereals can result in sufficient moisture through metabolism to allow further growth and mycotoxin formation. Consequently, the toxin may still be present in cereal products, thereby exposing human and animal populations to contamination [5].

The IARC has classified ochratoxin A as Group 2B, possibly carcinogenic to humans, based on certain indicators of carcinogenicity established in experimental animals [29].

Zearalenone (ZEN) is a mycotoxin produced by several *Fusarium* species, in particular *F. graminearum, F. culmorum, F. cerealis, F. equiseti, F. verticillioides* and *F. incarnatum* [30], which can cause several diseases in animals [31]. It is commonly found in maize but can also be found in other crops such as wheat, barley, sorghum and rye. In general, *Fusarium* specifically thrives and contaminates crops under wet and cold weather conditions. Although ZEN is primarily a contaminant of field crops, development of the toxin can also occur under inadequate storage conditions [32]. In the 2000s, the European Union's food safety policy was reformulated, in accordance with the approach of an integrative concept "from farm to fork", thus guaranteeing a high level of safety for food products in all stages of the production chain [33]; even feed mills, like food units, must have auto-control programs for contaminants [34,35].

Raw materials, such as cereals, oilseeds, legumes and, in particular, compound feed as complex matrices, are susceptible to contamination with bacterial or fungal mycotoxins [36]. For example, a study conducted in Poland shows that compound feeds for broilers are characterised by higher contamination with mycotoxins mainly belonging to the *trichothecenes* group; grower and finisher feeds are characterised by higher numbers of bacteria and fungi compared to starter feeds [37].

Considering the inclusion of compound feed production in the food chain, the current research highlights the presence of five mycotoxins such as AFT, DON, FUM, OTA and ZEN in samples taken over 5 years from a feed mill in the north of Romania. The results obtained were compared with those reported by other national and international researchers, and also the presence of more than one mycotoxin in the compound feed taken into study was identified and discussed. Although environmental factors (such as drought or heavy rainfall) and climate change have affected the geographical area indicated in this study, the mycotoxin values found in the samples did not exceed the legislative limits proposed by the European Commission. In addition to identifying potential new mycotoxins, future studies will examine how they affect animal health and how they co-occur in animal feed.

## 2. Materials and Methods

### 2.1. Feed Samples

Compound feeds are especially susceptible to contamination with multiple mycotoxins because they are a blend of multiple raw materials [10,38].

The broiler compound feed samples for starter, grower and finisher stages came from a Romanian feed mill that produces 85,000 t/year, which is representative of the country's feed production. In 2019, 284 samples of compound feed were analysed (92 starter, 79 grower and 113 finisher), in 2020, 241 samples were analysed (91 starter, 87 grower and 143 finisher), in 2021, 306 samples were analysed (98 starter, 82 grower and 126 finisher), in 2022, 333 samples were analysed (102 starter, 97 grower and 134 finisher) and in 2023, 350 samples were analysed (95 starter, 103 grower and 152 finisher). Samples of combined feeds were taken from the feed mill's storage bunkers and sent to the in-plant laboratory for analysis.

For the analysis of the compound feed, 4, 7, 11 and 14 incremental samples were manually sampled with a trowel from batches of 24, 48, 72 and 96 tonnes. The lot size dictated how many elementary samples were sampled overall, which were then separated and homogenised using a centrifugal mechanical divider to create the laboratory sample. The incremental sample size was a minimum of 3 kg, and the aggregate sample was made by reducing the incremental sample to a minimum of 0.5 kg. The feed samples were analysed as such. The results obtained were interpreted on a dry matter basis of 88% (12% moisture) in order to be compared with the maximum permissible limits established in the European Union legislation, presented in Table 2.

**Table 2.** Detection and quantification limits for five mycotoxins.

| Mycotoxins | Limits of Detection (LoD) (μg/kg) | Limits of Quantification (LoQ) (μg/kg) |
|---|---|---|
| AFT | 0.5 | 0.5 |
| DON | 50 | 50 |
| FUM | 50 | 50 |
| OTA | 0.5 | 0.5 |
| ZEN | 10 | 20 |

### 2.2. Equipment Used for Detection

The mycotoxin determination kits contained the following: microtiter plate spectrophotometer (450 nm); graduated cylinder (glass), 100 mL and 250 mL; glassware for preparing sample extract: filter funnel and 50 mL flask; 20 μL, 200 μL and 1000 μL micropipettes; 50 μL, 100 μL and 1000 μL micropipettes; filter paper: Whatman No. 1; scale (measurement range at least up to 50 g and precision of ±0.01 g); centrifuge (at least

3500× g) + centrifugal vials with cap (50 mL centrifuge tubes); vortex mixer 8-channel pipette for 50, 100 and 300 μL; grinder (mill); shaker; Ultra-Turrax.

### 2.3. Integrated Management System and Personnel Training

In the compound feed mill taken into study, the prevention of nonconformities in all phases under the control of the organisation is achieved by maintaining and continuously improving the effectiveness of an integrated management system in accordance with the requirements of the referenced standards: SR EN ISO 9001:2015, SR EN ISO 22000:2019 and SR ISO 45001:2018 [39–41]. In concordance with the above listed standards, the establishment is obliged to develop and comply with specific procedures on food quality and safety. In this regard, all team members who used the RIDASCREEN-FAST kit were trained by means of documented procedures (procedure on sampling techniques for laboratory examinations; procedure on quantitative determination of mycotoxins). The training included both theoretical aspects (technical principles of the kit and mycotoxins to be analysed) and practical aspects (correct use of the kit, handling of laboratory equipment, compliance with safety procedures). The training was an ongoing process, which included recap sessions and periodic staff performance evaluations, with the aim of ensuring that staff remained up to date with protocol updates.

### 2.4. Mycotoxins Analysis

The quantitative determination of mycotoxins was carried out according to the analytical method described in the RIDASCREEN®FAST enzyme immunoassay technical manuals provided by R-Biopharm AG, Darmstadt, Germany. The contamination levels of AFT are the sum of $AFB_1$, $AFB_2$, $AFG_1$ and $AFG_2$ and of DON and FUM are the sum of $FB_1$ and $FB_2$, and OTA and ZEN in the samples were measured using individual RIDASCREEN®FAST laboratory kits.

The extraction methods for mycotoxin determination are different depending on the specific toxin analysed. Total aflatoxins and fumonisins are extracted using 70% methanol, followed by mixing 5.0 g of ground sample with 25.0 mL of methanol. The extract is mixed thoroughly for three minutes and then filtered through filter paper. The resulting filtrate is then diluted with distilled water, containing 1.3 mL of fumonisin and 1 mL of aflatoxin. However, the extraction solvents are different for deoxynivalenol, ochratoxin A and zearalenone. DON is extracted by shaking 5.0 g of sample with 100 mL of distilled water, whereas ochratoxin A requires 50.0 mL of diluted ECO extractor. ZEN also uses 70% methanol, in the same way as AFT, but with a different dilution ratio.

In the case of incubation and washing, the test for AFT requires a 10 min incubation with the enzyme–antibody mixture at room temperature (20–25 °C), whereas DON, FUM and ZEN normally require a 5 min incubation. OTA requires a more complex extraction, with a 5 min mixing step followed by centrifugation, which differentiates it from other mycotoxins in terms of sample preparation.

Washing the wells is performed with 250 μL of buffer solution and by repeating the process twice, and it is similar for all mycotoxins determined. After washing, each well is treated with 100 μL of substrate/chromogen and incubated for varying periods: 3 min for DON, FUM and OTA and 5 min for AFT and ZEN, all at room temperature in the dark. After this incubation, 100 μL of stop solution is added to each well, and the absorbance is measured at 450 nm. For most mycotoxins, absorbance is normally read 10 min after the addition of the stop solution. In the case of OTA, the reading time is extended slightly, again requiring up to 15 min.

### 2.5. Statistical Analysis

The data obtained from the analyses were statistically processed and interpreted. The minimum and maximum values were determined, and the position and variance estimators, arithmetic mean ($\overline{x}$) and standard deviation (s) were calculated for the samples with positive results. Means and standard deviations were calculated using Microsoft Excel 2016 [42]. The statistical analysis of variance (ANOVA) was conducted using the GraphPad Prism (9.3.0) program to compare the levels of each mycotoxin across different feed types and the annual averages.

## 3. Results

The concentrations of AFT, DON, FUM, OTA and ZEN of the starter compound feed are presented in Table 3.

**Table 3.** Results of mycotoxicological assessment of starter compound feed.

| Mycotoxin | Year | No. of Samples | Positive Samples (μg/kg) | | | | | |
|---|---|---|---|---|---|---|---|---|
| | | | % | 1st Quartile | $\overline{x}$ | s | 3rd Quartile | Maximum |
| AFT | 2019 | 92 | 70.6 | 0.9 | 1.6 | 0.8 | 2 | 4.6 |
| | 2020 | 91 | 57.1 | 0.3 | 0.7 | 0.8 | 0.95 | 4.8 |
| | 2021 | 98 | 63.4 | 0.4 | 1.2 | 0.7 | 1 | 4.2 |
| | 2022 | 102 | 72.2 | 0.7 | 0.8 | 0.9 | 1.5 | 4.5 |
| | 2023 | 95 | 58.5 | 0.3 | 0.7 | 0.6 | 0.7 | 3.8 |
| DON | 2019 | 92 | 89.1 | 74 | 99.2 | 67.6 | 100 | 470 |
| | 2020 | 91 | 98.9 | 60 | 155.1 | 200.1 | 140 | 1120 |
| | 2021 | 98 | 92.3 | 57 | 136.3 | 85.7 | 230 | 670 |
| | 2022 | 102 | 97.4 | 50 | 98.4 | 128.2 | 120 | 910 |
| | 2023 | 95 | 79.6 | 47 | 87.2 | 78.1 | 100 | 730 |
| FUM | 2019 | 92 | 77.1 | 30 | 172.3 | 194.8 | 222 | 1230 |
| | 2020 | 91 | 51.6 | 10 | 65.3 | 97.5 | 65 | 460 |
| | 2021 | 98 | 65.2 | 50 | 82.4 | 113.8 | 270 | 820 |
| | 2022 | 102 | 87.2 | 65 | 142.7 | 189.6 | 300 | 1170 |
| | 2023 | 95 | 55.8 | 20 | 75.6 | 96.2 | 120 | 510 |
| OTA | 2019 | 92 | 82.6 | 0.62 | 1 | 0.7 | 1.3 | 3.59 |
| | 2020 | 91 | 61.5 | 0.1 | 0.5 | 1 | 0.5 | 7.4 |
| | 2021 | 98 | 72.3 | 0.1 | 1.2 | 0.5 | 1.7 | 8.2 |
| | 2022 | 102 | 85.3 | 0.5 | 1 | 0.8 | 1.4 | 6.8 |
| | 2023 | 95 | 57.8 | 0.2 | 0.5 | 1.2 | 0.9 | 4.7 |
| ZEN | 2019 | 92 | 94.5 | 11.2 | 19.7 | 10.4 | 25 | 49 |
| | 2020 | 91 | 90.1 | 5.1 | 11.9 | 11.2 | 12.6 | 57.4 |
| | 2021 | 98 | 95.8 | 8.3 | 20.8 | 13.7 | 32 | 45.2 |
| | 2022 | 102 | 86.5 | 12 | 13.9 | 12.4 | 27 | 67.4 |
| | 2023 | 95 | 97.4 | 6.5 | 21.4 | 10.2 | 15 | 52 |

$\overline{x}$—mean; s—standard deviation.

The most evident contamination in the case of AFT, according to the results obtained, occurred in the year 2022, the year in which most feed samples were tested, but the most significant contamination related to the number of samples tested and with a high maxima of 4.6 or even 4.8 was in the years 2019 and 2020, and also in 2020 we observed a massive contamination of samples with the DON mycotoxin, reaching a maximum of 1120 μg/kg. With the same approach of a ratio of positive samples to the number of samples studied, 2019 is the most significant year for FUM and OTA contamination of broiler starter feed.

Table 4 contains the results obtained for the grower feed with regard to the level of contamination with all mycotoxins investigated.

**Table 4.** Results of mycotoxicological assessment of grower compound feed.

| Mycotoxin | Year | No. of Samples | Positive Samples (μg/kg) | | | | | |
|---|---|---|---|---|---|---|---|---|
| | | | % | 1st Quartile | $\bar{x}$ | s | 3rd Quartile | Maximum |
| AFT | 2019 | 79 | 68.3 | 0.7 | 1.4 | 0.7 | 2 | 2.7 |
| | 2020 | 87 | 49.4 | 0.5 | 0.7 | 0.5 | 1.4 | 2.4 |
| | 2021 | 82 | 59.8 | 0.8 | 1.7 | 0.8 | 2.2 | 2.5 |
| | 2022 | 97 | 75.3 | 1 | 1.9 | 0.9 | 3.1 | 3.9 |
| | 2023 | 103 | 52.7 | 0.5 | 0.8 | 0.7 | 2.7 | 4.2 |
| DON | 2019 | 79 | 87.3 | 80 | 95.4 | 56.7 | 100 | 330 |
| | 2020 | 87 | 96.5 | 60 | 178.3 | 250.4 | 152.5 | 1290 |
| | 2021 | 82 | 77.6 | 100 | 98.5 | 327.2 | 370 | 1480 |
| | 2022 | 97 | 86.8 | 120 | 146.8 | 94.7 | 230 | 580 |
| | 2023 | 103 | 94.7 | 90 | 85.8 | 112.9 | 460 | 740 |
| FUM | 2019 | 79 | 69.6 | 110 | 156.5 | 176.2 | 440 | 912 |
| | 2020 | 87 | 43.6 | 65 | 56.3 | 72 | 70.5 | 300 |
| | 2021 | 82 | 51.3 | 90 | 87.6 | 162.7 | 300 | 500 |
| | 2022 | 97 | 86.8 | 60 | 122.8 | 134.6 | 290 | 972 |
| | 2023 | 103 | 42.7 | 50 | 54.2 | 82 | 160 | 420 |
| OTA | 2019 | 79 | 79.7 | 0.65 | 1.2 | 1.1 | 1.63 | 8.3 |
| | 2020 | 87 | 52.8 | 0.55 | 0.7 | 1.1 | 0.6 | 5.6 |
| | 2021 | 82 | 87.9 | 0.9 | 0.5 | 1.7 | 2.7 | 9.2 |
| | 2022 | 97 | 44.2 | 1.4 | 1.4 | 2.2 | 4.6 | 7.4 |
| | 2023 | 103 | 62.4 | 3.8 | 0.9 | 1.1 | 10.2 | 17.2 |
| ZEN | 2019 | 79 | 93.6 | 10.5 | 18.3 | 10.1 | 30 | 39.2 |
| | 2020 | 87 | 90.8 | 10.2 | 11.6 | 8.6 | 15 | 45 |
| | 2021 | 82 | 92.3 | 15.7 | 23.8 | 11.3 | 54.9 | 72.9 |
| | 2022 | 97 | 96.2 | 11.3 | 17.8 | 7.3 | 29.6 | 48 |
| | 2023 | 103 | 91.4 | 12.8 | 19.4 | 9.6 | 27.2 | 54.4 |

$\bar{x}$—mean; s—standard deviation.

Considering the results and applying the same rule of reporting the percentage of positive samples to the number of samples studied for the grower feed, we observe that the levels of AFT, DON and ZEN were the highest in the years 2019 and 2020. FUM and OTA recorded higher contamination levels in the years 2021 and 2022.

The results of the mycotoxicological assessment of the finisher compound feed are listed in Table 5. Analysing the results and applying the same rule of a ratio of the percentage of positive samples to the total number of samples studied for the finisher feed, we note a significant contamination with all mycotoxins in the years 2019 and 2020, with contamination reaching even maximum values of 1510 μg/kg for DON in 2020 or 1080 μg/kg for FUM in 2019.

The five-year investigation of mycotoxin contamination in broiler compound feed has revealed a complex pattern of co-occurrence between several mycotoxins, as can be seen in Figure 1. The most frequent co-occurrence involved four mycotoxins detected in 38.51% of the samples. The data highlight the risk of multiple mycotoxin contamination of broiler feed, with a wide range of co-occurrence combinations observed.

**Table 5.** Results of mycotoxicological assessment of finisher compound feed.

| Mycotoxin | Year | No. of Samples | Positive Samples (μg/kg) | | | | | |
|---|---|---|---|---|---|---|---|---|
| | | | % | 1st Quartile | $\bar{x}$ | s | 3rd Quartile | Maximum |
| AFT | 2019 | 113 | 64.9 | 0.9 | 1.5 | 0.7 | 2 | 4.6 |
| | 2020 | 143 | 51.04 | 0.7 | 0.8 | 0.7 | 1.2 | 3.9 |
| | 2021 | 126 | 69.2 | 1.3 | 1.3 | 0.8 | 3.7 | 4.8 |
| | 2022 | 134 | 49.3 | 1.8 | 0.7 | 0.6 | 4.2 | 5.2 |
| | 2023 | 152 | 53.9 | 1 | 1.9 | 0.7 | 2.2 | 3.7 |
| DON | 2019 | 113 | 91.07 | 70 | 90 | 54.1 | 90 | 321 |
| | 2020 | 143 | 98.6 | 60 | 161.7 | 219.1 | 160 | 1510 |
| | 2021 | 126 | 89.7 | 90 | 96 | 62.4 | 240 | 581 |
| | 2022 | 134 | 93.02 | 130 | 176.3 | 232.3 | 540 | 1740 |
| | 2023 | 152 | 95.4 | 80 | 128.5 | 119.8 | 470 | 842 |
| FUM | 2019 | 113 | 71.4 | 70 | 175.2 | 215.2 | 222 | 1080 |
| | 2020 | 143 | 67.1 | 60 | 95 | 156.5 | 100 | 900 |
| | 2021 | 126 | 61.8 | 90 | 150.4 | 162.3 | 320 | 870 |
| | 2022 | 134 | 78.2 | 130 | 184.7 | 226.3 | 280 | 1400 |
| | 2023 | 152 | 64.9 | 90 | 98 | 165.7 | 130 | 960 |
| OTA | 2019 | 113 | 82.1 | 0.5 | 1.4 | 2.2 | 1.6 | 18.5 |
| | 2020 | 143 | 65.03 | 0.5 | 0.8 | 1.2 | 0.9 | 8.5 |
| | 2021 | 126 | 72.2 | 1.5 | 1.2 | 2 | 4.2 | 9.4 |
| | 2022 | 134 | 63.5 | 0.8 | 0.9 | 1.1 | 2.8 | 13.7 |
| | 2023 | 152 | 84.2 | 1.3 | 1.9 | 2.4 | 3.7 | 21.4 |
| ZEN | 2019 | 113 | 91.9 | 11.6 | 19.3 | 14.3 | 25 | 78.7 |
| | 2020 | 143 | 95.8 | 10.3 | 13.4 | 10.2 | 17.4 | 51 |
| | 2021 | 126 | 89.5 | 12.4 | 17.6 | 15.6 | 19.2 | 87.9 |
| | 2022 | 134 | 93.2 | 19.6 | 23.8 | 13.4 | 35.7 | 89.4 |
| | 2023 | 152 | 94.7 | 11 | 14.2 | 11.4 | 18.9 | 62 |

$\bar{x}$—mean; s—standard deviation.

These compound feeds contain varying proportions of raw materials to provide different nutrient levels, with cereal grains always being the primary component. ANOVA results showed no significant differences ($p < 0.05$) in the levels of each mycotoxin across feed types and annual averages. The high proportion of cereals in all these compound feeds may account for the lack of differences. However, it is anticipated that starter compound feed would have less contamination than finisher compound feed, likely due to the lower amount of maize used in the formulation for young poultry, as maize is the main contributor to mycotoxin contamination.

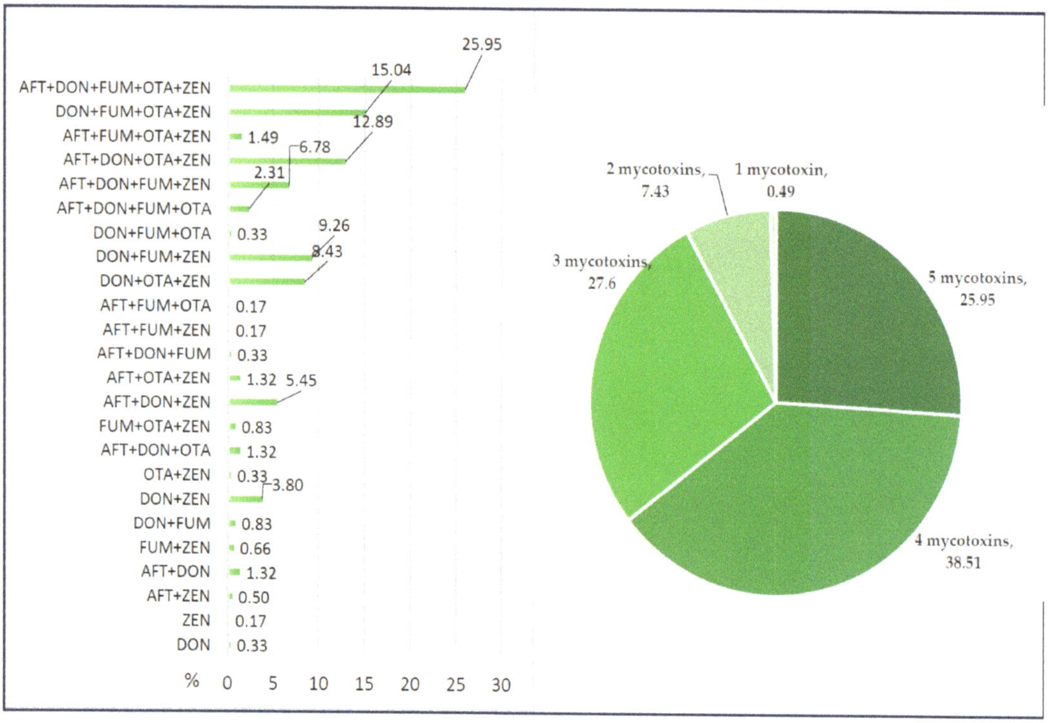


**Figure 1.** Co-occurrence of mycotoxins in broiler feed over the five-year study period.

## 4. Discussion

Broiler compound feeds are balanced feed formulas containing a variety of ingredients to ensure complete and healthy nutrition. Main feedstuffs include: maize and wheat flour—used to provide carbohydrates, protein and fibre; soy and sunflower meal—vegetable protein concentrates; vegetable oils (e.g., sunflower oil)—sources of fats, which are essential for energy and overall health of the chickens; salt—added to balance sodium levels; minerals (calcium, phosphorus, magnesium, etc. )—essential for bone and nervous system development; vitamins (A, D3, E, B)—help support the immune system and prevent deficiencies; plant fibre (e.g., wheat bran)—helps the digestive system function properly. The exact composition of compound feed varies depending on the age of the chick, stage of growth and production goals. Typically, broiler feeds are formulated to ensure fast and efficient growth while maintaining poultry health.

Mycotoxins in feed are a significant problem for animal and human safety, and recent studies have shown that food and feed contamination is the rule rather than the exception, impacting all segments of society, from farmers and feed producers to the general public [3,43–45]. Food and feed can be vectors for harmful bacteria, viruses or chemicals that are responsible for a wide range of human and animal health diseases [46,47]. Mycotoxins are secondary metabolites of filamentous fungi that cause massive losses to agriculture worldwide. AFT, OTA, DON, FUM, ZEN and trichothecenes are currently the most commonly tested in the food and feed safety industry [48].

In addition, the range of fungal species that produce these toxins is wide, including *Fusarium*, *Aspergillus* and *Penicillium* species. A two-year study assessed yeast and mould contamination of raw materials and compound feeds; in starter compound feeds, the genus *Aspergillus* was predominant in 2019 (46.6%), while in 2020, species of the genera *Penicillium* and *Cladosporium* were identified in the majority of samples (50%); for combined feeds for growing and finishing, the genus *Aspergillus* was predominantly identified in 2019 (60% and 72.2% of samples, respectively) and 2020 (61.5% and 46.6%, respectively) [49]. For the 60 most common mycotoxins found in feed, 48% were shown to be produced by the genus *Fusarium*, 13% by the genus *Aspergillus*, 8% by the genus *Penicillium* and 12% by the genus *Alternaria* [50].

In a study conducted by Shar et al. [51], it was found that the natural occurrence of toxins belonging to the genus *Fusarium* in compound feed was similar to that in the raw materials used in their production. The incidence of mycotoxins in feed followed the following order: ZEN > FUM > DON.

In the current research, the highest prevalence in feed was the mycotoxin ZEN (93%), followed by DON (91%) and then OTA, which was identified in 70% of the total samples studied. The incidence of mycotoxins in starter, grower and finisher compound feeds followed the following order: ZEN > DON > OTA.

Although there are hundreds of mycotoxins, regulatory limits or recommendations for maximum tolerated levels in food and feed have been established for only a small number of them [52]. The recognition that mycotoxins affect the health and productivity of poultry and pigs has led to the introduction of regulations setting maximum permissible limits for aflatoxins and guideline recommendations (recommended tolerance levels) for ochratoxins and a small number of fusariotoxins. The limits vary not only according to mycotoxin type, animal species, intended use, feed materials and feedstuffs but also according to regulatory organisation or country; the European Union has established guidelines for feed materials and feedstuffs, with differences depending on the age of the animal and the stage of production [53].

The main aflatoxins consist of aflatoxins $B_1$, $B_2$, $G_1$ and $G_2$ and can be produced by select isolates of *A. flavus* or *A. parasiticus* [25,54,55]. The Rapid Alert System for Food and Feed (RASFF) reported 5045 and 439 notifications of mycotoxin contamination of food and feed exported to European Union countries worldwide during 2010–2019, respectively, and approximately 89% of mycotoxin contamination notifications of food and 98.6% of feed contamination notifications were attributed to AFT contamination [56]. Averaging the results obtained in all five years taken in study for the feed tested from the feed mill, the maximum AFT content of the combined starter, grower and finisher feed samples was 2.4 μg/kg and reached 5.2 μg/kg for the finisher feed for the year 2022.

The European Commission has established a maximum level for aflatoxin B1 of 20 μg/kg for feed materials, 10 μg/kg for complementary and complete feedingstuffs, 5 μg/kg for compound feedingstuffs intended for chickens and young birds and 20 μg/kg for compound feedingstuffs for poultry (except young animals) [15]. For many years, these toxins were not considered a problem in European agricultural production until early 2013, when aflatoxins in maize for animal feed from the Balkan area caused serious problems in Europe [57,58]. In the present study, we did not identify AFT in concentrations above a maximum permitted level set by the European Commission.

In a study by Greco et al. [59], 44 out of 49 samples of compound feed for broiler chickens were contaminated with aflatoxins, with an average level of 2.685 μg/kg. Decastelli [60] analysed 616 feed samples, and AFT was found in 44 (7%) of the samples. Martins et al. [61] analysed poultry feed and found that 10% and 22%, respectively, were contaminated with $AFB_1$ at concentrations of 1–21 μg/kg. Šegvić Klarić et al. [62] determined AFT in feed in the range 4.2–10.3 μg/kg (mean 6.9 μg/kg) in 4 (31%) out of 13 samples. In our study, the mean levels of AFT contamination in the combined feed samples were lower than the results reported by previous studies.

Current research highlights a new way of minimising the presence of mycotoxins in complete animal feed; namely, in the case of AFT, various combinations of fruit pomace are used to minimise the number of mycotoxins, the most common being forest fruits [63].

DON is mainly produced by *F. graminearum* and, in some geographical areas, by *F. culmorum* [59]. The average results for DON in the current study were between 321 and 1740 μg/kg in all five years of the combined feed samples; these results were compared with the European limit values and did not exceed them, but in a recent study published by the EFSA regarding DON contamination in poultry feed, since 2017 the EFSA has recommended limits of 600 μg/kg for broilers and turkeys, with effects on gut health and growth suppression in broilers at concentrations of less than 1900 μg/kg DON in their feed [64]. While the current research results show that DON levels in broiler feed samples were below the European regulatory limits for compound feed, EFSA's recommendations underline the need for more stringent limits designed for the biological sensitivity of broilers. A study conducted by Greco et al. [59] shows that 44 out of 49 samples (90%) were contaminated with DON (median 222 μg/kg). In a study on feed, Cegielska-Radziejewska et al. [37] examined poultry feed samples and detected DON contamination in all samples in the range of 3.1–99.4 μg/kg (median 33.6 μg/kg). In another study, DON was found in 56% of poultry feeds, and the median concentration was 303 μg/kg [65]. Driehuis et al. [66] analysed 72 feed samples, and DON was found in 54% of the samples with a maximum concentration of 2408 μg/kg (mean 433 μg/kg). In our study, the average levels of DON contamination were higher than those found by Cegielska-Radziejewska et al. [37] and lower than the results obtained by Labuda et al. and Driehuis et al. [65,66].

FUMs are a group of mycotoxins (mainly $FB_1$, $FB_2$ and $FB_3$) produced mainly by *Fusarium verticillioides* and *F. proliferatum*, and maize is the main commodity affected by this group of toxins [25]. In the present study, the maximum FUM contents in the compound feed samples for starters, growers and finishers were between 300 μg/kg and 1400 μg/kg, with the highest values identified in 2022 in the compound feed for finishers. The European Commission established a guideline value of 60,000 μg/kg for the sum of $FB_1$ and $FB_2$ in maize feed materials and 20,000 μg/kg for complementary and complete feed for poultry [15]. In the present study, we did not detect FUMs at concentrations higher than the guideline values established by the European Commission.

A study conducted by Greco et al. [59] shows that fumonisins were detected in all samples analysed in a range of 222–6000 μg/kg, and Martins et al. [61] found that 1% of 337 poultry feed samples were contaminated with $FB_1$ in a range of 24–34 μg/kg. In another study, Almeida et al. [67] analysed 127 compound feed samples, and FUM was detected in 9% of the samples at a maximum content of 390 μg/kg (median 164 μg/kg). Another study found that FUM in compound feed for broilers had a mean concentration of 304 μg/kg (maximum 1160 μg/kg) in 49 out of 50 samples [66]. Zachariasova et al. [68] analysed 70 samples of poultry and pig feed, and FUM was detected at a maximum content of 10 μg/kg. In their study, Šegvić Klarić et al. [62] found that 7 out of 13 feed samples were contaminated with FUM, and their average was 2300 μg/kg, and the maximum content was 5000 μg/kg. In the present study, the average concentration of FUM in all the compound feed samples was lower than most of the results reported by previous studies.

OTA is a mycotoxin mainly produced by *P. verrucosum* and *A. ochraceus* [25]. In the current research, the maximum OTA content of combined starter, grower and finisher feed samples was determined, and the determined values ranged from 3.5 μg/kg to 21.4 μg/kg for all years included in this study. The European Commission has set a guideline limit value of 250 µg/kg for OTA in feed materials represented by cereals and cereal products and 100 µg/kg for complementary and compound feed for poultry [15]. In a study, Jaimez et al. [69] evaluated the occurrence of OTA in 22 samples of poultry feed; 43% of poultry feeds were contaminated with OTA at a mean content of 0.50 µg/kg. In a study about concentration of mycotoxins in broiler feed, Martins et al. [61] analysed 100 samples of poultry and pigs feed samples, and OTA was found in one sample at concentration of 4 µg/kg. In our study, the mean contamination levels of OTA in compound feed samples were higher than the results reported by Jaimez et al. [69] and lower than the results reported by Martins et al. [61].

ZEN is one of the most common mycotoxins, being mainly produced by *F. graminearum* and *F. culmorum* [25]. In this study, the maximum ZEN content found in the combined starter, grower and finisher feed samples was averaged over all years and ranged from 39.2 µg/kg to 89.4 µg/kg. The European Commission has set a guideline level of 2000 µg/kg for ZEN in feed materials from cereals and cereal products and 3000 µg/kg for maize by-products. The legal limits for ZEN for broilers are not found in the European Commission regulations; therefore, most research is based on limits based on clinical observations from studies conducted over time on poultry; therefore, the limits for ZEN in feed are very varied, ranging from 4 to 11,192 µg/kg [70]. In the current study, we did not detect ZEN at concentrations higher than the limit values observed in the clinical studies. Although the mycotoxins levels found in this study were below the regulatory limits set for broiler feed, these limitations are not always biologically safe. Chronic exposure to mycotoxins at subregulatory levels might have negative consequences, such as immunosuppression and impaired growth performance.

In a study by Greco et al. [59], 42 samples of compound feed for broiler chickens were contaminated with ZEN (median 50 µg/kg). In a study on mycotoxins in compound feed for broiler chickens, ZEN was found in 1 out of 22 samples at a low concentration of 0.5 µg/kg [69]. Labuda et al. [65] detected ZEN in 88% (44) of the samples with an average concentration of 21 µg/kg and a maximum content of 86 µg/kg. Driehuis et al. [66] analysed 72 samples of compound feed, and ZEN was identified in 28% of the samples with an average of 80 µg/kg and a maximum level of 363 µg/kg. In the compound feed samples analysed by Martins et al. [61], 13% of the samples were found to be contaminated with ZEN at a level between 104 and 356 µg/kg. Zachariasova et al. [68] analysed a total of 70 broiler compound feed samples, and ZEN was found at a maximum content of 104 g/kg. In our study, it was observed that the average levels of ZEN contamination in broiler compound feed samples were lower than those obtained in the studies by Driehuis et al., Martins et al. and Zachariasova et al. [61,66,68].

In northern Romania, according to the observations of the authors involved in this study, as well as local and national sources of monitoring of meteorological phenomena in the period 2019–2023, the increased humidity of 2019–2021 is highlighted as being a result of heavy rains with increased periodicity, leading to the occurrence of average humidity levels ranging between 74 and 76% compared to 2022 and 2023, which recorded an average humidity of 65%.

The grains that can contribute to mycotoxin pollution in compound feed are mainly those that are susceptible to infestation by mycotoxin-producing moulds or fungi. The most commonly affected are the following: maize—a staple in feeds and often affected by fungi of the genera Fusarium, Aspergillus and Penicillium, which produce mycotoxins such as AFT, FUM and DON; wheat—also vulnerable to mycotoxin contamination, especially DON; barley and rye can be contaminated especially with DON and ZEN.

Meteorological factors could explain the higher contamination of feed raw materials and later the complete compound feed for broilers, as observed in the results obtained for the years 2019 and 2020. However, we have to consider that some mycotoxins such as AFT, although they prefer high humidity, have been confirmed by some studies to also accumulate during periods of prolonged drought [71,72]. The natural protective mechanisms of forage plants are affected by long periods of drought, making it impossible for them to form protective structures against pathogens and thus to prevent the multiplication of mycotoxins.

*Co-Occurrence*

Co-occurrence in the case of compound feed is most often identified because the presence of mycotoxins in each of the raw materials used in the production of compound feed is very likely; therefore, current research is focused on the study of the synergistic or antagonistic effects of mycotoxins [10,73].

The data obtained in the current study highlight the co-occurrence of mycotoxins; thus, in 26% of the analysed samples, the concomitant presence of the five studied mycotoxins was confirmed. On the other hand, the combinations of AFT, FUM, OTA and AFT, FUM and ZEN were the least frequent, accounting for only 0.2% of cases in the compound feed samples analysed over the 5 years. In a 2019 investigation carried out in Spain, Arroyo-Manzanares et al. [74] demonstrated the co-occurrence of more than eight mycotoxins in 2.19% of pig compound feed samples; they also highlight that in 98.7% of the studied samples, they found at least two mycotoxins present simultaneously.

Also, in the case of the compound feed samples studied, we highlighted in Figure 1 that in most cases (38.51%) we identified the concomitant presence of four mycotoxins. Gruber Dominger et al. [75] explain that, in fact, the scenario of co-occurrence of mycotoxins in the feed is the most plausible.

## 5. Conclusions

Studies on the assessment of mycotoxin contamination in compound feed for broiler chickens in Romania are limited, and the high frequencies of contamination observed in this study emphasise the importance of improving mycotoxin control throughout the country. In the period 2019–2023, during which the analysed samples were collected from the combined feed factory in Romania, the samples showed contamination levels below the maximum limits allowed by the European Union; however, we consider it necessary to establish and apply biologically relevant regulatory thresholds and limits that could ensure the health of poultry. In addition, it has been shown that the distribution and presence of mycotoxins varied from one year to the next due to the changes in climatic conditions.

Co-occurrences of mycotoxins have been widely identified; even though levels have been low, there is a need to increase knowledge about their combined effects on animal and human health.

Concerning future research directions in Romania, these should be extended to more feed mills, feed materials and types of compound feed in order to develop a practical guide to provide a comprehensive overview of the risks in the food chain. In particular, it is also relevant to investigate the possibility of the transfer of mycotoxins into products of animal origin and to analyse the combined effects of mycotoxins on animal health, including synergistic or antagonistic effects between them.

**Author Contributions:** Conceptualisation, D.M.L. and I.M.P.; methodology, D.M.L. and S.-I.P.; software, D.M.L. and S.-I.P.; validation, D.M.L. and M.M.; formal analysis, D.M.L., C.-G.R.-R., S.-I.P. and I.M.P.; investigation, D.M.L., C.-G.R.-R. and M.M.; data curation, D.M.L. and I.M.P.; writing—original draft preparation, D.M.L. and S.-I.P.; writing—review and editing, D.M.L., S.-I.P., C.-G.R.-R. and I.M.P.; supervision, D.M.L. All authors have read and agreed to the published version of the manuscript.

**Funding:** This research received no external funding.

**Institutional Review Board Statement:** Not applicable.

**Data Availability Statement:** The data present in this study are available upon request from the corresponding author.

**Conflicts of Interest:** The authors declare no conflicts of interest.

## References

1. Lorenzo, J.M.; Munekata, P.; Domínguez, R.; Pateiro, M.; Saraiva, J.; Franco Ruiz, D. Chapter 3—Main Groups of Microorganisms of Relevance for Food Safety and Stability: General Aspects and Overall Description. In *Innovative Technologies for Food Preservation: Inactivation of Spoilage and Pathogenic Microorganisms*; Academic Press: Cambridge, MA, USA, 2018; pp. 53–107.
2. Marin, D.E.; Pistol, G.C.; Procudin, C.V.; Taranu, I. Co-Contamination of Food and Feed with Mycotoxin and Bacteria and Possible Implications for Health. *Agriculture* **2022**, *12*, 1970. [CrossRef]
3. Pleadin, J.; Frece, J.; Markov, K. Mycotoxins in food and feed. *Adv. Food Nutr. Res.* **2019**, *89*, 297–345. [CrossRef]
4. Zeece, M. Food additives. In *Introduction to the Chemistry of Food*; Academic Press: Cambridge, MA, USA, 2020; pp. 251–311. [CrossRef]
5. Richard, J.L. Mycotoxins—An overview. In *Romer Labs' Guide to Mycotoxins*; Binder, E.M., Krska, R., Eds.; Anytime Publishing Services: Leicestershire, UK, 2015; pp. 1–48.
6. Marroquín-Cardona, A.G.; Johnson, N.M.; Phillips, T.D.; Hayes, A.W. Mycotoxins in a changing global environment—A review. *Food Chem. Toxicol.* **2014**, *69*, 220–230. [CrossRef]
7. Eskola, M.; Kos, G.; Elliott, C.T.; Hajšlová, J.; Mayar, S.; Krska, R. Worldwide contamination of food-crops with mycotoxins: Validity of the widely cited 'FAO estimate' of 25%. *Crit. Rev. Food Sci. Nutr.* **2020**, *60*, 2773–2789. [CrossRef] [PubMed]
8. Alshannaq, A.; Yu, J.-H. Occurrence, Toxicity, and Analysis of Major Mycotoxins in Food. *Int. J. Environ. Res. Public Health* **2017**, *14*, 632. [CrossRef]
9. Anukul, N.; Vangnai, K.; Mahakarnchanakul, W. Significance of regulation limits in mycotoxin contamination in Asia and risk management programs at the national level. *J. Food Drug Anal.* **2013**, *21*, 227–241. [CrossRef]
10. Pinotti, L.; Ottoboni, M.; Giromini, C.; Dell'Orto, V.; Cheli, F. Mycotoxin Contamination in the EU Feed Supply Chain: A Focus on Cereal Byproducts. *Toxins* **2016**, *8*, 45. [CrossRef] [PubMed]
11. Matei, M.; Zaharia, R.; Petrescu, S.-I.; Radu-Rusu, C.G.; Simeanu, D.; Mierlita, D.; Pop, I.M. Persistent Organic Pollutants (POPs): A Review Focused on Occurrence and Incidence in Animal Feed and Cow Milk. *Agriculture* **2023**, *13*, 873. [CrossRef]
12. Probst, C.; Njapau, H.; Cotty, P.J. Outbreak of an acute aflatoxicosis in Kenya in 2004: Identification of the causal agent. *Appl. Environ. Microb.* **2007**, *73*, 2762–2764. [CrossRef] [PubMed]
13. Galarza-Seeber, R.; Latorre, J.D.; Bielke, L.R.; Kuttappan, V.A.; Wolfenden, A.D.; Hernandez-Velasco, X.; Merino-Guzman, R.; Vicente, J.L.; Donoghue, A.; Cross, D.; et al. Leaky gut and mycotoxins: Aflatoxin B1 does not increase gut permeability in broiler chickens. *Front. Vet. Sci.* **2016**, *3*, 10. [CrossRef] [PubMed]
14. D'Mello, J.P.E. Contaminants and toxins in animal feeds. In *Assessing Quality and Safety of Animal Feeds*; Jutzi, S., Ed.; Food and Agriculture Organization: Rome, Italy, 2004; Volume 160, pp. 107–128.
15. Commission Regulation (EU) No 574/2011 of 16 June 2011 amending Annex I to Directive 2002/32/EC of the European Parliament and of the Council as regards maximum levels for nitrite, melamine, *Ambrosia* spp. and carry-over of certain coccidiostats and histomonostats and consolidating Annexes I and II thereto. *Off. J. Eur. Union* **2011**, *L159*, 7–24.
16. Commission Recommendation of 17 August 2006 on the presence of deoxynivalenol, zearalenone, ochratoxin A, T-2 and HT-2 and fumonisins in products intended for animal feeding. *Off. J. Eur. Union* **2006**, *L229*, 7–9.
17. Reddy, K.R.N.; Salleh, B.; Saad, B.; Abbas, H.K.; Abel, C.A.; Shier, W.T. An overview of mycotoxin contamination in foods and its implications for human health. *Toxin Rev.* **2010**, *29*, 3–26. [CrossRef]
18. Ismail, A.; Gonçalves, B.L.; de Neeff, D.V.; Ponzilacqua, B.; Coppa, C.F.; Hintzsche, H.; Sajid, M.; Cruz, A.G.; Corassin, C.H.; Oliveira, C.A.F. Aflatoxin in foodstuffs: Occurrence and recent advances in decontamination. *Food Res. Int.* **2018**, *113*, 74–85. [CrossRef] [PubMed]
19. Rushing, B.R.; Selim, M.I. Aflatoxin B1: A review on metabolism, toxicity, occurrence in food, occupational exposure, and detoxification methods. *Food Chem. Toxicol.* **2019**, *124*, 81–100. [CrossRef]
20. Guo, H.; Wang, P.; Liu, C.; Chang, J.; Yin, Q.; Wang, L.; Jin, S.; Zhu, Q.; Lu, F. Compound Mycotoxin Detoxifier Alleviating Aflatoxin B1 Toxic Effects on Broiler Growth Performance, Organ Damage and Gut Microbiota. *Poult. Sci.* **2023**, *102*, 102434. [CrossRef] [PubMed]
21. Paneru, D.; Sharma, M.K.; Shi, H.; Wang, J.; Kim, W.K. Aflatoxin B1 Impairs Bone Mineralization in Broiler Chickens. *Toxins* **2024**, *16*, 78. [CrossRef] [PubMed]

22. Sobrova, P.; Adam, V.; Vasatkova, A.; Beklova, M.; Zeman, L.; Kizek, R. Deoxynivalenol and its toxicity. *Interdiscip. Toxicol.* **2010**, *3*, 94–99. [CrossRef] [PubMed]
23. Wan, S.; Sun, N.; Li, H.; Khan, A.; Zheng, X.; Sun, Y.; Fan, R. Deoxynivalenol damages the intestinal barrier and biota of the broiler chickens. *BMC Vet. Res.* **2022**, *18*, 311. [CrossRef]
24. Peng, W.X.; Marchal, J.L.M.; Van der Poel, A.F.B. Strategies to prevent and reduce mycotoxins for compound feed manufacturing. *Anim. Feed. Sci. Technol.* **2018**, *237*, 129–153. [CrossRef]
25. Richard, J.L. Some major mycotoxins and their mycotoxicoses—An overview. *Int. J. Food Microbiol.* **2007**, *119*, 3–10. [CrossRef]
26. Anttila, A.; Bhat, R.V.; Bond, J.A.; Borghoff, S.J.; Bosch, F.X.; Carlson, G.P.; Castegnaro, M.; Cruzan, G.; Gelderblom, W.C.A.; Hass, U.; et al. *International Agency for Research on Cancer (IARC) Monographs on the Evaluation of Carcinogenic Risks to Humans: Some Traditional Herbal Medicines, Some Mycotoxins, Naphthalene and Styrene*; IARC Press: Lyon, France, 2002; Volume 82, p. 367.
27. Völkel, I.; Schröer-Merker, E.; Czerny, C.P. The carry-over of mycotoxins in products of animal origin with special regard to its implications for the European food safety legislation. *Food Nutr. Sci.* **2011**, *2*, 852–867. [CrossRef]
28. El Khoury, A.; Atoui, A. Ochratoxin A: General overview and actual molecular status. *Toxins* **2010**, *2*, 461–493. [CrossRef] [PubMed]
29. World Health Organization (WHO); International Agency for Research on Cancer (IARC). *Some Naturally Occurring Substances: Food Items and Constituents, Heterocyclic Aromatic Amines and Mycotoxins*; IARC Monographs on the Evaluation of the Carcinogenic Risk of Chemicals to Humans; IARC Publications: Lyon, France, 1993; p. 56.
30. Gaffoor, I.; Trail, F. Characterization of two polyketide synthase genes involved in zearalenone biosynthesis in *Gibberella zeae*. *Appl. Environ. Microbiol.* **2006**, *72*, 1793–1799. [CrossRef]
31. Pop, I.M.; Halga, P.; Avarvarei, T. Animal nutrition and feeding. In *Nutritia si Alimentatia Animalelor*; Tipo Moldova: Iasi, Romania, 2006.
32. European Food Safety Authority (EFSA) Panel on Contaminants in the Food Chain. Scientific Opinion on the risks for public health related to the presence of zearalenone in food. *EFSA J.* **2011**, *9*, 2197. [CrossRef]
33. Pop, I.M. Food quality and safety management. In *Managementul Calitatii si Sigurantei Alimentelor*; Tipo Moldova: Iasi, Romania, 2022.
34. Lapusneanu, D.M.; Simeanu, D.; Radu-Rusu, C.; Pop, C.; Postolache, A.N.; Pop, I.M. Case study on the prevention and control of mycotoxin contamination in a feed mill unit. *J. Hyg. Eng. Des. JHED* **2020**, *33*, 41–47.
35. Regulation (EC) No 183/2005 of the European Parliament and of the Council of 12 January 2005 Laying down Requirements for Feed Hygiene. *Off. J. Eur. Union* **2005**, *L35*, 1–22.
36. Granados-Chinchilla, F.; Valenzuela-Martínez, C.; García-Murillo, B.; Aguilar-Madrigal, D.; Redondo-Solano, M.; Molina, A. Microbiological safety and presence of major mycotoxins in animal feed for laboratory animals in a developing country: The case of Costa Rica. *Animals* **2021**, *11*, 2389. [CrossRef]
37. Cegielska-Radziejewska, R.; Stuper-Szablewska, K.; Szablewski, T. Microflora and mycotoxin contamination in poultry feed mixtures from western Poland. *Ann. Agric. Environ. Med. AAEM* **2013**, *20*, 30–35.
38. Smith, M.-C.; Madec, S.; Coton, E.; Hymery, N. Natural Co-Occurrence of Mycotoxins in Foods and Feeds and Their in vitro Combined Toxicological Effects. *Toxins* **2016**, *8*, 94. [CrossRef]
39. *SR EN ISO 9001:2015*; Quality Management Systems—Requirements. Romanian Standards Association: Bucharest, Romania, 2015.
40. *SR EN ISO 22000:2019*; Food Safety Management Systems—Requirements for Any Organization in the Food Chain. Romanian Standards Association: Bucharest, Romania, 2019.
41. *SR ISO 45001:2018*; Occupational Health and Safety Management Systems—Requirements with Guidance for Use. Romanian Standards Association: Bucharest, Romania, 2018.
42. Rayat, C.S. Applications of Microsoft Excel in Statistical Methods. In *Statistical Methods in Medical Research*, 1st ed.; Rayat, C.S., Ed.; Springer: Singapore, 2018; pp. 139–146.
43. Abdolshahi, A.; Behdad Shokrollahi, Y. Chapter 2—Food contamination. In *Mycotoxins and Food Safety*; Suna, S., Ed.; IntechOpen: Rijeka, Croatia, 2020.
44. Guerre, P. Mycotoxin and Gut Microbiota Interactions. *Toxins* **2020**, *12*, 769. [CrossRef] [PubMed]
45. Stanley, D.; Bajagai, Y.S. Feed Safety and the Development of Poultry Intestinal Microbiota. *Animals* **2022**, *12*, 2890. [CrossRef] [PubMed]
46. Todd, E.C.D. Foodborne Diseases: Overview of Biological Hazards and Foodborne Diseases. In *Encyclopedia of Food Safety*; Academic Press: Cambridge, MA, USA, 2014; pp. 221–242. [CrossRef]
47. Cimbalo, A.; Alonso Garrido, M.; Font, G.; Manyes, L. Toxicity of mycotoxins in vivo on vertebrate organisms: A review. *Food Chem. Toxicol.* **2020**, *137*, 111161. [CrossRef] [PubMed]

48. Abrunhosa, L.; Morales, H.; Soares, C.; Calado, T.; Vila-Cha, A.S.; Pereira, M.; Venancio, A. A Review of Mycotoxins in Food and Feed Products in Portugal and Estimation of Probable Daily Intakes. *Crit. Rev. Food Sci. Nutr.* **2016**, *56*, 249–265. [CrossRef] [PubMed]
49. Lăpușneanu, D.M.; Simeanu, D.; Radu-Rusu, C.-G.; Zaharia, R.; Pop, I.M. Microbiological Assessment of Broiler Compound Feed Production as Part of the Food Chain—A Case Study in a Romanian Feed Mill. *Agriculture* **2023**, *13*, 107. [CrossRef]
50. Khoshal, A.K.; Novak, B.; Martin, P.G.P.; Jenkins, T.; Neves, M.; Schatzmayr, G.; Oswald, I.P.; Pinton, P. Co-Occurrence of DON and Emerging Mycotoxins in Worldwide Finished Pig Feed and Their Combined Toxicity in Intestinal Cells. *Toxins* **2019**, *11*, 727. [CrossRef]
51. Shar, Z.H.; Shar, H.H.; Jatoi, A.; Sherazi, S.T.H.; Mahesar, S.A.; Khan, E.; Phanwar, Q.K. Natural co-occurrence of *Fusarium* toxins in poultry feed and its ingredients. *J. Consum. Prot. Food Saf.* **2020**, *15*, 341–350. [CrossRef]
52. Bennett, J.W.; Klich, M. Mycotoxins. *Clin. Microbiol. Rev.* **2003**, *16*, 497–516. [CrossRef] [PubMed]
53. Guerre, P. Worldwide Mycotoxins Exposure in Pig and Poultry Feed Formulations. *Toxins* **2016**, *8*, 350. [CrossRef]
54. Rawal, S.; Kim, J.E.; Coulombe, R., Jr. Aflatoxin B1 in poultry: Toxicology, metabolism and prevention. *Res. Vet. Sci.* **2010**, *89*, 325–331. [CrossRef] [PubMed]
55. Ochieng, P.E.; Scippo, M.; Kemboi, D.C.; Croubels, S.; Okoth, S.; Kang'ethe, E.K.; Doupovec, B.; Gathumbi, J.K.; Lindahl, J.F.; Antonissen, G. Mycotoxins in Poultry Feed and Feed Ingredients from Sub-Saharan Africa and Their Impact on the Production of Broiler and Layer Chickens: A Review. *Toxins* **2021**, *13*, 633. [CrossRef]
56. Alshannaq, A.; Yu, J.-H. Analysis of E.U. Rapid Alert System (RASFF) Notifications for Aflatoxins in Exported U.S. Food and Feed Products for 2010–2019. *Toxins* **2021**, *13*, 90. [CrossRef]
57. De Rijk, T.C.; Van Egmond, H.P.; Van der Fels-Klerx, H.J.; Herbes, R.; De Nijs, M.; Samson, R.A.; Slate, A.B.; Van der Spiegel, M. A study of the 2013 Western European issue of aflatoxin contamination of maize from the Balkan area. *World Mycotoxin J.* **2015**, *8*, 641–651. [CrossRef]
58. Kosicki, R.; Błajet-Kosicka, A.; Grajewski, J.; Twarużek, M. Multiannual mycotoxin survey in feed materials and feeding stuffs. *Anim. Feed Sci. Technol.* **2016**, *215*, 165–180. [CrossRef]
59. Greco, M.V.; Franchi, M.L.; Rico Golba, S.L.; Pardo, A.G.; Pose, G.N. Mycotoxins and mycotoxigenic fungi in poultry feed for food-producing animals. *Sci. World J.* **2014**, *2014*, 968215. [CrossRef] [PubMed]
60. Decastelli, L.; Lai, J.; Gramaglia, M.; Monaco, A.; Nachtmann, C.; Oldano, F.; Ruffer, M.; Sezian, A.; Bandirola, C. Aflatoxins occurrence in milk and feed in Northern Italy during 2004–2005. *Food Control* **2007**, *18*, 1263–1266. [CrossRef]
61. Martins, H.M.; Marques, M.; Almeida, I.; Guerra, M.M.; Bernardo, F. Mycotoxins in feedstuffs in Portugal: An overview. *Mycotoxin Res.* **2008**, *24*, 19–23. [CrossRef] [PubMed]
62. Šegvić Klarić, M.; Cvetnić, Z.; Pepeljnjak, S.; Kosalec, I. Co-occurrence of Aflatoxins, Ochratoxin A, Fumonisins, and Zearalenone in Cereals and Feed, Determined by Competitive Direct Enzyme-Linked Immunosorbent Assay and Thin-Layer Chromatography. *Arch. Ind. Hyg. Toxicol.* **2009**, *60*, 427–433. [CrossRef]
63. Mierlita, D.; Teușdea, A.C.; Matei, M.; Pascal, C.; Simeanu, D.; Pop, I.M. Effect of Dietary Incorporation of Hemp Seeds Alone or with Dried Fruit Pomace on Laying Hens' Performance and on Lipid Composition and Oxidation Status of Egg Yolks. *Animals* **2024**, *14*, 750. [CrossRef] [PubMed]
64. EFSA Panel on Contaminants in the Food Chain (CONTAM); Schrenk, D.; Bignami, M.; Bodin, L.; Chipman, J.K.; Del Mazo, J.; Grasl-Kraupp, B.; Hogstrand, C.; Hoogenboom, L.R.; Leblanc, J.C.; et al. Risk assessment of aflatoxins in food. *EFSA J.* **2020**, *18*, e06040. [CrossRef]
65. Labuda, R.; Parich, A.; Berthiller, F.; Tančinová, D. Incidence of trichothecenes and zearalenone in poultry feed mixtures from Slovakia. *Int. J. Food Microbiol.* **2005**, *105*, 19–25. [CrossRef] [PubMed]
66. Driehuis, F.; Spanjer, M.C.; Scholten, J.M.; Te Giffel, M.C. Occurrence of mycotoxins in feedstuffs of dairy cows and estimation of total dietary intakes. *J. Dairy Sci.* **2008**, *91*, 4261–4271. [CrossRef] [PubMed]
67. Almeida, I.; Martins, H.M.; Santos, S.; Costa, J.M.; Bernardo, F. Co-occurrence of mycotoxins in swine feed produced in Portugal. *Mycotoxin Res.* **2011**, *27*, 177–181. [CrossRef]
68. Zachariasova, M.; Dzuman, Z.; Veprikova, Z.; Hajkova, K.; Jiru, M.; Vaclavikova, M.; Zachariasova, A.; Pospichalova, M.; Florian, M.; Hajslova, J. Occurrence of multiple mycotoxins in European feedingstuffs, assessment of dietary intake by farm animals. *Anim. Feed Sci. Technol.* **2014**, *193*, 124–140. [CrossRef]
69. Jaimez, J.; Fente, C.A.; Franco, C.M.; Cepeda, A.; Vazquez, B.I. A survey of the fungal contamination and presence of ochratoxin A and zearalenone on Spanish feed and raw materials. *J. Sci. Food Agric.* **2004**, *84*, 832–840. [CrossRef]
70. Liu, J.; Applegate, T. Zearalenone (ZEN) in Livestock and Poultry: Dose, Toxicokinetics, Toxicity and Estrogenicity. *Toxins* **2020**, *12*, 377. [CrossRef] [PubMed]
71. Liu, C.; Van der Fels-Klerx, H.J. Quantitative Modeling of Climate Change Impacts on Mycotoxins in Cereals: A Review. *Toxins* **2021**, *13*, 276. [CrossRef]

72. Medina, A.; Gilbert, M.K.; Mack, B.M.; OBrian, G.R.; Rodríguez, A.; Bhatnagar, D.; Payne, G.; Magan, N. Interactions between water activity and temperature on the Aspergillus flavus transcriptome and aflatoxin B1 production. *Int. J. Food Microbiol.* **2017**, *256*, 36–44. [CrossRef] [PubMed]
73. Muñoz-Solano, B.; González-Peñas, E. Co-Occurrence of Mycotoxins in Feed for Cattle, Pigs, Poultry, and Sheep in Navarra, a Region of Northern Spain. *Toxins* **2023**, *15*, 172. [CrossRef] [PubMed]
74. Arroyo-Manzanares, N.; Rodríguez-Estévez, V.; Arenas-Fernández, P.; García-Campaña, A.M.; Gámiz-Gracia, L. Occurrence of Mycotoxins in Swine Feeding from Spain. *Toxins* **2019**, *11*, 342. [CrossRef]
75. Gruber-Dorninger, C.; Jenkins, T.; Schatzmayr, G. Global Mycotoxin Occurrence in Feed: A Ten-Year Survey. *Toxins* **2019**, *11*, 375. [CrossRef] [PubMed]

agriculture

MDPI

Article

# Genetic Diversity and Phylogenetic Analysis of the Endangered Transylvanian Pinzgau Cattle: A Key Resource for Biodiversity Conservation and the Sustainability of Livestock Production

**Mădălina-Alexandra Davidescu [1], Claudia Pânzaru [2], Bianca-Maria Mădescu [1], Răzvan-Mihail Radu-Rusu [2], Marius Gheorghe Doliș [2], Cristina Simeanu [2], Alexandru Usturoi [1,*], Andrei Ciobanu [2] and Ștefoil Creangă [1]**

1 Department of Control, Expertise and Services, Faculty of Food and Animal Sciences, "Ion Ionescu de la Brad" Iasi University of Life Sciences, 700489 Iasi, Romania; madalina.davidescu@iuls.ro (M.-A.D.); bianca.madescu@iuls.ro (B.-M.M.); steofil.creanga@iuls.ro (Ș.C.)

2 Department of Animal Resources and Technologies, Faculty of Food and Animal Sciences, "Ion Ionescu de la Brad" Iasi University of Life Sciences, 700489 Iasi, Romania; claudia.panzaru@iuls.ro (C.P.); razvan.radu@iuls.ro (R.-M.R.-R.); marius.dolis@iuls.ro (M.G.D.); cristina.simeanu@iuls.ro (C.S.); andrei.ciobanu@iuls.ro (A.C.)

* Correspondence: alexandru.usturoi@iuls.ro

**Abstract:** Animal biodiversity is essential for maintaining the functionality of local food systems and ensuring sustainable livelihoods. Starting in 2000, the Food and Agriculture Organization of the United Nations (F.A.O.) has drawn attention to the decline in cattle populations, including the Transylvanian Pinzgau breed from Romania. Renowned for its hardiness, adaptability, and enhanced resistance to diseases and climate change, the Transylvanian Pinzgau is regarded as an important genetic asset for advancing livestock production. As a result, tracking genetic diversity has become a key focus in breeding programs, particularly for small, endangered local populations that play a vital role in regional agro-ecological systems. This research paper sought to assess the genetic diversity of a group of 24 head of cattle from the Transylvania region by analyzing two mtDNA markers, *cytochrome b* and *D-loop* sequences, both widely recognized for their relevance and importance in the analysis of genetic diversity of cattle and phylogenetic studies. The findings, derived through statistical analysis of nucleotide sequences using specialized software, indicated that the analyzed cattle are part of the ancestral T haplogroup, with a direct lineage tracing back to Bos taurus. This information can aid in developing crossbreeding programs focused on conserving essential genetic resources, improving other cattle breeds, and protecting biodiversity.

**Keywords:** animal production; biodiversity; genetic diversity; phylogeny; Pinzgau; mitochondrial DNA

## 1. Introduction

Human societies have primarily concentrated on cattle husbandry in the agricultural sector, considering it an essential activity for supporting rural economies [1,2]. This practice was closely linked to the needs of communities, providing not only food sources through meat and milk production but also resources for agricultural labor, transportation, and construction, thereby contributing to the social and economic development of society [1–3].

The domestication of cattle, dating back to the dawn of civilization, has represented a crucial influence on agricultural practices and the development of societies throughout history. Indigenous cattle breeds belong to a significant social and cultural identity, highly valued for their remarkable resilience to harsh environmental conditions and climatic challenges, innate resistance to diseases and ectoparasites, and rapid recuperative capacity post-illness [1,2,4–8].

In recent decades, there has been a marked global decline in livestock genetic diversity [8,9], primarily driven by production specialization that prioritizes the most widespread, highly productive breeds. Notably, the rigorous selection process of "elite sires" for the purpose of artificial insemination in extensive cattle populations has further reduced genetic diversity within breeds [6,8,10–13].

Among the oldest European cattle breeds is the Transylvanian Pinzgau, whose historical origins remain incompletely understood [1,13–15]. The Romanian Pinzgau breed, also known as the Transylvanian Pinzgau after its location area in the Transylvania region, is an alpine cattle breed, adapted to growth and exploitation in mountainous altitudes [12,16,17]. This breed was developed between 1670 and 1740 through the interbreeding of native cattle from the Pinzgau region (in the Salzburg area of Austria) with the Bernese spotted cattle (imported from Switzerland, specifically from Tux and Zillertal, between 1690 and 1740). Key attributes of this breed include robust resistance to pathogens and environmental stressors, extended productive longevity, and superior mobility compared to improved breeds. These morphological and functional attributes are crucial for the breed's persistence under the specific demands of high-altitude regions [1,4,7,12]. Consequently, it is essential to prioritize the preservation and growth of the current Pinzgau nucleus, in addition to preserving the breed's genetic diversity, especially considering accelerating global climate change [15–18].

Since 2000, reports from the FAO organization have indicated that the Transylvanian Pinzgau cattle breed faces a risk of extinction due to a significant population decline [19,20].

The genetic composition and lineage relationships between the Transylvanian Pinzgau and other varieties from the Pinzgau breed worldwide reveal varying degrees of kinship and evolutionary divergence [16]. Comparative genetic studies have shown that, while the Transylvanian Pinzgau shares ancestral genetic markers with other Pinzgau populations in Europe, distinct genetic drift and adaptation to local environmental conditions have shaped its unique genetic profile over time [16,17]. By analyzing molecular markers and constructing phylogenetic trees, researchers can assess the genetic distances between these populations, which provide insights into their common heritage and the evolutionary processes that have led to regional genetic variations. Such analyses are essential for understanding both the conservation needs and the genetic uniqueness of the Transylvanian Pinzgau within the broader context of the Pinzgau breed globally (Austrian Pinzgau, German Pinzgau, Slovak Pinzgau, Swiss Pinzgau, etc.) [18]. The common lineage of this cattle breed can be traced to the wild progenitor *Bos taurus*, which was considered extinct by the 16th century [21–23]. From this ancestral stock, cattle have inherited numerous valuable traits [2].

Numerous studies focused on the origin of cattle [14,21,24–26], based on mtDNA analysis, have established that all taurine descend from a common ancestor in the area of Near East approximately 10,000 years ago, coinciding with the shift to the Neolithic era. Recent investigations into the mitochondrial genome of *Bos taurus* have identified macro-haplogroup T [21,27], which comprises two sibling clades, T1′2′3 and T5. The T1′2′3 clade includes the initially characterized haplogroups T1, T2, T3, and T4, with T4 being a subgroup of T3. All T haplogroups are believed to have likely originated in the Fertile Crescent, where domestication occurred. From this region, they disseminated alongside the movement of domestic cattle herds of *Bos taurus*. The mitochondrial DNA of contemporary cattle is not wholly identified by haplogroups T1–T5 [21,28–30].

Presently, the Research and Development Station for Cattle Breeding in Târgu-Mureș, Romania, houses 24 Pinzgau individuals included in a nationwide biodiversity conservation program. The cows are not known to be closely related. They were identified in the Transylvanian region and selected based on their phenotypic traits characteristic of the Transylvanian Pinzgau breed, ensuring the representation of the breed's diversity. This population serves as the biological material utilized for ongoing research efforts, aimed at protecting biodiversity and the key genetic materials of this cow, which represents a sustainable alternative to the increasingly pronounced climate changes to which modern improved cattle are increasingly susceptible [4,12,31].

The population trend of this breed demonstrates a decline from approximately 5% of Romania's total cattle population in 1937 to 1.35% by 1995 [32–35]. This breed is distinguished by its unique coat pattern, which is a deep cherry color marked with characteristic white patterns (Figure 1).

**Figure 1.** The Transylvanian Pinzgau cattle from the Research and Development Station for Cattle Breeding, Târgu-Mureș, Romania (original photograph).

The Transylvanian Pinzgau reaches an average weight up to 900 kg and a height of approximately 134 cm in males, while females can weigh up to 500 kg and reach a height of 127 cm [35]. Morphologically, Transylvanian Pinzgau cattle are considered descendants of *Bos taurus*, with cranial features that exhibit notable similarities to certain Spanish breeds, making it valuable in terms of a genetic perspective [34].

Studying the phylogeny of the Transylvanian Pinzgau is crucial for several reasons. First, understanding its genetic lineage and evolutionary history can provide insights into the breed's unique adaptations and resilience to local environmental conditions [12,31]. Additionally, phylogenetic analysis can reveal the relationships between the Transylvanian Pinzgau and other Pinzgau populations worldwide, enhancing our comprehension of genetic diversity within the breed [10,14,31]. Moreover, examining the phylogeny of the Transylvanian Pinzgau can aid in identifying genetic markers linked to disease resistance and overall fitness, thereby contributing to improved management practices [4,5,18]. Ultimately, this research is essential not only for the conservation of this specific breed but also for the continued sustainability of livestock farming in Romania and beyond.

This study sought to assess the diversity of genetic resources of the Transylvanian Pinzgau cattle population utilizing mitochondrial genes (*cytochrome b* and *D-loop*) that are pertinent to investigations of variability, phylogeny lineage, and genetic mapping of phylogeography. In mitochondrial DNA (mtDNA), *cytochrome b* (*cyt b*) is a component of the *cytochrome bc1* complex (also known as Complex III) in the mitochondrial electron transport chain [17,21,24,29]. This protein plays a crucial role in cellular respiration by facilitating electron transfer and contributing to the generation of the proton gradient used for ATP synthesis. *Cytochrome b* is encoded by the mitochondrial DNA and is highly conserved across species, making it a popular target in studies of genetic variation, phylogenetics, and species identification [21,24,30]. The *D-loop* is a non-coding region known for its role in the onset of DNA replication and transcription. It contains control elements essential for regulating the replication and expression of the mitochondrial genome. This region is often used in genetic studies because it has a high mutation rate, making it useful for analyzing genetic diversity, maternal lineage, and evolutionary studies [24,30].

Yan et al., in 2019, assessed the genetic diversity and phylogenetic relationships of Chinese cattle using mtDNA, specifically the *cytochrome b* marker. The researchers found that the analysis of these mitochondrial markers clarified the evolutionary relationships among different cattle breeds and was essential for the safeguarding and oversight of these populations of animals [11].

Another study, published by Prihandini et al. (2020) on the genetic diversity of the mitochondrial *cytochrome b* gene in a native Indonesian cattle nucleus, demonstrates that the mitochondrial *cytochrome b* marker can be effectively used to understand genetic structure and identify the main haplogroups that define different cattle breeds [15].

A study published by Davidescu et al. (2022) focusing on the phylogenetic analysis of endangered Grey Steppe cows shows that *cytochrome b* was the mt marker highly skilled for accurately assessing the phylogenetic lineage between the studied breed and other cattle breeds in Europe, demonstrating importance for identifying genetic differences. Meanwhile, the *D-loop* mt gene provided the most reliable topological support [13].

This research aims to investigate the genetic composition and evolutionary links of the endangered Transylvanian Pinzgau cattle, a key resource for biodiversity conservation and the sustainability of livestock production.

## 2. Materials and Methods

### 2.1. Collection of Blood Samples

The initial phase in fulfilling the research objectives involved obtaining blood specimens from 24 female Transylvanian Pinzgau cattle, a population housed at the Research and Development Station for Cattle Breeding in Târgu-Mureș, within Romania's Transylvania region. The average age of the analyzed cows was 52 months, with limits between 12 and 90 months, respectively. Blood samples were acquired via jugular venipuncture, using 2 mL vacutainer tubes containing EDTA (ethylene-diamine-tetra-acetic acid) to prevent blood clotting. The collected blood samples were numbered from tp_01 to tp_24, "tp" being the abbreviation from the name of breed, namely Transylvanian Pinzgau. The blood samples were preserved under ideal storage conditions in a freezer at −20 °C before the stage of DNA isolation and analysis.

### 2.2. Extraction and Quantification of Genomic DNA from Blood Samples

The DNA was isolated from blood through an automated protocol using the Maxwell™ 16 and 16 MDx systems, with a specific kit supplied by the Promega distributor (Promega Corporation, Madison, WI, USA): the Maxwell 16 LEV Blood DNA Kit (catalog code AS1290) [36,37]. This kit includes 50 LEV cleansing cartridges, 50 collection tubes for elution, 20 mL of cell lysis reagent, two 1 mL solutions of K proteinase, and 20 mL of washing buffer. The Maxwell™ 16 system is capable of processing 16 samples in 40 min, and the extracted DNA can be applied to several techniques, including PCR and agarose gel electrophoresis.

The DNA quantification was conducted with the spectrophotometer Nanodrop ASP-3700 (ACTGene Inc., Princeton, NJ, USA), measuring absorbance at absorbance wavelengths of A260 nm and A280 nm to determine DNA extraction yield. This spectrophotometric technique is guided by the principle that most biological substances exhibit a characteristic absorbance rate within the ultraviolet (UV) radiation spectrum. Specifically, an absorption rate at 260 nm corresponds to nucleic acids (DNA/RNA), 280 nm to proteins, and 230 nm to various contaminants. Beer–Lambert's law establishes a linear correlation between the concentration of a compound and its absorbance at a specific wavelength. This principle underlies the measurement of DNA concentration, enabling assessments of purity in relation to protein content [37,38].

### 2.3. Primer Design, PCR Amplification, and Gene Sequencing

The amplification of mitochondrial markers, specifically *cytochrome b* and the *D-loop*, was conducted through polymerase chain reaction (PCR), a molecular technique used to efficiently generate millions to billions of copies of targeted DNA segments using specific amplification primers. In this study, PCR amplification of *cytochrome b* and the *D-loop* genes was achieved with two primer pairs (forward primer and reverse primer, detailed in Table 1) designed based on the *Bovine Reference Sequence* (BRS; GenBank V00654) [39]. The mitochondrial genome in *Bos taurus* measures 16,341 base pairs (bp) in total length. The *cytochrome b* marker, approximately 1140 bp in length, and the mitochondrial *D-loop* control region, about 910 bp long (Figure 2), were amplified using the PCR technique.

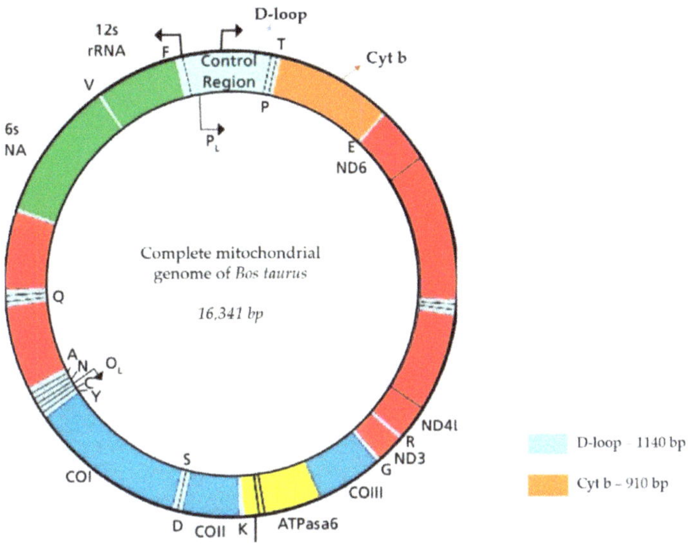


**Figure 2.** Total mitochondrial genome of *Bos taurus*: focus on mt markers analyzed—*cytochrome b* (1140 bp); *D-loop*-(910 bp) [13,40–42].

**Table 1.** Amplification primer pair characteristics for gene sequence analyzed.

| Primer Selection | Primer Specificity Gene | Primer Sequence (5′-3′) | Base Composition G[1] + C[2](%) | GenBank Accession No. and Genome Position [43] | Amplicon Length (bp) |
|---|---|---|---|---|---|
| BCYT | *cytochrome b* | Forward sequence primer: TTCTTACATGGAATCTAACCATGA | 33.3 | *V00654.1* 14,443–14,466 | 1140 |
| | *cytochrome b* | Reverse sequence primer: GGGAGGTTAGTTGTTCTCCTTCTC | 50.0 | *V00654.1* 473–497 | |
| BRS | *D-loop* | Forward sequence primer: CCTAAGACTCAAGGAA-GAAACTGC | 45.8 | *V00654.1* 15,718–15,741 | 910 |
| | *D-loop* | Reverse sequence primer: CAGTGAGAATGCCCTCTAGGTT | 50.0 | *V00654.1* 496–517 | |

"G[1]"—guanine. "C[2]"—cytosine.

The DNA extracts that were isolated and purified were processed through amplification using polymerase chain reaction (PCR). The total reaction volume for PCR was set at 25.5 µL, comprising 2 µL of DNA extract, 12.5 µL of GoTaq® Green Master Mix (Promega, Madison, WI, USA), 1.5 µL of both primers, and 7.5 µL of deionized water without nucleases. The reaction was further enhanced with 0.5 µL of $MgCl_2$. For *cytochrome b* fragment amplification, the primer annealing temperature was adjusted to 62 °C, while for *D-loop* fragment amplification, it was set to 60 °C. The PCR protocol for amplifying *cytochrome b* and *D-loop* sequences comprised 35 cycles, with conditions detailed in Figure 3a,b.

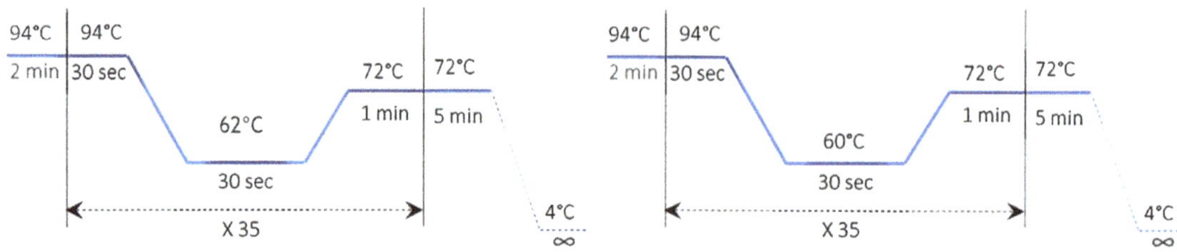


(**a**) (**b**)

**Figure 3.** PCR_Program: (**a**) *cytochrome b*; (**b**) *D-loop*.

The analyzed markers were successfully replicated, after which the PCR products were submitted to Macrogen Europe Genomic Sequencing Lab in Amsterdam, The Netherlands, for sequencing genes. Sanger sequencing was employed to determine the genetic code sequences of markers, using the primers specified in Table 2.

### 2.4. Validation of Amplified PCR Products in Agarose Gel Electrophoresis

To assess the degree of purity, molecular size, and potential nonspecific contaminants, the amplicons resulting from the amplification of the gene sequences (*cyt b* and *D-loop*) were validated using agarose gel electrophoresis (Figure 4a,b). The PCR amplicons were validated using a 1% agarose gel with 0.5× TBE buffer solution, followed by electrophoretic migration conducted at 100 volts for 30 min. To approximate the length of the PCR fragments, a 100-base-pair molecular weight marker was used. The sequences of interest were effectively amplified, with the specific primers showing strong specificity for the targeted mtDNA markers. The PCR amplicons had lengths of around 1140 base pairs for *cytochrome b* and about 910 base pairs for the *D-loop*, aligning with the reported values for the complete *Bos taurus* mitochondrial genome *(GenBank accession no.: V00654)*.

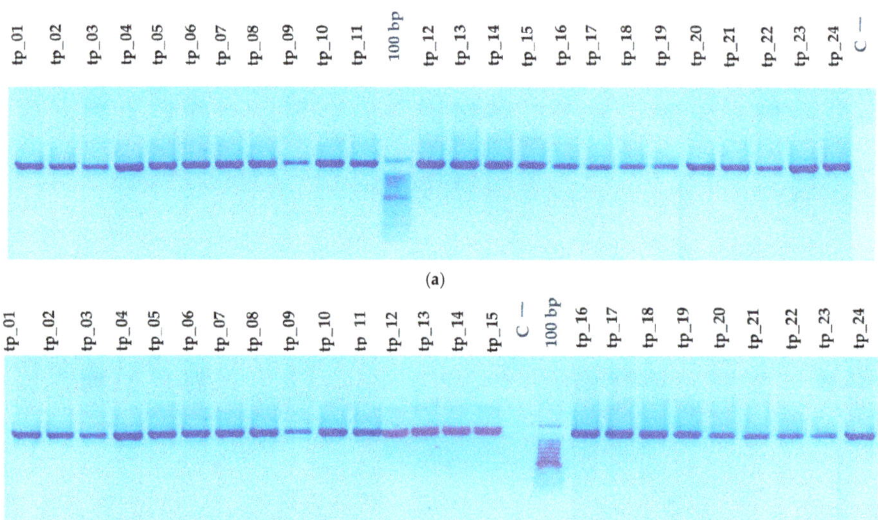


(**b**)

**Figure 4.** PCR amplification of mtDNA genes: (**a**) Sequence *cytochrome b* (M, 100 bp molecular marker; C—negative sample control; tp_1–tp_24—sample numbers). (**b**) Sequence *D-loop* (M, 100 bp molecular marker; C—negative sample control; tp_1–tp_24—sample numbers).

*2.5. Quantitative Assessment of Nitrogenous Base Proportions in Cyt b and D-loop Sequences from Transylvanian Pinzgau*

The complete mitochondrial DNA (mtDNA) *cytochrome b* sequence, measuring 1140 base pairs [24], was successfully obtained for all 24 samples analyzed. The nucleotide composition was determined to be as follows: 30.5% adenine, 25.4% thymine, 29.9% cytosine, and 14.2% guanine. In addition, the nucleotide composition of the *D-loop* marker, which spans 910 base pairs [24], was characterized as follows: 32.6% adenine, 29.1% thymine, 24.5% cytosine, and 13.8% guanine. These results demonstrate the success of the sequencing (Figure 5).

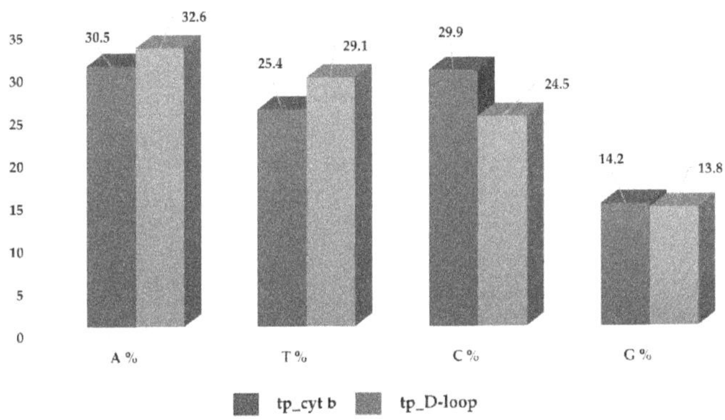


**Figure 5.** Frequency distribution of DNA nucleotides (A: adenine; T: thymine; C: cytosine; G: guanine) in the sequences of *cytochrome b* and the *D-loop*.

The values are very similar to those obtained in analyses of sequences specific to the Pinzgau breed from other studies, with only slight variation [16,17]. This indicates a balanced proportion among the four DNA nucleotides.

*2.6. The Coefficient of Specificity*

The specificity coefficient was determined using the frequencies of the nucleotide bases, represented by the A + T/C + G ratio, which illustrates the variations among individuals of a species. This parameter is utilized in molecular phylogenetic research to assess variability in nucleotide structure. Typically, it exhibits values ranging from 1.2 to 1.5 [21], which are characteristic of most animal species. When considering the nitrogenous dominant bases in frequency, two distinct types of DNA can be identified: AT-type DNA (where A + T > C + G) and GC-type DNA (where A + T < C + G). In the analysis of the *cytochrome b* nucleotide sequences, all individuals examined displayed a characteristic AT-type DNA profile (A + T > C + G). A comparable conclusion was reached regarding the nucleotide sequences of the *D-loop* region. The coefficient of specificity for the two combined nucleotide sequences averaged 1.43, a value consistent with the typical range observed in animal organisms.

*2.7. Data Analysis*

Data analysis was conducted using a series of specialized software programs (Table 2). Initial analysis of the sequences, including fluorogram generation and correction, was performed with DNABaser (DNA assembly programs). To examine the phylogenetic lineage between the Transylvanian Pinzgau cattle and their wild ancestor, *Bos taurus*, we aligned the total mitochondrial genome sequences of the species, alongside the specific *cytochrome b* genes and mitochondrial fragments (*D-loop*) of the Pinzgau cows, obtained from database GenBank-NCBI [43]. For every sample, the sequences forward and reverse were combined based on the sequences generated by the two primers, then trimmed to produce a single, unique sequence. ClustalW was used to align all sequences from the samples [44], implemented in the MegaAlignment module. Sequence alignment and phylogenetic tree generation were performed using the MEGA X 11.0.10 software (Center for Evolutionary Medicine and Informatics, Tempe, AZ, USA, providing Molecular Evolutionary Genetics Analysis).

**Table 2.** Data analysis tools and software used.

| Software No. | Specification of the Software Program Utilized | Type of Data Analysis | Reference |
|---|---|---|---|
| 1. | CEQ™ 8000 BECKMAN COULTER Life Sciences macrogen EUROPE | PCR_amplicon sequencing<br>✓→*Sanger sequencing* | [45] |
| 2. | DNA BASER sequence assembly software | Chromatogram alignment and adjustment of sequences<br>✓→*DNA Baser 5.15* | [46] |
| 3. | MEGA Molecular Evolutionary Genetics Analysis | DNA sequence alignment<br>✓→*Mega X 11.0.10* | [46,47] |
| 4. | jModelTest | Evaluation of the optimal substitution model<br>✓→*jModelTest 2.1.10* | [48,49] |
| 5. | PopART | Assembly of the haplotype network<br>✓→*PopArt 1.7* | [49–51] |
| 6. | SEA- PhyML | Phylogenetic tree build<br>✓→*SeaView/PhyML 5.0.5* | [52–54] |
| 7. | DnaSP | Evaluation of nucleotide sequence diversity<br>✓→*DnaSP 5.10.1* | [51,55,56] |

The sequence diversity under examination was assessed using LaunchDnaSP version 4.50.3 program [52]. Median-joining network analysis was conducted utilizing the PopART 1.7 software. The neighbor-joining tree was constructed employing the Kimura 2-parameter model, incorporating the following parameters: 1000 bootstrap replicates, a gamma distribution (+G) utilizing five rate categories and an evolutionary invariability model (+I).

## 3. Results

### *3.1. Dynamics of the Evolutionary Rate of mtDNA Markers*

To evaluate variation in the genetic makeup of *D-loop* sequences, the analysis included aligning the 24 sequences to identify haplotypes and nucleotide variation sites, using the reference sequence encoded V00654 in the GenBank database. The haplogroup T3 exhibited the most frequent occurrence, consistent with predictions for European cattle [57,58]. Also, the observed genetic variation in haplotypes was measured at $0.908 \pm 0.005$. Utilizing the gene sequences of the mitochondrial markers studied, the demographic and territorial expansion of the Transylvanian Pinzgau cattle was evaluated through the model for calculating mismatch distribution [59,60]. This model illustrates the distribution of nucleotide differences observed among haplotype pairs. Generally, the distribution can take a unimodal form, indicative of populations that have experienced recent expansion in population or area, characterized by significant migration, or multimodal, typical of populations in a stable demographic equilibrium. Concerning the population investigated, the analysis of both mitochondrial markers yielded a multimodal distribution (Figure 6a,b).

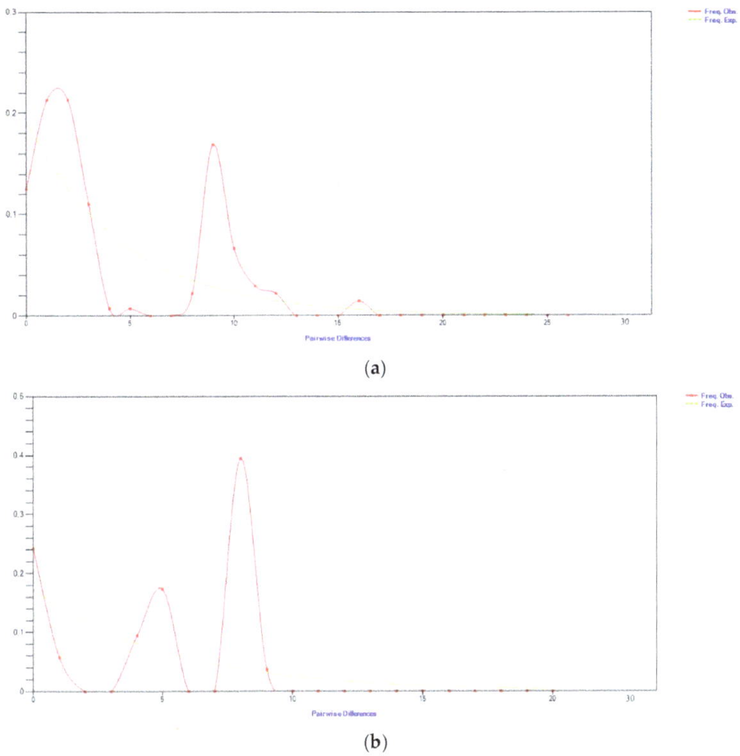


**Figure 6.** Demographic expansion of the Transylvanian Pinzgau cattle: (**a**) *cytochrome b*; (**b**) *D-loop*.

The gene sequences were aligned using the ClustalW method via the MegAlign module. This process involved merging both forward and reverse sequences for each animal through alignment with the respective primer sequences, followed by the separation of the two strands and their subsequent combination into a singular sequence. A total of 23 variable sites (1.98%) were identified, among which 17 were classified as informative sites (1.51%). In the analysis of *D-loop* sequences, 22 variable sites (2.44%) were recorded, with 15 of these being informative sites (1.61%).

The occurrence rates of the nitrogenous bases were 30.5% A, 25.4% T, 14.3% G, and 29.9% C (*cytochrome b*). A<->G- and T<->C-type transversions exhibited values of 4.6481 and 0.5065, respectively (Figure 7a), with their proportion being 0.637 (Table 3).

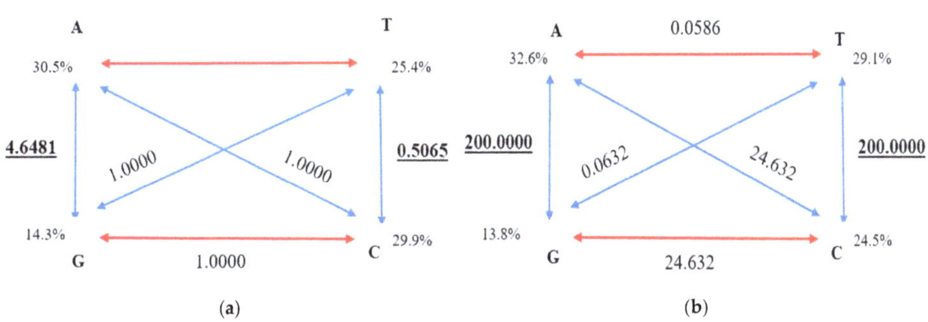


**Figure 7.** Rates of nucleotide substitution calculated using jModelTest: (**a**) sequences of *cytochrome b*; (**b**) sequences of the *D-loop*.

**Table 3.** Count of nucleotide sites and Ti/Tv ratio for *cyt b* and *D-loop* mtDNA characteristic of the Transylvanian Pinzgau.

| Sequences | Total of Informative Sites | % | Total of Variable Sites | % | Tr/Tv [1] Ratio |
|---|---|---|---|---|---|
| *cyt b* | 17 | 1.51 | 23 | 1.98 | 0.637 |
| *D-loop* | 15 | 1.61 | 22 | 2.44 | 5.267 |

[1] ti/tv: transition/trasversion ratio.

For the *D-loop* sequences, the distribution of nitrogenous bases was as follows: 32.6% A, 29.1% T, 24.5% C, and 13.8% G. The Ti/Tv ratio exhibited a value of 5.267 (Table 3 and Figure 7b).

### 3.2. Haplotype Frequency Evaluation

Through the analysis and interpretation of the DNA sequences of the two mitochondrial markers (*cyt b* and *D-loop*), three haplotypes with varying frequencies were identified (T1, T2, and T3), into which the 24 cows examined were grouped (Figure 8). The molecular diversity assessment revealed that contemporary taurine mitochondrial genomes cluster within several closely related lineages, designated as T haplogroup, respectively T1, T2, T3, and T4, exhibiting distinct geographical structuring: T1 is predominantly located in Africa; T2 is believed to have originated in the Near East and Western Asia; T3 is primarily found in Europe and is originating from the growth of a small cattle population domesticated in the Middle East; T4 was identified as a derived subclade within T3, likely disseminated throughout East Asia due to a founder effect resulting from the eastward migration of cattle [21,40].

The T3 haplotype had the highest frequency, occurring in 18 of the analyzed animals (75%). The T2 haplotype was found in five animals, representing 21% and one animal individual was included in the T1 haplotype, which showed the lowest occurrence, at only 4% (Table 4).

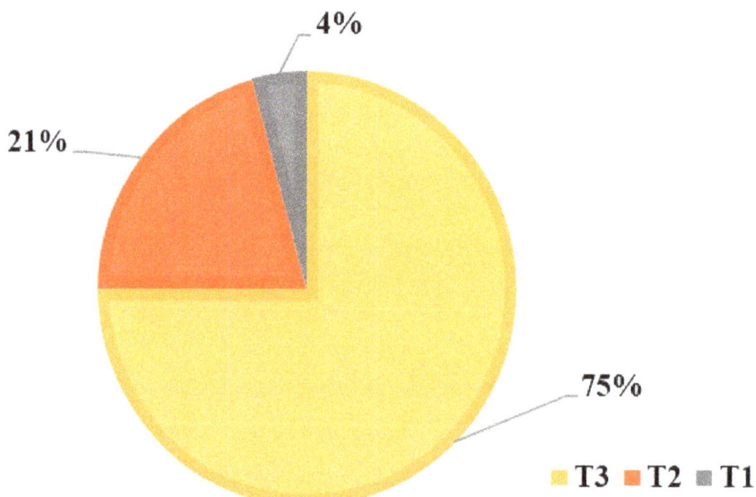


**Figure 8.** Distribution of haplotype frequencies within the sampled animals of the Transylvanian Pinzgau population.

**Table 4.** Identified haplotypes and their representative individuals within the Transylvanian Pinzgau population.

| Identified Haplotypes | Representative Individuals of Haplotype | Total No. of Individuals/Haplotype |
|---|---|---|
| T1 | tp_09 | 1 |
| T2 | tp _01; tp _11; tp _12; tp _20; tp _21 | 5 |
| T3 | tp _02; tp _03; tp _04; tp _05; tp _06; tp _07; tp _08; tp _10; tp _13; tp _14; tp _15; tp _16; tp _17; tp _18; tp _19; tp _22; tp _23; tp _24 | 18 |

### 3.3. *Analysis of Haplotype Networks and Construction of Phylogenetic Trees*

The gene sequences were examined by utilizing the Network 10.2.0.0 software, producing the haplotype design (Figure 9). Three primary haplotypes (T1, T2, and T3) were identified, each with distinct connection patterns. All haplotypes were associated with a specific number of animals.

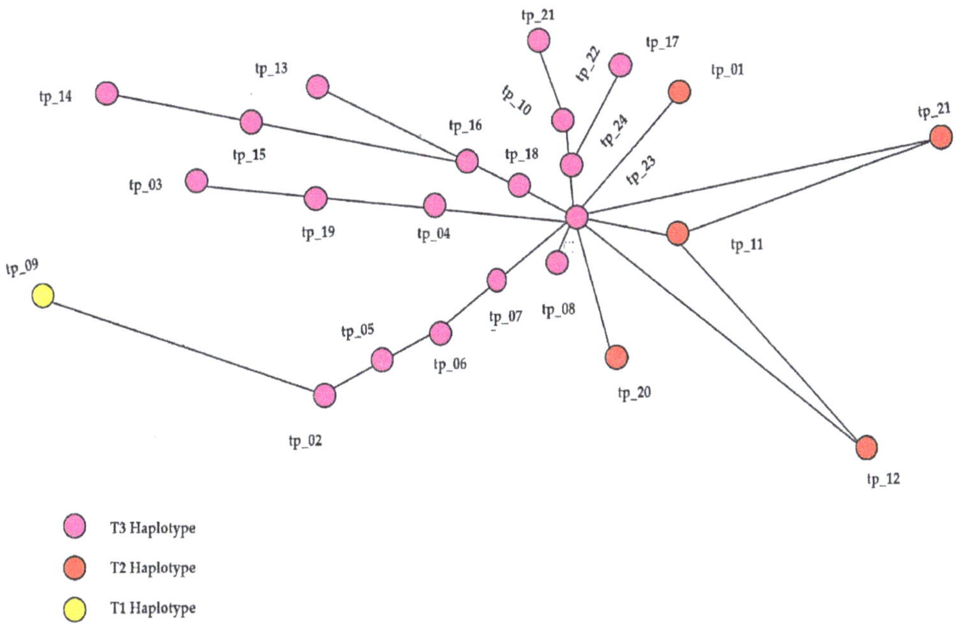


**Figure 9.** Haplotype network of Transylvanian Pinzgau cattle.

The analysis of the distribution of the three haplotypes revealed that the T3 haplotype had the highest frequency, found in 18 of the 24 individuals studied. The genetic divergence between haplotypes T1, T2, and T3 ranged from one to four nucleotide sites. The identification of T-derived haplotypes indicates that the population has undergone demographic expansion and confirms that the cattle analyzed are directly descended from *Bos taurus*. This is a crucial finding for genetic conservation programs aimed at preserving this valuable breed (Figure 10).

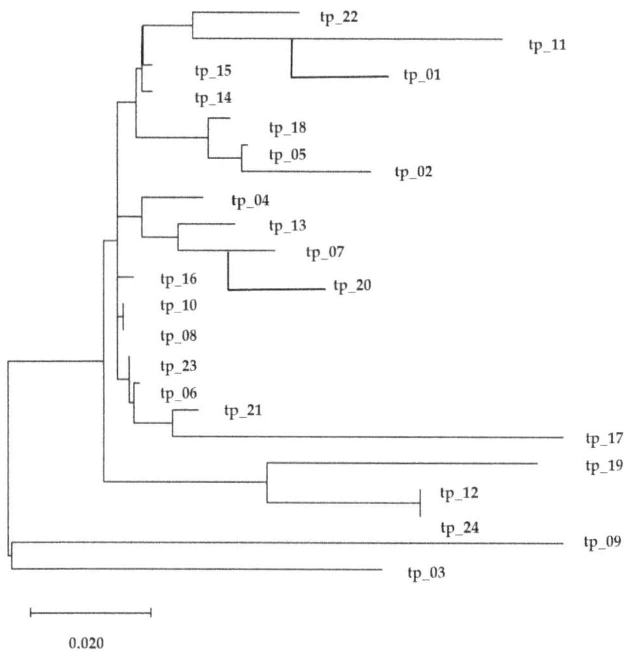


**Figure 10.** Phylogenetic tree constructed based on the sequence analysis of the Transylvanian Pinzgau cattle.

## 4. Discussion

Until now, only a limited number of studies have been conducted regarding the genetic structure of endangered Romanian cattle breeds in comparison to other cattle breeds globally. Most of these studies have been restricted to a limited number of breeds and primarily concentrated on the nuclear DNA. This research presents, for the first time, the genetic diversity and phylogenetic characteristics of native Transylvanian Pinzgau cattle, derived from the sequencing of mitochondrial DNA genes (*cytochrome b* and *D-loop*).

Numerous studies show that the haplotypes identified in the Transylvanian Pinzgau were found to be identical to those documented in Austrian Pinzgau, German Pinzgau, Slovak Pinzgau, and Swiss Pinzgau populations [17,57,58].

Recent investigations [17,55,57,61–63] have demonstrated that nearly all taurine cattle are assigned to the T macro-haplogroup, with a predicted divergence time of approximately 16,000 years, suggesting a significant evolutionary bottleneck in the taurine lineage within the *Bos taurus* genus. The T macro-haplogroup is further subdivided into two related subclades (T5 and T1/T2/T3), with T1/T2/T3 being the most common subclade. Over time, T4 has been incorporated into T3 [17,64].

The phylogeny of the Transylvanian Pinzgau cattle breed, when analyzed alongside other Pinzgau breeds in Europe and beyond, reveals both shared ancestry and unique genetic distinctions shaped by regional breeding practices and geographic isolation. Mitochondrial DNA (mtDNA) studies, often leveraging markers like the *D-loop* region, trace maternal lineage and suggest that the Transylvanian Pinzgau shares a close ancestry with other European Pinzgau populations, particularly the Austrian Pinzgau [18,32,34,35]. This connection reflects historical breed movements across central and eastern Europe, although some genetic markers show divergence, possibly due to limited gene flow in isolated regions such as Transylvania [18,32,34,64–66]. Genomic studies indicate that while Transylvanian Pinzgau cattle maintain a unique genetic profile, they are also closely linked to broader European cattle populations, shaped by shared domestication and migration patterns [64–66]. Understanding these phylogenetic relationships is crucial for conservation, as it highlights the variability of genetic traits within the breed and underscores the need to preserve this unique lineage.

Current research in bovine genomics emphasizes the importance of maintaining genetic diversity in isolated breeds like the Transylvanian Pinzgau to ensure the resilience and sustainability of global cattle genetics [23,24,67].

The Transylvanian Pinzgau cattle, like many European cattle breeds, are primarily associated with mitochondrial haplogroup T3, which is common across Europe and reflects maternal lineage tracing back to Near Eastern domestication events that later spread into Europe. Haplogroup T3 is widely prevalent in central and eastern European breeds, including the Pinzgau lineage, due to shared ancestry and historical cattle movement within Europe [40,64,68]. Although rarer, minor occurrences of other haplogroups such as T1—more common in African cattle—have occasionally been identified in European cattle, likely due to ancient trade and migration influences, though it is less common in the Pinzgau [21,69]. These haplogroups, studied through mitochondrial DNA analysis, highlight the genetic distinctions and evolutionary history of the Transylvanian Pinzgau relative to other Pinzgau and European breeds, offering valuable insights for conservation and genetic diversity assessments [21,24,67].

The Transylvanian Pinzgau cattle population analyzed shows a strong presence of haplogroup T3 (18 out of 24 individuals), which is typical of European cattle breeds and mirrors findings in other Pinzgau populations, such as the Austrian Pinzgau, where T3 also predominates due to shared ancestry from early European domesticated cattle [21,67]. The presence of haplogroups T2 (in five individuals) and T1 (in one individual) adds diversity to the Transylvanian Pinzgau's lineage, with T2 being relatively common in Near Eastern and European breeds and reflecting ancient domestication patterns that reached Europe. The single T1 individual is intriguing, as T1 is typically associated with African cattle and likely entered Europe through historic trade or migration routes, as occasionally seen in Mediterranean breeds [21]. These findings align with genomic studies on European cattle that show clustering but with distinct genetic variations based on geographic isolation, as seen in the Transylvanian Pinzgau, which may have undergone genetic bottlenecks due to isolation in Transylvania [23,70]. This diversity highlights the conservation value of the Transylvanian Pinzgau, as preserving unique haplotypes within isolated populations contributes to overall genetic resilience in cattle breeds [7].

Studies of mitochondrial DNA (mtDNA) across various Pinzgau cattle populations, including Austrian, German, Slovak, and Swiss breeds, reveal a strong predominance of haplogroup T3, which is the dominant lineage in European cattle [23,70]. In Austrian Pinzgau, haplogroup T3 accounts for nearly all identified mtDNA haplotypes, consistent with its ancestry from early domestication events in the Near East. Similarly, German Pinzgau cattle predominantly exhibit haplogroup T3, with occasional instances of haplogroup T2, reflecting historical breeding practices and limited genetic mixing from neighboring regions. The Slovak Pinzgau also primarily features haplogroup T3, with rare occurrences of T2 and even more infrequently, T1, which may indicate low-level gene flow from ancient trade routes [71–77]. Likewise, Swiss Pinzgau cattle show a similar pattern, predominantly carrying haplogroup T3, with T2 appearing sporadically [78]. Collectively, these findings highlight the genetic homogeneity within Pinzgau breeds, largely shaped by shared ancestry and geographical isolation, with haplogroup T3 serving as a marker of their common European lineage [79].

The occurrence of haplotypes derived from haplogroup T indicates a demographic expansion model for these individuals and signifies their lineage within *Bos taurus*, along with domestication in the Fertile Crescent during the Neolithic era [80–82].

## 5. Conclusions

This research highlights the importance of conserving the genetic diversity and unique phylogenetic lineage of the endangered Transylvanian Pinzgau cattle. Using mitochondrial

DNA markers, we identified strong genetic links to other European Pinzgau populations, alongside distinct variations shaped by geographical isolation. The detection of haplogroups T3 and T1 emphasizes shared ancestry and ancient gene flow, contributing to the breed's genetic richness.

These findings underscore the need for targeted conservation efforts to preserve isolated breeds like the Transylvanian Pinzgau, which enhance global cattle resilience. Additionally, maintaining unique haplotypes supports sustainable breeding practices, ensuring adaptability to future environmental and agricultural challenges.

**Author Contributions:** Conceptualization, M.-A.D.; methodology, M.-A.D., C.P., R.-M.R.-R. and B.-M.M.; software, M.-A.D., Ș.C. and M.G.D.; validation, M.-A.D. and C.S.; formal analysis, M.-A.D., Ș.C. and A.C.; investigation, M.-A.D., A.U., R.-M.R.-R. and B.-M.M.; data curation, C.P.; writing—original draft preparation, M.-A.D.; writing—review and editing, M.-A.D., A.U., M.G.D. and A.C.; supervision, M.-A.D. All authors have read and agreed to the published version of the manuscript.

**Funding:** This research received no external funding.

**Institutional Review Board Statement:** In this research, blood samples were obtained from 24 head of cattle, in compliance with the EU regulation UE 2016/679 concerning the "general data protection regulation." The samples were collected via jugular vein puncture, following the procedures outlined in the EU Directive 2010/63/EU regarding the "protection of animals used for scientific purposes." No experimental procedures were applied to the animals.

**Data Availability Statement:** The data present in this study are available on request from the first author and corresponding authors.

**Conflicts of Interest:** The authors declare no conflicts of interest.

## References


1. Medugorac, I.; Medugorac, A.; Russ, I.; Veit-Kensch, C.E.; Taberlet, P.; Luntz, B.; Mix, H.M.; Förster, M. Genetic diversity of European cattle breeds highlights the conservation value of traditional unselected breeds with high effective population size. *Mol. Ecol.* **2009**, *18*, 3394–3410. [CrossRef] [PubMed]
2. Koyun, H.; Koncagül, S.; Karakuş, K. Significance of Genetic Diversity in Farm Animal Production. *Türk Bilimsel Derlemeler Derg.* **2016**, *1*, 82–83.
3. Simeanu, D.; Radu-Rusu, R.M. Animal nutrition and productions. *Agriculture* **2023**, *13*, 943. [CrossRef]
4. Maciuc, V. *Cattle Breeding Management*; Alfa: Iasi, Romania, 2006; pp. 76–78. ISBN 973-8953-18-9.
5. Kasarda, R.; Jamborová, L.; Moravcíková, N. Genetic diversity and production potential of animal food resources. *Acta Fytotech. Zootech.* **2020**, *23*, 102–108. [CrossRef]
6. Mădescu, B.M.; Lazăr, R.; Ciobanu, M.M.; Boișteanu, P.C. Morpfo-Productive Characteristics of Aubrac Cattle Breed: A Sistematic Review. *Sci. Pap. Ser. D Anim. Sci.* **2021**, *64*, 260–265.
7. Groeneveld, L.F.; Lenstra, J.A.; Eding, H.; Toro, M.A.; Scherf, B.; Pilling, D. Genetic diversity in farm animals—A review. *Anim. Genet.* **2010**, *11*, 6–31. [CrossRef]
8. Usturoi, A.; Avarvarei, B.V.; Nistor, C.E.; Simeanu, C.; Davidescu, M.A.; Usturoi, M.G. The Quality of Some Acid Dairy Products Obtained in the Traditional System. *Sci. Pap. Anim. Sci. Biotechnol.* **2024**, *57*, 187–193.
9. Kasarda, R.; Vostrý, L.; Vostrá-Vydrová, H.; Candráková, K.; Moravčíková, N. Food Resources Biodiversity: The Case of Local Cattle in Slovakia. *Sustainability* **2021**, *13*, 1296. [CrossRef]
10. Barnes, K.; Collins, T.; Dion, S.; Reynolds, H.; Riess, S.; Stanzyk, A.; Wolfe, A.; Lonergan, S.; Boettcher, P.; Charrondiere, U.R.; et al. Importance of cattle biodiversity and its influence on the nutrient composition of beef. *Anim. Front.* **2012**, *2*, 54–60. [CrossRef]
11. Yan, L.; She, Y.; Elzo, M.A.; Zhang, C.; Fang, X.; Chen, H. Exploring genetic diversity and phylogenic relationships of Chinese cattle using gene mtDNA 16S rRNA. *Arch. Anim. Breed.* **2019**, *62*, 325–333. [CrossRef]
12. Davidescu, M.A.; Grădinaru, A.C.; Creangă, Ș. Endangered romanian cattle breeds–between traditional breeding and genetic conservation. *Sci. Pap. Anim. Sci. Ser. Sci. Pap. Anim. Husb. Ser.* **2021**, *75*, 66–75.
13. Davidescu, M.-A.; Simeanu, D.; Gorgan, D.-L.; Ciorpac, M.; Creanga, S. Analysis of Phylogeny and Genetic Diversity of Endangered Romanian Grey Steppe Cattle Breed, a Reservoir of Valuable Genes to Preserve Biodiversity. *Agriculture* **2022**, *12*, 2059. [CrossRef]
14. Upadhyay, M.R.; European Cattle Genetic Diversity Consortium; Chen, W.; Lenstra, J.A.; Goderie, C.R.J.; MacHugh, D.E.; Park, S.D.E.; Magee, D.A.; Matassino, D.; Ciani, F.; et al. Genetic origin, admixture and population history of aurochs (*Bos primigenius*) and primitive European cattle. *Heredity* **2017**, *118*, 169–176. [CrossRef] [PubMed]
15. Prihandini, P.W.; Primasari, A.; Luthfi, M.; Efendy, J.; Pamungkas, D. Genetic Diversity of Mitochondrial DNA *Cytochrome b* in Indonesian Native and Local Cattle Populations. *J. Ilmu Ternak Dan Vet.* **2020**, *25*, 39–47. [CrossRef]

16. Küttel, L.; Letko, A.; Häfliger, I.M.; Signer-Hasler, H.; Joller, S.; Hirsbrunner, G.; Mészáros, G.; Sölkner, J.; Flury, C.; Leeb, T.; et al. A complex structural variant at the KIT locus in cattle with the Pinzgauer spotting pattern. *Anim. Genet.* **2019**, *50*, 423–429. [CrossRef]
17. Kasarda, R.; Moravcikova, N.; Sidlova, V.; Krupova, Z.; Krupa, E.; Kadlecik, O. Progress in evaluation of Diversity in Pinzgau cattle based on molecular markers. *Arch. Zootech.* **2016**, *19*, 37–44.
18. Ciocan-Alupii, M.; Radu-Rusu, R.M.; Pânzaru, C.; Nistor-Anton, M.; Bilkevich, V.; Maciuc, V. Study of productive performance in the Pinzgau breed exploited in the Dornelor Basin, Suceava County. *Sci. Pap. Ser. D Anim. Sci.* **2022**, *LXV*, 298–303.
19. Food and Agriculture Organization of the United Nations. Available online: https://www.fao.org (accessed on 12 January 2024).
20. Scherf, B.D. *World Watch List for Domestic Animal Diversity*, 3rd ed.; FAO: Rome, Italy, 2000; Available online: https://www.fao.org/docrep/009/x8750e/x8750e00.htm (accessed on 11 October 2023).
21. Bonfiglio, S.; Ginja, C.; De Gaetano, A.; Achilli, A.; Olivieri, A.; Colli, L.; Tesfaye, K.; Agha, S.H.; Gama, L.T.; Cattonaro, F.; et al. Origin and spread of *Bos taurus*: New clues from mitochondrial genomes belonging to haplogroup T1. *PLoS ONE* **2012**, *7*, e38601. [CrossRef]
22. Scheu, A.; Powell, A.; Bollongino, R.; Vigne, J.-D.; Tresset, A.; Çakırlar, C.; Benecke, N.; Burger, J. The genetic prehistory of domesticated cattle from their origin to the spread across Europe. *BMC Genet.* **2015**, *16*, 54. [CrossRef]
23. Decker, J.E.; McKay, S.D.; Rolf, M.M.; Kim, J.; Alcalá, A.M.; Sonstegard, T.S.; Hanotte, O.; Götherström, A.; Seabury, C.M.; Praharani, L.; et al. Worldwide patterns of ancestry, divergence, and admixture in domesticated cattle. *PLoS Genet.* **2014**, *10*, e1004254. [CrossRef]
24. Achilli, A.; Bonfiglio, S.; Olivieri, A.; Malusà, A.; Pala, M. The multifaceted origin of taurine cattle reflected by the mitochondrial genome. *PLoS ONE* **2009**, *4*, e5753. [CrossRef]
25. Gurke, M.; Vidal-Gorosquieta, A.; Pajimans, J.L.A.; Węcek, K.; Barlow, A.; González-Fortes, G.; Hartmann, S.; Grandal-d'Anglade, A.; Hofreiter, M. Insight into the introduction of domestic cattle and the process of Neolithization to the Spanish region Galicia by genetic evidence. *PLoS ONE* **2021**, *16*, e0249537. [CrossRef] [PubMed]
26. Cubric-Curik, V.; Novosel, D.; Brajkovic, V.; Stabelli, O.R.; Krebs, S.; Sölkner, J.; Salamon, D.; Ristov, S.; Berger, B.; Trivizaki, S.; et al. Large-scale mitogenome sequencing reveals consecutive expansions of domestic taurine cattle and supports sporadic aurochs introgression. *Evol. Appl.* **2022**, *15*, 663–678. [CrossRef]
27. Bollongino, R.; Edwards, C.J.; Alt, K.W.; Burger, J.; Bradley, D.G. Early history of European domestic cattle as revealed by ancient DNA. *Biol. Lett.* **2006**, *2*, 155–159. [CrossRef] [PubMed]
28. Janák, V.; Novák, K.; Kyselý, R. Late History of Cattle Breeds in Central Europe in Light of Genetic and Archaeogenetic Sources—Overview, Thoughts, and Perspectives. *Animals* **2024**, *14*, 645. [CrossRef]
29. Kukučková, V.; Kasarda, R.; Žitný, J.; Moravčíková, N. Genetic markers and biostatistical methods as appropriate tools to preserve genetic resources. *AGROFOR Intern. J.* **2018**, *3*, 41–48. [CrossRef]
30. Stock, F.; Edwards, C.J.; Bollongino, R.; Finlay, E.K.; Burger, J.; Bradley, D.G. *Cytochrome b* sequences of ancient cattle and wild ox support phylogenetic complexity in the ancient and modern bovine populations. *Anim. Genet.* **2009**, *40*, 694–700. [CrossRef]
31. Necula, D.; Ciupe, S.; Tamas-Krumpe, O.; Todoran, D.; Ognean, L. Analysis of the Current Opportunities for Valorization and Conservation of the Main Autochthonous Cattle Breeds in the Conditions of the Carpathian Mountain Areas. *Vet. Med.* **2024**, *81*, 1–9. [CrossRef]
32. Popa, R.; Popa, D.; Maftei, M.; Dronca, D.; Băcilă, V. Animal biodiversity conservation, a key of sustainable agriculture. Case study: The Romanian Pinzgau breed in Transylvania region. *Sci. Pap. Anim. Sci. Ser. D* **2012**, *LV*, 25–29.
33. Maciuc, V.; Pânzaru, C.; Ciocan-Alupii, M.; Radu-Rusu, C.-G.; Radu-Rusu, R.-M. Comparative Assessment of the Nutritional and Sanogenic Features of Certain Cheese Sorts Originating in Conventional Dairy Farms and in "Mountainous" Quality System Farms. *Agriculture* **2024**, *14*, 172. [CrossRef]
34. Cărătuș, N.; Vidu, L.; Mărginean, G.E. Study on the exploitation of cattle in Transylvania. *Ann. "Valahia" Univ. Targoviste* **2019**, *13*, 57–62. [CrossRef]
35. Necula, D.; Cătună-Boca, C.; Tamas-Krumpe, O.M.; Ognean, L. Evolution of the Transylvanian Pinzgau cattle breed in its natural habitat in the Carpathian mountain areas (A short review). *ABAH Bioflux* **2022**, *14*, 96–101.
36. PCR Master Mix—Promega Corporation. PCR Master Mix. Available online: https://www.promega.ro/products/pcr/taq-polymerase/master-mix-pcr/?catNum=M7502 (accessed on 29 March 2024).
37. Minhas-Khan, A.; Ghafar-Zadeh, M.; Shaffaf, T.; Forouhi, S.; Scime, A.; Magierowski, S.; Ghafar-Zadeh, E. UV-Vis Spectrophotometric Analysis of DNA Retrieval for DNA Storage Applications. *Actuators* **2021**, *10*, 246. [CrossRef]
38. Koetsier, G.; Cantor, E. A Practical Guide to Analyzing Nucleic Acid Concentration and Purity with Microvolume Spectrophotometers. In *Technical Note*; New England Biolabs Inc.: Ipswich, MA, USA, 2019; pp. 1–8.
39. GenBank Overview-NCBI. Bos taurus Complete Mitochondrial Genome. Available online: https://www.ncbi.nlm.nih.gov/nuccore/V00654 (accessed on 9 December 2023).
40. Seroussi, E.; Yakobson, E. Bovine mtDNA *D-loop* haplotypes exceed mutations in number despite reduced recombination: An effective alternative for identity control. *Animal* **2010**, *4*, 1818–1822. [CrossRef]
41. Arbizu, C.I.; Ferro-Mauricio, R.D.; Chávez-Galarza, J.C.; Vásquez, H.V.; Maicelo, J.L.; Poemape, C.; Gonzales, J.; Quilcate, C.; Corredor, F.-A. The Complete Mitochondrial Genome of a Neglected Breed, the Peruvian Creole Cattle (*Bos taurus*), and Its Phylogenetic Analysis. *Data* **2022**, *7*, 76. [CrossRef]

42. Hiendleder, S.; Lewalski, H.; Janke, A. Complete Mitochondrial Genomes of *Bos taurus* and *Bos indicus* Provide New Insights into Intra-Species Variation, Taxonomy and Domestication. *Cytogenet. Genome Res.* **2008**, *120*, 150–156. [CrossRef]
43. UCSC Genome Browser. Available online: https://genome.ucsc.edu/ (accessed on 9 September 2023).
44. Thompson, J.D.; Higgins, D.G.; Gibson, T.J. CLUSTAL W: Improving the sensitivity of progressive multiple sequence alignment through sequence weighting, position-specific gap penalties and weight matrix choice. *Nucleic Acids Res.* **1994**, *22*, 4673–4680. [CrossRef] [PubMed]
45. Furutani, S.; Furutani, N.; Kawai, Y.; Nakayama, A.; Nagai, H. Rapid DNA Sequencing Technology Based on the Sanger Method for Bacterial Identification. *Sensors* **2022**, *22*, 2130. [CrossRef]
46. Li, M.; Nordborg, M.; Li, L.M. Adjust quality scores from alignment and improve sequencing accuracy. *Nucleic Acids Res.* **2004**, *32*, 5183–5191. [CrossRef]
47. Kumar, S.; Stecher, G.; Li, M.; Knyaz, C.; Tamura, K. MEGA X: Molecular Evolutionary Genetics Analysis across Computing Platforms. *Mol. Biol. Evol.* **2018**, *35*, 1547–1549. [CrossRef]
48. Chen, W.; Kenney, T.; Bielawski, J.; Gu, H. Testing adequacy for DNA substitution models. *BMC Bioinform.* **2019**, *20*, 329. [CrossRef] [PubMed]
49. Sinding, M.H.S.; Gilbert, M.T.P. The Draft Genome of Extinct European Aurochs and its Implications for De-Extinction. *Open Quat.* **2016**, *2*, 7. [CrossRef]
50. Gouy, M.; Tannier, E.; Comte, N.; Parsons, D.P. Seaview version 5: A multiplatform software for multiple sequence alignment, molecular phylogenetic analyses, and tree reconciliation. *Mult. Seq. Alignment Methods Protoc.* **2021**, *2231*, 241–260. [CrossRef]
51. Rozas, J.; Ferrer-Mata, A.; Sánchez-DelBarrio, J.C.; Guirao-Rico, S.; Librado, P.; Ramos-Onsins, S.E.; Sánchez-Gracia, A. DnaSP 6: DNA Sequence Polymorphism Analysis of Large Data Sets. *Mol. Biol. Evol.* **2017**, *34*, 3299–3302. [CrossRef] [PubMed]
52. Rozas, J.; Sánchez-DelBarrio, J.C.; Messeguer, X.; Rozas, R. DnaSP, DNA polymorphism analyses by the coalescent and other methods. *Bioinformatics* **2003**, *19*, 2496–2497. [CrossRef] [PubMed]
53. Tarekegn, G.M.; Ji, X.Y.; Bai, X.; Liu, B.; Zhang, W.; Birungi, J.; Djikeng, A.; Tesfaye, K. Variations in mitochondrial *cytochrome b* region among Ethiopian indigenous cattle populations assert *Bos taurus* maternal origin and historical dynamics. *Asian-Australas. J. Anim. Sci.* **2018**, *31*, 1393–1400. [CrossRef] [PubMed]
54. Yang, S.L.; Lin, R.Y.; Xu, L.X.; Cheng, L. Analysis of polymorphisms of mitochondrial DNA *D-loop* and Mc1R gene in Chinese Wuchuan Black cattle. *J. Appl. Anim. Res.* **2014**, *42*, 487–491. [CrossRef]
55. Kukučková, V.; Moravčíková, N.; Ferenčaković, M.; Simčič, M.; Mészáros, G.; Sölkner, J.; Trakovická, A.; Kadlečík, O.; Curik, I.; Kasarda, R. Genomic characterization of Pinzgau cattle: Genetic conservation and breeding perspectives. *Conserv. Genet.* **2017**, *18*, 893–910. [CrossRef]
56. Singhla, T.; Boonyayatra, S.; Chulakasian, S.; Lukkana, M.; Alvarez, J.; Sreevatsan, S.; Wells, S.J. Determination of the sensitivity and specificity of bovine tuberculosis screening tests in dairy herds in Thailand using a Bayesian approach. *BMC Vet. Res.* **2019**, *15*, 149. [CrossRef]
57. Kasarda, R.; Moravčí-ková, N.; Trakovická, A.; Mészáros, G.; Kadlečí-k, O. Genome-wide selection signatures in Pinzgau cattle. *Potravin. Slovak J. Food Sci.* **2015**, *9*, 268–274. [CrossRef]
58. Miluchová, M.; Gábor, M.; Gašper, J. Analysis of the Genetic Structure of Slovak Holstein Cattle Using Seven Candidate Genes Related to Milk Quality. *Diversity* **2022**, *14*, 989. [CrossRef]
59. Smitz, N.; Berthouly, C.; Cornélis, D.; Heller, R.; Van Hooft, P.; Chardonnet, P.; Caron, A.; Prins, H.; van Vuuren, B.J.; De Iongh, H.; et al. Pan-African genetic structure in the African buffalo (*Syncerus caffer*): Investigating intraspecific divergence. *PLoS ONE* **2013**, *8*, e56235. [CrossRef]
60. Li, M.H.; Zerabruk, M.; Vangen, O.; Olsaker, I.; Kantanen, J. Reduced genetic structure of north Ethiopian cattle revealed by Y-chromosome analysis. *Heredity* **2007**, *98*, 214–221. [CrossRef]
61. Felius, M.; Koolmees, P.A.; Theunissen, B.; European Cattle Genetic Diversity Consortium; Lenstra, J.A. On the Breeds of Cattle—Historic and Current Classifications. *Diversity* **2011**, *3*, 660–692. [CrossRef]
62. Dorji, J.; Vander-Jagt, C.; Chamberlain, A.; Cocks, B.; MacLeod, I.; Daetwyler, H. Cattle Maternal Diversity Inferred from 1,883 Taurine and Indicine Mitogenomes-Preprint. *Res. Sq.* **2021**, 1–35. [CrossRef]
63. Dorji, J.; Vander Jagt, C.J.; Chamberlain, A.J.; Cocks, B.G.; MacLeod, I.M.; Daetwyler, H.D. Recovery of mitogenomes from whole genome sequences to infer maternal diversity in 1883 modern taurine and indicine cattle. *Sci. Rep.* **2022**, *12*, 5582. [CrossRef]
64. Liu, H.; Zhai, J.; Wu, H.; Wang, J.; Zhang, S.; Li, J.; Niu, Z.; Shen, C.; Zhang, K.; Liu, Z.; et al. Diversity of Mitochondrial DNA Haplogroups and Their Association with Bovine Antral Follicle Count. *Animals* **2022**, *12*, 2350. [CrossRef] [PubMed]
65. Baumung, R.; Sölkner, J. Analysis of pedigrees of Tux-Zillertal, Carinthian Blond and Original Pinzgau cattle population in Austria. *J. Anim. Breed. Genet.* **2002**, *119*, 175–181. [CrossRef]
66. Matiuti, M.; Bogdan, A.T.; Crainiceanu, E. A case study of Transylvanian Pinzgau in Banat areal. *Sci. Pap. Anim. Sci. Biotechnol.* **2009**, *42*, 169–174.
67. Mannen, H.; Kohno, M.; Nagata, Y.; Tsuji, S.; Bradley, D.G.; Yeo, J.S.; Nyamsamba, D.; Zagdsuren, Y.; Yokohama, M.; Nomura, K.; et al. Independent mitochondrial origin and historical genetic differentiation in North Eastern Asian cattle. *Mol. Phylogenet Evol.* **2004**, *32*, 539–544. [CrossRef]

68. Kasarda, R.; Moravčíková, N.; Lehocká, K.; Olšanská, B.; Prišťák, J.; Trakovická, A.; Kadlečík, O. Haplotype block structure in the genome of Slovak Pinzgau cattle. In Proceedings of the X International Scientific Agricultural Symposium Agrosym, Jahorina, Bosnia and Herzegovina, 3–6 October 2019; pp. 1440–1445.
69. Šidlová, V.; Kasarda, R.; Moravčíková, N.; Trakovická, A.; Kadlečík, O. Microsatellite analysis of population structure in Slovak Pinzgau cattle. *Acta Agrar. Kaposváriensis* **2014**, *18*, 24–29.
70. Mang, N. Genetic Structure of Quantitative Characters in Pinzgau of Transilvania Breed from Hateg Region. *Sci. Pap. Anim. Sci. Biotechnol.* **2011**, *44*, 432–434.
71. Brenig, B. Endangered Pinzgauer cattle subtype Jochberger Hummel are genetically distinct. *Anim. Genet.* **2020**, *51*, 590–594. [CrossRef] [PubMed]
72. Kadlečík, O.; Kasarda, R.; Pavlík, I.; Hazuchová, E. Pedigree Analysis of Slovak Pinzgau Breed. *Agric. Conspec. Sci.* **2011**, *76*, 165–168.
73. Kasarda, R.; Kadlecik, O.; Mészáros, G. Trends of endangered population of Pinzgau Cattle in Slovakia. *Arch. Zootech.* **2008**, *11*, 82–87.
74. Šidlová, V.; Moravčíková, N.; Trakovická, A.; Ferenčaković, M.; Curik, I.; Kasarda, R. Production type of Slovak Pinzgau cattle in respect of related breeds. *Acta Fytotechn. Zootechn.* **2015**, *18*, 25–29. [CrossRef]
75. Kukučková, V.; Moravčíková, N.; Trakovická, A.; Kadlečík, O.; Kasarda, R. Genetic differentiation of Slovak Pinzgau, Simmental, Charolais and Holstein cattle based on the linkage disequilibrium, persistence of phase and effective population size. *Acta Argic. Slov.* **2016**, *5*, 37–40. [CrossRef]
76. Melus, V.; Kasarda, R.; Kadlecik, O.; Trakovicka, A. Breeding potential of the Slovak Pinzgau cattle: Seeking biochemical and molecular biologic traits. *Arch. Zootech.* **2009**, *12*, 34–37.
77. Mészáros, G.; Fuerst, C.; Fuerst-Waltl, B.; Kadlečík, O.; Kasarda, R.; Sölkner, J. Genetic evaluation for length of productive life in Slovak Pinzgau cattle. *Arch. Anim. Breed.* **2008**, *51*, 438–448. [CrossRef]
78. Taberlet, P.; Coissac, E.; Pansu, J.; Pompanon, F. Conservation genetics of cattle, sheep, and goats. *Comptes Rendus Biol.* **2011**, *334*, 247–254. [CrossRef]
79. Pavlík, I.; Sölkner, J.; Kadlečík, O.; Kasarda, R.; Mészáros, G.; Fuerst, C.; Fuerst-Waltl, B. Joint genealogical analysis as a tool for diversity evaluation in Pinzgau cattle populations. *Arch. Anim. Breed.* **2014**, *57*, 14. [CrossRef]
80. Magee, D.A.; MacHugh, D.E.; Edwards, C.J. Skip Nav Destination Interrogation of modern and ancient genomes reveals the complex domestic history of cattle. *Anim. Front.* **2014**, *4*, 7–22. [CrossRef]
81. Zhang, Y.; Wei, Z.; Zhang, M.; Wang, S.; Gao, T.; Huang, H.; Zhang, T.; Cai, H.; Liu, X.; Fu, T.; et al. Population Structure and Selection Signal Analysis of Nanyang Cattle Based on Whole-Genome Sequencing Data. *Genes* **2024**, *15*, 351. [CrossRef]
82. Verdugo, M.P.; Mullin, V.E.; Scheu, A.; Mattiangeli, V.; Daly, K.G.; Delser, P.M.; Hare, A.J.; Burger, J.; Collins, M.J.; Kehati, R.; et al. Ancient cattle genomics, origins, and rapid turnover in the Fertile Crescent. *Science* **2019**, *365*, 173–176. [CrossRef] [PubMed]

agriculture

MDPI

Article

# The Influence of Season and Age on Specific Semen Traits and Reproductive Behavior in Carpatina Breed Bucks

**Constantin Pascal [1,*], Claudia Pânzaru [1], Răzvan-Mihail Radu-Rusu [1], Vasile Maciuc [1], Alexandru Marian Florea [2] and Ionică Nechifor [2,*]**

[1] Faculty of Food and Animal Sciences, "Ion Ionescu de la Brad" Iasi University of Life Sciences, 700490 Iasi, Romania; claudia.panzaru@iuls.ro (C.P.); razvan.radu@iuls.ro (R.-M.R.-R.); vasile.maciuc@iuls.ro (V.M.)

[2] Research and Development Station for Sheep and Goat Breeding, Popăuți-Botoșani, 710004 Botoșani, Romania; florea.marian@karakul-moldoovis.ro

* Correspondence: constantin.pascal@iuls.ro (C.P.); nechifor.ionica@karakul-moldoovis.ro (I.N.)

**Abstract:** This study explores how age and seasonal changes impact semen characteristics and reproductive behavior in Carpatina breed bucks. Males were divided into three age groups: young (14–23 months; L14), adult (3–4 years; L34), and older (5–6 years; L56). Scrotal biometry was determined using a measuring tape, and testicular volume was evaluated by fully submerging the testes in a water-filled container and measuring the displaced water. Semen analysis was conducted on samples collected each season, with volume, color, and acidity being assessed. The evaluation of specific semen characteristics (motility, sperm concentration, normal spermatozoa) was conducted using a Computer-Assisted Semen Analysis (CASA) system, and testosterone levels were measured in blood samples collected at the start of each season. Behavior and sexual reflexes were evaluated based on mating desire and the bucks' reaction to the presence of females. Key findings indicate that testicular volume varies significantly with both age and season, with the most pronounced differences between younger bucks and the older groups, especially during autumn. Semen quality parameters such as ejaculate volume, sperm concentration, and motility also showed seasonal fluctuations, with younger bucks having lower sperm concentrations. Testosterone levels were observed to increase with age, peaking in autumn. Behavioral observations revealed that younger bucks exhibited less intense sexual activity, although this improved during autumn. Additionally, a significant correlation was identified between body weight and testicular volume in adult bucks ($R = 0.942$, $p$-value = 0.016 for L34; $R = 0.797$, $p$-value = 0.022 for L56), suggesting that age plays a crucial role in reproductive potential. Our findings highlight that, while bucks are capable of year-round reproduction, autumn provides optimal conditions for semen quality and reproductive performance. This research has valuable implications for optimizing breeding programs, contributing to genetic advancement, and improving management strategies in goat farming, especially within temperate continental climates.

**Keywords:** buck semen; semen traits; testicular size; Carpatina goat breed; sexual behavior

## 1. Introduction

In the area where this study was conducted, goat farming has seen significant growth recently. The Carpatina breed is well adapted to the local environment, demonstrating exceptional organic resistance to microclimatic factors such as temperature fluctuations, humidity, and variations in feed availability. This breed is known for its hardiness, allowing it to thrive in rugged, mountainous areas. It is primarily raised for both milk and meat production, with its milk being valued for its high nutritional content, including good fat and protein levels, making it suitable for dairy products. Additionally, the meat of the Carpatina breed is appreciated for its tenderness and flavor, contributing to its value in local markets. The breed's adaptability, combined with its dual-purpose nature, makes it an integral part of the local farming systems in the Carpathian region [1].

Male goats, known as bucks, play an essential role in producing future generations of offspring. Effective management in goat farming includes situations where bucks are utilized for breeding outside their natural reproductive season. In these instances, certain semen traits may deviate from refqce values, potentially leading to negative effects on reproductive success [1]. Another important aspect of goat farm management is the timing of using young bucks for breeding. Although bucks typically reach puberty between five and eight months of age [2,3], they are generally not used for mating until the next breeding season in most seasonal breeds [4]. Delaying their use until they reach 12–15 months results in lost time and resources, impacting not only the generation interval but also genetic progress in the herd [5–7]. This effect is primarily due to annual changes in photoperiod, which vary according to latitude and create distinct reproductive and non-reproductive seasons worldwide [8]. Other external factors that vary with the season, such as temperature, relative humidity, rainfall, and nutrition, also affect the reproductive function of bucks [1,9].

Understanding the effects of these influencing factors is essential, as bucks play a key role in enhancing the genetic quality of the local populations from which they originate. The breeding buck is considered to be half of the herd, as he is responsible for the conception rate and the degree of genetic improvement in the kids [10,11]. Buck fertility is a critical factor in goat farm management, as a single buck is typically used to breed with numerous does. Consequently, any issues affecting a buck's fertility can have significant implications for the overall productivity and success of the breeding program [7]. Most studies examining factors that influence bucks' breeding quality have focused on populations in Western countries, where data are more readily available. However, research conducted in Central and Eastern European countries remains limited, and findings from these regions show considerable variability. This lack of consistent data highlights the need for more region-specific studies to better understand the factors affecting bucks' fertility in different climates and management systems. In this context, the aim of this research was to analyze how age and season affect the semen characteristics, as well as the morphology of the testes, sexual behavior, and reflexes, in Carpatina breed bucks. These studies are both useful and necessary, because goat farming in many parts of the world is in a process of development and modernization. Additionally, the research is justified by the fact that, in this part of the European Union, goat populations are increasing, and in some countries the annual growth rate exceeds 1% [12].

Knowing seasonal variations in some semen characteristics provides useful information for farmers who want to plan breeding activities outside the natural period [1]. By collecting semen, followed by conditioning and storing it through cryopreservation, it becomes possible to use it in other seasons or periods of the year, with positive effects on improving reproduction rates, as evidenced in several studies and research papers [6,8,9,13–15].

The purpose of this research was to determine how the age and season influence key semen characteristics, as well as testicular morphology, sexual behavior, and sexual reflexes, in Carpatina breed bucks. Additionally, this study assessed the correlation between these variables and the main semen quality parameters. This information aims to optimize the selection of reproductive bucks with desirable semen traits, leading to a reduction in the number of bucks needed for breeding, and lowering maintenance costs.

## 2. Materials and Methods

### *2.1. Ethical Approval*

The present study was approved by the Scientific Research Ethics Committee of the Research and Development Station for Sheep and Goat Breeding, Popăuți-Botoșani (registration number: 278, date: 10 March 2023). Furthermore, during both the handling and sampling of biological materials, all ethical requirements were observed, ensuring that the animals undergoing experimental procedures did not experience painful treatments or go through periods of stress and discomfort.

### 2.2. The Research Area and Climatological Conditions

The geographical area where the research was conducted is located in the northeastern part of Romania, with the geographical coordinates 47°47′45″ N and 26°40′43″ E [16]. The natural vegetation is characteristic of the forest–steppe zone, mainly consisting of agricultural land and ancient grasslands that have replaced former forests.

The climate is temperate and continental, strongly influenced by air masses of Asian origin. The winters in the area are harsh and cold, and strong winds predominate from the northeast. Summers, by contrast, are extremely hot, with temperatures often exceeding 35 °C and a relative humidity of around 65%. Precipitation varies throughout the year, with the highest rainfall occurring during spring and autumn. In spring, temperatures fluctuate between 9 °C and 25 °C, while in autumn they range from 8 °C to 20 °C, with relative humidity levels exceeding 75% [17].

### 2.3. Animal Selection and Management

The research animals were 18 Carpathian breed bucks, all meeting the necessary conditions for reproductive use. To assess the influence of two variables—age and season—on semen traits and other reproductive functions, the bucks were divided into three groups (L14, L34, and L56). Each buck was identified and underwent a thorough veterinary examination, including an assessment of genital integrity. The grouping was based on age at the start of the study, with each group containing six bucks. Group L14 consisted of young bucks aged 14 to 23 months, Group L34 included adult bucks aged between three and four years, and Group L56 comprised bucks aged five to six years.

### 2.4. Experimental Design

The experimental period lasted 12 months, from March 2023 to March 2024, and covered the four annual seasons typical of the Northern Hemisphere, located at the intersection of the 45° northern latitude and 25° eastern longitude meridian: spring (21 March–20 June), summer (21 June–20 September), autumn (21 September–21 December), and winter (21 December–20 March) [18]. Throughout the research, the bucks were housed under identical conditions, with each group having free access to an outdoor paddock for exercise. During the entire study, sanitary/veterinary and animal welfare regulations were strictly observed. The feeding of the three groups was carried out using forage rations that met the nutritional requirements recommended by the HYBRIMIN diet optimization program. For each group, the nutritional requirements were determined based on body weight as follows: for the group of bucks under two years of age (L14), the requirements were 10 MJ of metabolizable energy (ME) for ruminants, 170 g of crude fiber (CF), and 80 g of crude protein (CP); for adult bucks aged between three and four years (L34), the requirements were 11.3 MJ of ME for ruminants, 190 g of CF, and 90 g of CP; for bucks aged five to six years (L56), the requirements were 12.7 MJ of ME for ruminants, 220 g of CF, and 100 g of CP. Clean water was provided from sanitary sources and available at all times, and salt licks were placed in the housing areas assigned to each group.

### 2.5. The Scrotal Biometry

Measurements related to testicular circumference (TC), testicular length (TL), and testicular volume (TV) were taken individually at the beginning of each season over three consecutive days. The testicular circumference was determined by gently pulling the testicles downward toward the lower part of the scrotum and then placing a measuring tape around the widest point (Figure 1a), with the result expressed in centimeters. Testicular length (TL) was measured by determining the distance between the point of insertion near the abdominal attachment and the end of the third distal portion of the scrotum (Figure 1b), also expressed in centimeters. Testicular volume (TV) was determined using the water displacement method (Figure 1c), expressed in cubic centimeters ($cm^3$). For this measurement, the testicles were completely submerged in a container filled with water, and the residual water displacement was used to calculate the desired volume.

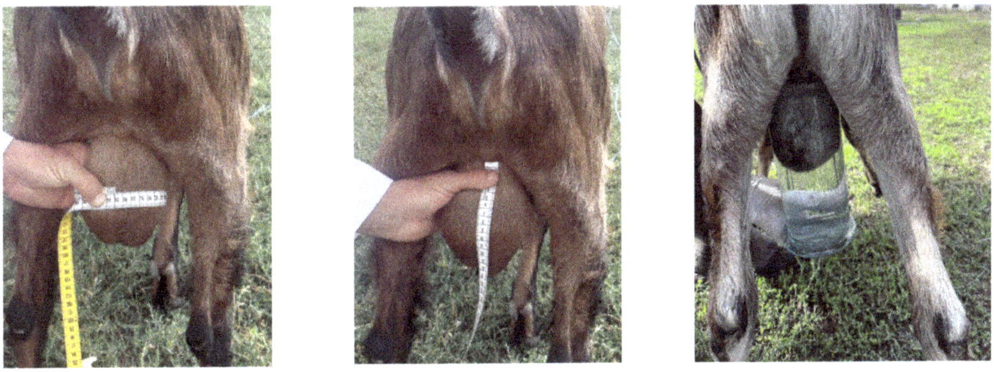

(**a**) (**b**) (**c**)

**Figure 1.** The determination of scrotal circumference (**a**), testicular length (**b**), and testicular volume (**c**).

### 2.6. *Semen Analysis*

The evaluation of semen was carried out on samples collected from each buck immediately after the start of each season. The collection was performed by experienced personnel using the artificial vagina method. To avoid contamination of the semen, the prepuce was sanitized before collection. Collection took place once per day for four consecutive days. To ensure a uniform interval between collections, the semen was collected at 8 a.m. every day. The collected samples were stored in a water bath at a temperature of 35–37 °C and transported to the laboratory within a maximum of ten minutes for evaluation.

#### 2.6.1. The Volume, Color, and pH of the Ejaculate

The volume of semen collected was measured directly using the graduated container used during collection. The color of the seminal fluid was assessed visually and assigned a score ranging from 1 to 4 according to a model presented by Jha et al. [19], as follows: 1 = watery; 2 = milky; 3 = yellowish-white; 4 = cream-white. The degree of semen acidity (pH) was estimated from the collected samples using a colorimetric method, using indicator paper (phenolphthalein paper) and comparing it with a color scale.

#### 2.6.2. The Evaluation of Specific Characteristics of Semen

For the analysis of semen-specific traits, a CASA analyzer was used (CEROS II CASA, IMV Technologies, L'Aigle, France), endorsed with a 10× objective for negative and positive phases, and compatible with specimen chambers of 20 μm depth. Motile sperm and total concentration were analyzed using WHO 5-phase slides, equipped with Animal Breeders II Software (version 1.13.7; Hamilton Thorne, Beverly, MA, USA). The system also included a Trinocular Zeiss Axiolab A5 microscope (Carl Zeiss Microscopy GmbH, Jena, Germany) with 100× magnification. Additionally, the used slides, coverslips, and chamber were pre-warmed to 37 °C for at least 10 min before the sample analysis. The initial sperm concentration was measured by diluting each sperm sample 1:100 in a 0.025% glutaraldehyde solution. Before assessing motility or viability, $200 \times 10^6$ spermatozoa/mL semen samples were carefully mixed and re-diluted to a final concentration of $50 \times 10^6$ spermatozoa/mL using TRIS extenders for motility and viability evaluation. The analysis was conducted on 3 μL samples loaded onto Leja 4-chamber 20 μm slides. To these samples, 1 μL of propidium iodide and 1 μL of acridine orange were added. After application, the samples were gently mixed for homogenization. The homogenized mixture was covered with a coverslip, and images were captured under the microscope. This procedure resulted in the differentiation of dead and live cells, which were highlighted in red and green, respectively, through staining.

2.6.3. Assessment of Blood Testosterone Levels

The level of testosterone was determined from blood samples collected on the same day that semen collection was performed (during the first four days of each season at 8 AM). The collection was preceded by a needle incision into the jugular vein at an angle of approximately 20 to 30 degrees relative to the skin surface, to collect blood from the jugular vein of each male, with a volume of 10 mL being collected each time. After collection, the samples were stored in individual tubes and transported to the laboratory. There, they were centrifuged (4000 rpm), and the resulting serum was frozen at −20 °C until the testosterone levels were determined. The determination of testosterone was performed using the radioimmunoassay (RIA) method, utilizing TESTO-RIA-CT (DIAsource ImmunoAssays, Ottignies-Louvain-la-Neuve, Belgium) and a commercial kit (KIP1709) that facilitates the quantitative measurement of testosterone (T) protein in serum and plasma, with a detection range from 0.1 ng/mL to 20 ng/mL and a sensitivity of 0.05 ng/mL.

*2.7. The Sexual Behavior and Reflexes*

To properly assess the sexual behavior and reflexes of bucks, they were isolated from females throughout the experimental period. Contact with females occurred only at the beginning of each season. The females used in this study were clinically healthy but infertile and, for certain reasons, had not become pregnant in the previous breeding season.

Behavior was analyzed using a method described by Goshme et al. [20], focusing on the following aspects: the reaction and desire of the buck to mate, the time taken to approach the female, the attitude adopted in the presence of the female, and the reaction time of the bucks upon the introduction of the female into a pen measuring 5 $\times$ 6 m. The assessment of the behavior manifested and how the bucks reacted in the presence of infertile females was based on assigning scores according to the following types of manifestation:

Excellent (5): Typical behavior observed in the buck when introduced into the testing pen includes a high level of eagerness to mount, restlessness, and provocative movements. The buck shows an intense desire to mate immediately, appearing agitated and difficult to control.

Very good (4): Sexual reflexes are considered optimal when the buck shows a strong desire to mate within three minutes of introduction to the testing environment. The buck is often restless, aggressive, and difficult to control, indicating heightened sexual arousal and readiness to mate.

Good (3): The buck typically sniffs the females and expresses a desire to mate within the first five minutes of being introduced to them. Although it may be agitated during this time, it remains relatively easy to control compared to a more intense response.

Weak (2): The time required for the buck to initiate mating extends to approximately ten minutes. During this period, the buck remains docile and is relatively easy to handle, indicating a more calm state of sexual arousal compared to more aggressive or impatient individuals.

Very weak (1): The male buck exhibits delayed and incomplete sexual reflexes, with the time required to display a desire to mate extending beyond ten minutes. This extended response time suggests a lowered level of sexual arousal or potential issues with reproductive readiness, as the buck's response to the presence of females is slower and less intense compared to more responsive individuals.

The manifestations of erection and ejaculation reflexes were evaluated according to the technical standards presented by the National Agency for Zootechnics in Bucharest [21], with points awarded as follows: 1 = lack of erection or incomplete erection of the penis; 2 = satisfactory erection, not seeking the vulva, ejaculation in the furrow; 3 = well-defined erection, evident seeking of the vulva, no ejaculation; 4 = well-defined erection, energetic seeking of the vulva, delayed ejaculation; 5 = immediate erection of the penis in the presence of the female, energetic seeking of the vulva, rapid ejaculation with obvious spurting.

*2.8. Statistical Data Processing*

Experimental data were input in a column-type database and processed within the GraphPad Prism v. 9.4.1. software (Palo Alto, CA, USA) to obtain statistical descriptive values (mean, standard error) and to compare the three groups' performances, using the ANOVA single-factor test followed by Tukey post hoc processing. The Mann–Whitney U test for independent groups was used to compare scorings of semen color, sexual behavior, and sexual reflexes (discontinued variables), and one-to-one comparisons were carried out (within each season, between the age groups). Also, Pearson correlation coefficients and the level of significance for the correlations were computed using the same software package. Correlations were calculated within groups, between body weight and certain testicular traits, and between the ejaculate volume and certain quality traits of the semen.

## 3. Results

*3.1. Body Weight and Scrotal Measurements*

Body weight (BW) is an important indicator that has a direct relationship with many traits affecting reproductive activity in males. In all established groups, the evolution of both body weight and scrotal measurements was monitored each season (Table 1). Therefore, at the beginning of each season, both BW and the changes occurring at the testicular level were determined, specifically concerning testicular circumference (TC), testicular length (TL), and testicular volume (TV).

**Table 1.** Mean (±SE) body weight and different measurements of the testicles in relation to age and season.

| Season | Traits | Animal Group | | |
|---|---|---|---|---|
| | | L14 | L34 | L56 |
| Spring | BW (kg) | $49.58^{a} \pm 1.53$ | $63.72^{dx} \pm 1.41$ | $74.78^{dz} \pm 0.88$ |
| | TC (cm) | $25.90^{a} \pm 0.82$ | $31.12^{d} \pm 0.37$ | $32.71^{d} \pm 0.63$ |
| | TL (cm) | $9.21^{a} \pm 0.06$ | $10.73^{d} \pm 0.22$ | $10.56^{c} \pm 0.29$ |
| | TV ($cm^3$) | $209.70^{a} \pm 0.44$ | $219.48^{b} \pm 2.99$ | $225.66^{d} \pm 1.91$ |
| Summer | BW (kg) | $51.26^{a} \pm 1.48$ | $67.38^{dx} \pm 1.70$ | $75.38^{dy} \pm 0.92$ |
| | TC (cm) | $26.16^{a} \pm 0.19$ | $33.72^{d} \pm 0.42$ | $33.90^{d} \pm 0.23$ |
| | TL (cm) | $9.41^{a} \pm 0.10$ | $10.90^{d} \pm 0.31$ | $10.60^{c} \pm 0.14$ |
| | TV ($cm^3$) | $212.70^{a} \pm 1.52$ | $226.58^{c} \pm 3.27$ | $227.46^{c} \pm 1.12$ |
| Autumn | BW (kg) | $54.46^{a} \pm 1.34$ | $68.38^{dx} \pm 1.68$ | $76.01^{dw} \pm 0.81$ |
| | TC (cm) | $26.62^{a} \pm 0.40$ | $34.66^{d} \pm 0.38$ | $34.58^{d} \pm 0.37$ |
| | TL (cm) | $9.50^{a} \pm 0.17$ | $10.96^{c} \pm 0.35$ | $10.57^{b} \pm 0.12$ |
| | TV ($cm^3$) | $215.30^{a} \pm 0.93$ | $230.98^{d} \pm 2.03$ | $229.8^{d} \pm 0.25$ |
| Winter | BW (kg) | $53.56^{a} \pm 1.14$ | $67.68^{dx} \pm 1.33$ | $74.98^{dy} \pm 0.79$ |
| | TC (cm) | $25.64^{a} \pm 0.23$ | $33.54^{d} \pm 0.27$ | $34.16^{d} \pm 0.37$ |
| | TL (cm) | $9.20^{a} \pm 0.23$ | $10.58^{c} \pm 0.21$ | $10.30^{c} \pm 0.14$ |
| | TV ($cm^3$) | $209.18^{a} \pm 0.51$ | $220.48^{c} \pm 2.51$ | $224.26^{d} \pm 1.47$ |

Notes: BW—body weight; TC—testicular circumference; TL—testicular length; TV—testicular volume. Statistically different: $^{ab}$ for $p < 0.05$; $^{ac}$ for $p < 0.01$; $^{ad}$ for $p < 0.001$ between L14 and L34, L56; $^{xw}$ for $p < 0.05$; $^{xy}$ for $p < 0.01$; $^{xz}$ for $p < 0.001$ between L34 and L56.

BW showed values that varied according to the age of the bucks and the season in which this trait was evaluated. The differences between two-year-old bucks (L14) and those older than three years (L34 and L56) were highly significant in each season ($p < 0.001$). Additionally, the difference recorded in the analysis of BW specific to bucks from the L34 and L56 groups indicated very significant differences in spring ($p < 0.001$), distinct significant differences in summer and winter ($p < 0.01$), and significant differences at the beginning of autumn ($p < 0.05$).

In contrast, for TC and TL, the differences recorded between L34 and L56 were not significant, indicating that, after the age of three years, neither season nor age influences testicular dimensions anymore.

As a result of the intensification of spermatogenesis, at the beginning of autumn, testicular volume (TV) records its highest value. The differences recorded between L34 and L56 were minimal and statistically non-significant ($p \geq 0.05$). Between L14 and L34, the differences were significant in spring ($p < 0.05$), distinctly significant in summer and winter, and very significant in autumn ($p < 0.001$). Additionally, the differences in TV between L14 and L56 were very significant ($p < 0.001$) in each season, except for the difference recorded in the summer season, which was distinctly significant ($p < 0.01$).

### 3.2. *Characteristics of Semen Quality in Bucks*

For the semen analysis, various macroscopic and microscopic analyses were performed. The values obtained for ejaculate volume (EV), semen concentration (SC), live spermatozoa (LS), normal spermatozoa (NS), semen motility (MS), and color of semen (CS), based on age and season, are presented in Table 2.

**Table 2.** Mean (±SE) differences in the quality traits of the semen of Carpatina bucks.

| Season | Traits | Animal Group | | |
|---|---|---|---|---|
| | | L14 | L34 | L56 |
| Spring | EV (mL) | 0.84 $^a$ ± 0.03 | 1.22 $^d$ ± 0.01 | 1.19 $^d$ ± 0.02 |
| | SC ($10^9$/mL) | 3.85 $^a$ ± 0.04 | 4.12 $^c$ ± 0.06 | 4.15 $^c$ ± 0.06 |
| | LS (%) | 79.01 ± 1.14 | 79.60 ± 0.50 | 80.20 ± 0.20 |
| | NS (%) | 82.40 $^a$ ± 1.17 | 89.01 $^d$ ± 0.31 | 89.80 $^d$ ± 0.20 |
| | MS (%) | 77.60 $^a$ ± 0.72 | 81.20 $^d$ ± 0.20 | 80.40 $^c$ ± 0.24 |
| | CS (points) | 2.40 $^a$ ± 0.24 | 3.20 $^b$ ± 0.30 | 3.00 ± 0.31 |
| Summer | EV (mL) | 0.91 $^a$ ± 0.42 | 1.27 $^d$ ± 0.05 | 1.25 $^d$ ± 0.01 |
| | SC ($10^9$/mL) | 3.91 $^a$ ± 0.04 | 4.24 $^b$ ± 0.03 | 4.37 $^c$ ± 0.11 |
| | LS (%) | 81.00 $^a$ ± 0.54 | 85.4 $^d$ ± 0.24 | 86.00 $^d$ ± 0.31 |
| | NS (%) | 84.80 $^a$ ± 0.37 | 90.40 $^d$ ± 0.40 | 91.00 $^d$ ± 0.54 |
| | MS (%) | 78.40 $^a$ ± 0.08 | 81.64 $^c$ ± 0.18 | 81.03 $^c$ ± 0.01 |
| | CS (points) | 2.80 ± 0.21 | 3.60 ± 0.34 | 3.80 ± 0.44 |
| Autumn | EV (mL) | 1.11 $^a$ ± 0.02 | 1.33 $^d$ ± 0.06 | 1.30 $^d$ ± 0.01 |
| | SC ($10^9$/mL) | 4.01 $^a$ ± 0.17 | 4.39 $^b$ ± 0.13 | 4.43 $^c$ ± 0.04 |
| | LS (%) | 82.20 $^a$ ± 0.20 | 85.20 $^d$ ± 0.20 | 86.00 $^d$ ± 0.31 |
| | NS (%) | 84.15 $^a$ ± 0.22 | 88.20 $^d$ ± 0.20 | 88.25 $^d$ ± 0.15 |
| | MS (%) | 80.20 ± 1.62 | 86.20 ± 0.80 | 86.40 ± 0.74 |
| | CS (points) | 3.20 ± 0.20 | 3.80 ± 0.20 | 3.80 ± 0.20 |
| Winter | EV (mL) | 0.96 $^a$ ± 0.01 | 1.19 $^c$ ± 0.06 | 1.22 $^c$ ± 0.02 |
| | SC ($10^9$/mL) | 3.83 $^a$ ± 0.59 | 4.17 $^b$ ± 0.03 | 3.97 ± 0.10 |
| | LS (%) | 79.80 ± 0.37 | 81.20 ± 0.20 | 80.80 ± 0.20 |
| | NS (%) | 83.60 $^a$ ± 0.68 | 89.60 $^d$ ± 0.68 | 89.80 $^c$ ± 0.73 |
| | MS (%) | 71.80 $^a$ ± 0.37 | 81.20 $^b$ ± 0.19 | 80.80 ± 0.20 |
| | CS (points) | 2.80 ± 0.20 | 3.40 ± 0.24 | 3.20 ± 0.20 |

Notes: EV—ejaculate volume; SC—semen concentration; LS—live spermatozoa; NS—normal spermatozoa; SM—sperm motility; Sc—semen color. Statistically different: $^{ab}$ for $p < 0.05$; $^{ac}$ for $p < 0.01$; $^{ad}$ for $p < 0.001$ between L14, L34, and L56.

The statistical analysis of the data for EV, CS, LS, NS, and MS indicated that, with respect to the season, there were significant differences ($p < 0.001$; $p < 0.01$) between the group of young bucks (L14) and the two groups of adult bucks of different ages (L34 and L56). In the evaluation of Cs, the differences recorded were not significant ($p > 0.05$), except for the scores assigned to samples collected at the beginning of summer, where a statistically significant difference appeared between L14 and L56 ($p < 0.05$).

The semen traits related to live spermatozoa (LS), normal spermatozoa (NS), and semen motility (MS) were not affected by age ($p > 0.05$). For bucks in groups L14 and L34, the season did not represent an important influencing factor, as all mean values remained statistically insignificant ($p > 0.05$). In contrast, semen concentration (SC) was strongly influenced by age and season. Younger bucks (<2 years) had lower semen cell concentrations than those older than 2 years (Table 2). The color of semen was statistically different (U = 4 and $p$-value = 0.0476) between L14 and L34 samples ($p < 0.05$) only in spring, according to the Mann–Whitney U test.

### *3.3. Variations in Testosterone, pH, and Sexual Behavior and Reflexes in Relation to the Season and Bucks' Age*

Testosterone (T4) is an important sexual hormone with a major role in the body. The average value for T4 showed increasing levels from the beginning of spring across all groups. Regarding age, the seasonal differences between L14 vs. L34 and L14 vs. L56 were distinctly significant in spring ($p < 0.01$) and highly significant ($p < 0.001$) at the beginning of summer, autumn, and winter (Table 3). The differences between L34 and L56 were non-significant ($p > 0.05$). According to these values, it can be concluded that the level of T4 in the blood influences the reproductive activity with respect to the age of the bucks and the season.

**Table 3.** Mean (±SE) of differences for testosterone, semen acidity, sexual behavior, and sexual reflexes in Carpatina bucks.

| Season | Traits | Animal Group | | |
|---|---|---|---|---|
| | | L14 | L34 | L56 |
| Spring | T4 (ng/mL) | $2.05^{a} \pm 0.06$ | $2.42^{c} \pm 0.06$ | $2.45^{c} \pm 0.07$ |
| | pH (°T) | $6.42 \pm 0.24$ | $6.52 \pm 0.06$ | $6.57 \pm 0.11$ |
| | SB (points) | $2.20^{a} \pm 0.20$ | $3.60^{b} \pm 0.24$ | $3.40^{b} \pm 0.24$ |
| | SR (points) | $2.40^{a} \pm 0.24$ | $3.60^{b} \pm 0.24$ | $3.2^{b} \pm 0.20$ |
| Summer | T4 (ng/mL) | $2.50^{a} \pm 0.11$ | $4.20^{d} \pm 0.06$ | $4.02^{d} \pm 0.19$ |
| | pH (°T) | $6.45^{a} \pm 0.24$ | $6.82^{b} \pm 0.25$ | $6.57 \pm 0.27$ |
| | SB (points) | $2.40^{a} \pm 0.24$ | $3.40^{b} \pm 0.24$ | $3.20^{b} \pm 0.20$ |
| | SR (points) | $2.60 \pm 0.24$ | $3.40 \pm 0.24$ | $3.20 \pm 0.20$ |
| Autumn | T4 (ng/mL) | $4.89^{a} \pm 0.31$ | $8.98^{d} \pm 0.37$ | $8.17^{d} \pm 0.42$ |
| | pH (°T) | $6.53^{a} \pm 0.2$ | $6.82^{b} \pm 0.35$ | $6.77 \pm 0.22$ |
| | SB (points) | $2.80^{a} \pm 0.20$ | $4.60^{b} \pm 0.24$ | $4.40^{c} \pm 0.40$ |
| | SR (points) | $2.80^{a} \pm 0.20$ | $4.2^{b} \pm 0.37$ | $3.80^{b} \pm 0.37$ |
| Winter | T4 (ng/mL) | $2.26^{a} \pm 0.08$ | $4.42^{d} \pm 0.11$ | $4.23^{d} \pm 0.26$ |
| | pH (°T) | $6.51 \pm 0.03$ | $6.64 \pm 0.08$ | $6.58 \pm 0.02$ |
| | SB (points) | $2.60^{a} \pm 0.02$ | $4.40^{b} \pm 0.02$ | $4.20^{b} \pm 0.03$ |
| | SR (points) | $2.60 \pm 0.24$ | $3.20 \pm 0.20$ | $3.00 \pm 0.01$ |

Notes: T—testosterone; pH—sperm acidity; SB—sexual behavior; SR—sexual reflexes. Statistically different: $^{ab}$ for $p < 0.05$; $^{ac}$ for $p < 0.01$; $^{ad}$ for $p < 0.001$ between L14 and L34/L56, respectively.

According to the results obtained, it can be specified that the pH of the semen has specific values indicative of semen quality in all groups (Table 3). There were also some significant differences ($p < 0.05$) only between L14 and L34, and only in the analysis of samples collected in summer and autumn.

The sexual behavior and sexual reflexes of bucks are very important aspects that influence the fertilization capacity of the does. In adult bucks from groups L34 and L56, the average scores obtained indicated a seasonal influence on mating behavior. The highest values for sexual behavior (SB) were recorded in autumn, exceeding 4 points in both groups. Young bucks (L14) displayed less intense SB, with a lower score in spring (2.20 ± 0.20) and maximum values in the autumn season (2.80 ± 0.20). Although they expressed a desire to mate, the young bucks became very agitated, and the time taken to mount was significantly prolonged. Significant differences ($p < 0.05$) related to sexual behavior were mostly recorded between younger bucks (L14) and the older ones (L 34 and L56).

Sexual reflexes (SR) are an extremely important trait for males intended for reproduction. Among the three groups, bucks aged between three and four years (L34) exhibited the most intense SR in the presence of females. The differences recorded between L14 and the older bucks in L34 and L56 were significant in spring and autumn ($p < 0.05$) and not significant during summer and winter ($p > 0.05$).

### 3.4. Variation in Seasonal Relationships Between Different Reproductive Traits in Relation to the Age of Carpatina Breed Bucks

The inclusion among the research objectives of determining the correlations established between BW in each season and some testicular traits (TC, TL, TV, and EV), as well as between EV determined in each season and the main semen traits (CS, LS, NS, MS), was motivated by the desire to conduct a comprehensive evaluation of all aspects that can be used in assessing the quality of breeding bucks. The results obtained show that both negative and positive correlations were established between the BW of the bucks determined at the beginning of each calendar season and some specific values of the main testicular dimensions (Table 4). The data indicate that, for BW recorded in the bucks from the L34 and L56 groups at the beginning of spring, a positive and significant relationship ($p < 0.05$) with TV was observed. When analyzing the correlations between BW at the start of autumn and the values obtained from testicular measurements, the relationships were positive but not significant ($p > 0.05$).

**Table 4.** The relationship between body weight estimated in each season and certain testicular traits.

| Animal Group | BW | Testicular Circumference | | Testicular Length | | Testicular Volume | | Ejaculate Volume | |
|---|---|---|---|---|---|---|---|---|---|
| | | R | *p*-Value | R | *p*-Value | R | *p*-Value | R | *p*-Value |
| Spring | | | | | | | | | |
| L14 | 49.58 | −0.201 | 0.729 NS | 0.151 | 0.807 NS | 0.472 | 0.105 NS | 0.549 | 0.549 NS |
| L34 | 63.72 | −0.654 | 0.230 NS | 0.256 | 0.362 NS | 0.942 | 0.016 * | 0.327 | 0.590 NS |
| L56 | 74.78 | 0.763 | 0.133 NS | 0.550 | 0.336 NS | 0.797 | 0.022 * | 0.276 | 0.652 NS |
| Summer | | | | | | | | | |
| L14 | 51.26 | 0.472 | 0.431 NS | 0.519 | 0.369 NS | 0.863 | 0.072 NS | 0.524 | 0.364 NS |
| L34 | 67.38 | −0.166 | 0.789 NS | 0.279 | 0.648 NS | 0.819 | 0.087 NS | 0.226 | 0.971 NS |
| L56 | 75.38 | 0.758 | 0.137 NS | 0.695 | 0.192 NS | 0.060 | 0.922 NS | 0.306 | 0.615 NS |
| Autumn | | | | | | | | | |
| L14 | 54.46 | 0.244 | 0.691 NS | 0.056 | 0.924 NS | 0.624 | 0.260 NS | 0.873 | 0.053 NS |
| L34 | 63.38 | 0.382 | 0.524 NS | −0.854 | 0.065 NS | 0.446 | 0.451 NS | −0.351 | 0.356 NS |
| L56 | 76.01 | −0.667 | 0.208 NS | −0.378 | 0.529 NS | 0.332 | 0.584 NS | 0.572 | 0.313 NS |
| Winter | | | | | | | | | |
| L14 | 53.56 | 0.190 | 0.758 NS | −0.545 | 0.341 NS | −0.738 | 0.154 NS | 0.700 | 0.187 NS |
| L34 | 67.68 | 0.018 | 0.976 NS | 0.603 | 0.280 NS | 0.862 | 0.059 NS | −0.130 | 0.834 NS |
| L56 | 74.98 | 0.760 | 0.134 NS | 0.148 | 0.812 NS | 0.061 | 0.921 NS | −0.635 | 0.249 NS |

Notes: * $p < 0.05$; NS = non-significant.

Regarding the results related to the correlations established between EV collected in each season and sperm traits (CS, LS, NS, and MS), both positive and negative correlations were observed (Table 5). For semen collections performed in the spring from the three groups of bucks, it was found that only the relationship established between EV and CS in L14 was significant ($p < 0.05$).

**Table 5.** The relationship between ejaculate volume determined in each season and certain semen quality traits.

| Animal Group | EV | Spermatozoa Concentration | | Live Spermatozoa | | Normal Spermatozoa | | Spermatozoa Motility | |
|---|---|---|---|---|---|---|---|---|---|
| | | R | *p*-Value | R | *p*-Value | R | *p*-Value | R | *p*-Value |
| Spring | | | | | | | | | |
| L14 | 0.84 | 0.933 | 0.0205 * | −0.567 | 0.318 NS | −0.774 | 0,131 NS | −0.330 | 0.587 NS |
| L34 | 1.22 | −0.397 | 0.507 NS | 0.430 | 0.469 NS | −0.554 | 0.331 NS | 0.000 | 0.099 NS |
| L56 | 1.19 | 0.109 | 0.860 NS | −0.311 | 0.601 NS | 0.415 | 0.486 NS | 0.847 | 0.069 NS |
| Summer | | | | | | | | | |
| L14 | 0.91 | 0.515 | 0.373 NS | −0.658 | 0.226 NS | −0.727 | 0.163 NS | −0.812 | 0.094 NS |
| L34 | 1.27 | −0.339 | 0.576 NS | −0.038 | 0.951 NS | −0.306 | 0.615 NS | 0.431 | 0.468 NS |
| L56 | 1.25 | −0.363 | 0.547 NS | 0.110 | 0.860 NS | 0.318 | 0.602 NS | 0.422 | 0.574 NS |
| Autumn | | | | | | | | | |
| L14 | 1.11 | −0.889 | 0.043 * | −0.322 | 0.579 NS | −0.516 | 0.373 NS | −0.587 | 0.114 NS |
| L34 | 1.33 | −0.259 | 0.673 NS | 0.534 | 0.353 NS | 0.900 | 0.036 NS | −0.200 | 0.746 NS |
| L56 | 1.30 | −0.532 | 0.355 NS | −0.337 | 0.579 NS | −0.380 | 0.527 NS | −0.661 | 0.224 NS |
| Winter | | | | | | | | | |
| L14 | 0.96 | −0.487 | 0.404 NS | 0.674 | 0.211 NS | −0.133 | 0.830 NS | 0.625 | 0.259 NS |
| L34 | 1.19 | 0.890 | 0.042 * | 0.033 | 0.957 NS | −0.715 | 0.174 NS | −0.061 | 0.923 NS |
| L56 | 1.22 | 0.021 | 0.813 NS | 0.399 | 0.252 NS | 0.529 | 0.163 NS | 0.565 | 0.245 NS |

Notes: * $p < 0.05$; NS = non-significant.

For semen collections performed in L34 at the beginning of summer, negative but non-significant correlations ($p > 0.05$) were found between EV and other semen quality traits—with CS (r = −0.339), LS (r = −0.038), and NS (r = −0.306)—and a positive correlation only with MS (r = 0.468).

## 4. Discussion

### *4.1. Seasonal Variations in Body Weight and Scrotal Measurements in Relation to Age in Bucks*

In all experimental groups, body weight (BW) showed a trend of increasing from spring to autumn. The decrease in live weight was primarily due to the mobilization of body reserves as a result of the effort expended during mating. This weight loss fell within normal limits, knowing that a ram or a buck can lose up to 25% of its body weight during a breeding season [1,22–25]. In Group L14, it was observed that, at the beginning of autumn, BW was 8.96% higher than at the beginning of spring, with average values of 49.58 ± 1.53 kg and 54.46 ± 1.34 kg, respectively. The fact that young bucks (L14) had a BW in autumn representing over 70% of the BW recorded for L34 and L56 indicates that the requirement for body maturity was met, and bucks younger than two years can be safely used for reproduction.

In the adult groups L34 vs. L56, the increase in BW was due to body recovery, as well as the establishment of subcutaneous fat reserves necessary to support the effort exerted during the breeding season. At the beginning of winter, BW in both groups decreased very slightly (<1%) compared to the BW recorded in autumn. The results obtained are similar to other values cited in the specialized literature [22–25], demonstrating that young bucks of the Carpatina breed can be utilized for mating or semen collection before reaching two years of age. Based on these results, it can be stated that both age and season had a major and significant influence ($p < 0.001$, $p < 0.001$, and $p < 0.05$) on the seasonal increase or decrease in BW.

For bucks used in breeding, understanding the changes in testicular dimensions during both the growth phase and the sexually active or resting adult stages provides objective indicators of reproductive readiness. In bucks younger than two years (L14), monitoring changes in testicular size can help determine the optimal age at which they should be introduced into the breeding group. Furthermore, observing the moment when morphological changes in the testes occur, indicated by an increase in reference dimensions, also signals the onset of active spermatogenesis. This physiological shift indicates that the buck is becoming sexually mature and ready for breeding. Tracking these testicular dimensions allows for more precise management of bucks in breeding programs, ensuring that they are introduced at the right age to optimize reproductive success [26,27]. All of the obtained results align with other values cited in the literature. Many scientific articles highlight that changes in specific testicular dimensions are closely related to body weight (BW) and the degree of body development [28–31]. Body size and testicular measurements are proven to be important parameters for assessing reproductive soundness. For the adult groups of bucks older than two years (L34 and L56), the data confirm that the average values of testicular circumference (TC) showed an evolution influenced by both age and season. Thus, starting from the minimum TC values recorded at the beginning of spring ($31.12 \pm 0.37$ cm for L34 and $32.71 \pm 0.63$ cm for L56), maximum values were reached at the beginning of autumn ($34.66 \pm 0.38$ cm for L34 and $34.58 \pm 0.37$ cm for L56). The data indicate that body maturity was achieved in the groups with adult bucks, and the recorded evolutions were limited, close, and statistically non-significant ($p > 0.05$). The statistical analysis of the TC data indicates that there were very significant differences ($p < 0.001$) between L14 vs. L34 and L14 vs. L56 at the beginning of each season. The fact that, at the beginning of autumn, TC for L14 showed an average value of $26.62 \pm 0.04$ cm indicates good development of this organ. In other studies on different breeds raised under different conditions, the evolution recorded for TC was similar in level and direction. Many scientific articles highlight a trend of increasing heritability for TC with age, suggesting that environmental factors influencing traits become less significant in older animals [9,14,32].

The testicular length (TL) values were close and statistically non-significant ($p \geq 0.05$) in each season between L34 and L56. In Group L14, the average value increased by approximately 3% from spring to the end of summer, from $9.21 \pm 0.06$ cm to $9.50 \pm 0.17$ cm. The differences between L14 vs. L34 were very significant ($p < 0.001$) in the spring and summer seasons and distinctively significant ($p < 0.01$) in autumn and winter. Between L14 vs. L56, it was noted that the differences in TL were distinctly significant ($p < 0.01$), except for the difference recorded at the beginning of autumn ($p < 0.05$). The obtained values suggest that both age and season are influential factors for some testicular dimensions, including TL. The relationship between testicular size and BW in young individuals undergoing growth is synthetically presented in many scientific articles [33,34]. Some research indicates that the emergence of statistical differences ($p < 0.05$) in TL is mainly due to the age of the bucks. Thus, Abba et al. (2015) [35] confirmed that the average values of TL increased from $8.39 \pm 0.68$ cm in bucks that had reached 18 months to $9.89 \pm 0.61$ cm at the age of 30 months.

Testicular volume (TV) is an important indicator, as it is associated with semen production. In all experimental groups of bucks, it was observed that TV was higher at the beginning of autumn, which marks the onset of the natural breeding season for small ruminants in the research area. In Group L14, the average values of TV were lower during the seasons outside the natural breeding period, measuring $209.70 \pm 0.44$ cm$^3$ in spring, $212.07 \pm 1.52$ cm$^3$ in summer, and $209.18 \pm 0.51$ cm$^3$ in winter. At the beginning of autumn, which coincides with the start of the breeding season, TV was 2.60% higher compared to spring and 1.39% higher than the level recorded in summer. The obtained results support the idea that both age and season have a significant influence on TV. The greater increase in size and volume in L14 may also be attributed to the period of growth in body weight (BW) and body dimensions. In the groups of adult bucks (L34 and L56), the increase in TV may have resulted from physiological changes as well as the growth of the testicular parenchyma, which includes increases in the length, width, and thickness of the testes—an aspect highlighted by other bibliographic sources [6,15,36,37].

### *4.2. Variations in the Quantity and Quality of Semen in Relation to the Age of Bucks and the Season*

In regions with a temperate continental climate, seasonal variations in conception and birth rates have long been linked to seasonal changes in semen parameters. It has been observed that both the season and the age of male sheep and goats affect the values influencing semen quality. Ejaculate volume (EV) represents one of the key traits used to express semen quality and contributes to the fertility assessment of bucks. Based on samples collected at the beginning of each season, it was found that EV showed some differences among the groups (Table 2). Additionally, the results indicate that age is a factor affecting EV in bucks. This assertion is supported by the fact that the volume differences between L14 vs. L34 and L14 vs. L56 were very significant ($p < 0.001$) at the beginning of each season, except for the difference determined from samples collected at the onset of winter ($p < 0.01$). In L14, it was observed that, with the intensification of spermatogenesis, EV increased from $0.84 \pm 0.03$ mL in spring to $1.11 \pm 0.02$ mL at the beginning of autumn. Since no statistically significant differences were recorded between L34 and L56, it can be said that age influences this trait to a limited extent. The differences registered between the group of young bucks under two years old (L14) and those aged 2–4 years (L34) and 5–6 years (L56) were due to the fact that testicular development in the young bucks was not yet complete. Bucks from L14 can be used for breeding activities, since they have a volume that allows for both natural mating and the fractionation of seminal fluid into straws for artificial insemination. Information in these scientific articles confirms that EV varies throughout the year, being lower in the off-season (0.44 mL) compared to other times of the year (0.86 mL). Higher values recorded during the peak sexual activity season are a result of intensified spermatogenesis, which is influenced by various environmental factors. When these factors reach certain thresholds, they positively affect spermatogenesis, leading to an increase in testicular volume (EV), a phenomenon noted in several specialized studies. Additionally, semen concentration (SC) is a crucial factor when objectively assessing semen quality, as it provides key insights into the reproductive potential of bucks. Monitoring both EV and SC offers valuable indicators for evaluating breeding readiness and fertility levels in bucks.

The obtained values are used for certifying fertility, diagnosing and/or predicting reproductive disorders, processing insemination doses, characterizing seminal samples for trade, and evaluating the effects of treatments regarding semen production (for example, toxicology and nutrition studies). According to the obtained data, no significant differences were found between the adult bucks aged three to four years (L34) and those aged five to six years (L56). This aspect confirms that age has less influence on this trait among sexually active adult bucks.

Between the young bucks in L14 and those in L34, the differences were statistically significant in spring ($p < 0.01$) and significant in summer, autumn, and winter ($p < 0.05$). Between L14 and L56, the differences were statistically significant in spring ($p < 0.01$), while those at the beginning of winter were not significant ($p \geq 0.05$), confirming that age has some influence on the CS of the semen. Since the highest values in all three groups were recorded in autumn ($4.01 \pm 0.17 \times 10^9$ in L14, $4.39 \pm 0.13 \times 10^9$ in L34, and $4.43 \pm 0.04 \times 10^9$ in L56), it can be said that CS is also influenced by the season. For CS in Carpatina breed bucks, the highest values were recorded during the breeding season, which aligns with the findings of a couple of other author collectives [38–41]. CS tended to increase in both breeds at the end of winter and at the beginning of summer and autumn, while it decreased in spring and winter.

Live spermatozoa (LS) is a very important trait, included in any quality analysis of semen and fertility of bucks [1,42–44]. Based on the recorded values, LS has direct effects on the economic and productive activity [44] in goat farms. For LS, it was noted that between L14 vs. L34 and L14 vs. L56, the differences recorded in summer and autumn had a high degree of statistical significance ($p < 0.001$), while they became non-significant in samples collected in spring and winter ($p > 0.05$).

Data processing for NS highlighted some very significant differences ($p < 0.001$) between groups, with the exception of the difference between L14 and L56 recorded at the beginning of winter, which had a different level of statistical significance ($p < 0.01$). According to these values, it can be stated that semen production and its quality are correlated with testicular dimensions, testis weight, and scrotal circumference. The same conclusion was reached in studies conducted by other research groups [8,11,25,45].

Semen motility (SM) is a critical factor that influences the ability of sperm to navigate through the female reproductive tract after sexual contact. In cases where average motility values are low, fertility tends to decrease across various mammalian species, including bucks. Reduced motility impairs the sperm's ability to fertilize the egg, thus lowering reproductive success. This trait is closely linked to semen quality and is a key indicator of fertility. Low motility can be influenced by various factors, including environmental stress, nutrition, and overall health, all of which can negatively impact reproductive outcomes in bucks [43]. According to the obtained data, we observed that motility had very good values in all seasons and in all groups of bucks. Statistical differences were noted between L14 and L34 only in the seasons outside the natural breeding period, being very significant in spring ($p < 0.001$), distinctly significant in summer and winter ($p < 0.01$), and non-significant in autumn ($p \geq 0.05$).

The appearance of these differences between groups shows that the season represents a factor with a greater influence on MS. In L14, the average values for MS reached their highest level at the beginning of autumn (82.20 $\pm$ 1.62%), while the lowest level of motility was found during winter (71.80 $\pm$ 0.37%). The fact that the average motility in L14 exceeded 80% during the autumn season supports the use of young bucks for reproduction without reservations. This high motility rate indicates that the sperm are highly active and capable of successfully fertilizing eggs, making young bucks a viable option for breeding during this period. In 2011, Rachmawati et al. [46] reported much lower sperm motility in Kacang bucks (42.22 $\pm$ 16.41%) and Etawa bucks (65.56 $\pm$ 9.17%).

The *color of semen* (Cs) is influenced by SC, EV, and other factors. In our three groups, it was noted that the season influenced Cs less. Between L14 and L34/L56, significant differences in Cs appeared ($p < 0.05$) only in semen samples collected in spring and summer.

According to the obtained data, we can affirm that the physical aspect of the ejaculates varied from a yellow or yellowish-white color in the summer–autumn period to a cream-white color in other seasons of the year. Consequently, the score assigned was higher for samples collected in autumn (3.20 $\pm$ 0.20 in L14, 3.80 $\pm$ 0.2 in L34 and L56). At the same expression level for this trait, other research groups that analyzed the color and aspect of the ejaculate collected in different seasons and from other breeds of bucks also reached the same level [47–49].

### 4.3. The Testosterone Levels, Semen Acidity, and Reproductive Behavior in Carpatina Bucks

The presence of testosterone in males is crucial, as it regulates sexual appetite (libido), bone mass, fat distribution, muscle mass, red blood cell production, and spermatogenesis. In bucks located in temperate continental climates, testosterone levels increase and peak in autumn, promoting intense sexual activity and high sperm production. This seasonal fluctuation supports the idea that testosterone levels are influenced by both age and environmental factors such as season. Research has shown that seasonal variations, particularly in the autumn, align with heightened reproductive activity in bucks, confirming this hormone's role in regulating these processes. The average values obtained for T4 in blood serum indicated a favorable disposition for sexual activity in all groups, including the young bucks (L14). This is due to the fact that, in bucks found in areas with a temperate climate, sexual behavior is seasonal, accentuating in autumn and early winter precisely because of the increase in testosterone levels. Compared to the determination of T4 in samples collected in spring, it was noted that, in L14 at the beginning of autumn, there was an increase from 2.05 ± 0.06 ng/mL to 4.89 ± 0.31 ng/mL. In the natural mating season, T4 reaches values > 8 ng/mL in adult bucks. In accordance with these values, we can specify that both age and season are important factors for the variation in T4 levels in blood.

The obtained values are similar to other data cited in several scientific articles. For example, the concentration of T4 in some white goats was 4.30 ± 0.47 ng/mL [50], in other hybrid goats resulting from crossing with Etawah bucks it was 6.82 ± 4.18 ng/mL, and in Kejobong goats the level reached was 12.00 ± 6.56 ng/mL [51]. The onset of increased behavioral manifestations typical of reproduction occurs after an increase in testosterone levels from a minimum value of 2 ng/mL to values even reaching 20 ng/mL [52].

As the pH increases and reaches its maximum values at the beginning of autumn (in L14 it was 6.41 ± 0.2, in L34 it was 7.38 ± 0.35, and in L56 it was 7.14 ± 0.22), this shows a slight influence of the season and less of the age of the reproducers. If pH values < 6.5 are recorded in bucks used for breeding, both motility and sperm metabolism gradually reduce. For sperm quality, a stable pH is essential. Since some specialized articles indicate that the seminal fluid collected from bucks has a pH between 6.4 and 7.2, or an average of 6.8 [53,54], we can conclude that, in our three groups of Carpatina breed bucks, the acidity of the sperm ranged within limits associated with an optimal pH. In conclusion, the small variations in pH between seasons, as well as between groups, cannot be attributed exclusively to the influence exerted by the season or age.

To stimulate their sexual behavior and reflexes, the bucks from the three groups were completely separated from the females. Contact occurred only on planned days for studying sexual behavior and reflexes. Differences in the scores resulting from the evaluation of sexual behavior (SB) between L14 and L34 were significant in summer ($p < 0.05$) and distinctly significant ($p < 0.01$) in autumn, winter, and spring. Between L14 and L56, statistically significant differences ($p < 0.01$) were recorded in all seasons except summer ($p > 0.05$). In the presence of females, they exhibited complete erection and immediately showed mating desire by seeking the vulva; however, they had a faster ejaculation in autumn and a slightly delayed one in the other seasons. Consequently, the score assigned for sexual reflexes (SR) had an upward trend, from 2.60 ± 0.24 points in the winter season for L14 to 2.80 ± 0.20 points in the autumn season. In the case of adult bucks from L34 and L56, it was found that erection was rapid in the presence of females; the bucks immediately sought the vulva, and ejaculation occurred after the jump, resulting in an average score of more than 4 points. In contrast, in summer, under the influence of high temperatures, the average score assigned was 3.40 ± 0.24 points for L34, 3.20 ± 0.20 points for L56, and 2.60 ± 0.24 points for L14. Based on the scores obtained for SB and SR, as well as the recorded differences in approach and behavior, it can be confirmed that both age and season are factors that influence both the behavior and the reflexes of the bucks.

Studies on goat populations in different regions have shown varied behaviors among bucks due to multiple influencing factors, particularly climate and photoperiod. In temperate and subtropical climates, bucks typically undergo a period of sexual rest, with significant changes in behavior influenced by seasonal shifts in light exposure. These behaviors include actions such as nudging, anogenital sniffing, mounting attempts, self-urination, mounting with or without intromission, flehmen responses, and vocalizations. These actions are directly linked to the changes in photoperiod, which regulate sexual activity and reproductive behaviors. Research has demonstrated that these behaviors are crucial indicators of the buck's sexual readiness and can vary significantly depending on environmental factors like climate and seasonal changes. These behavioral shifts are also important in managing breeding programs, as understanding seasonal patterns can help optimize breeding success [55,56]. The same behavior is observed in goats raised in Central Europe, which modify their sexual behavior during a sexual rest period from January to May. In contrast to this type of behavior, the literature mentions that male sheep and goats from tropical regions exhibit active behavior throughout the year [8,29]. This phenomenon is explained by the fact that the main influencing factor is not photoperiod but, rather, nutrition, high ambient temperatures, relative humidity, and atmospheric precipitation [57–59].

### *4.4. The Relationship Between Seasonal Body Weight and Testicular Measurements, and Between Ejaculate Volume and Sperm Quality Traits, in Carpatina Bucks*

Body weight (BW), particularly that of bucks intended for mating, plays a significant role in their fertility, and this is especially important at the start of the natural mating season. Maintaining optimal body condition at this time is crucial, as it supports the physical demands associated with reproduction. Bucks with adequate body condition are better equipped to handle the physical exertion required for mating, which can positively influence their fertility and overall reproductive performance. Research indicates that bucks in poor body condition at the onset of the breeding season may experience reduced libido, lower semen quality, and lower reproductive success. Conversely, bucks in optimal condition are more likely to exhibit strong sexual drive, better semen quality, and higher mating success rates. Proper nutrition and management of body weight prior to the breeding season are therefore essential for maximizing fertility and ensuring efficient reproduction [42,43]. Many studies conducted on various small ruminant populations highlight the role of BW and testicular traits, emphasizing that they represent important criteria in evaluating male fertility in animal production [25,27,36,60]. According to the obtained results, significant and positive correlations ($p < 0.05$) between BW and testicular volume (TV) were found only in groups of adult males over three years old (L34 and L56). This significant correlation suggests that, as BW increases, body development also progresses, leading to an increase in TV. In the case of seasonal BW and scrotal circumference (SC), testicular length (TL), and ejaculate volume (EV), the established correlations were not significant ($p > 0.05$). All of these data are consistent with the findings of Waheed et al. (2011) [61] and Mathapo et al. (2022) [62], contradicting the results of Gemeda et al. (2017) [63], who found that that BW was strongly and significantly correlated with TL in some indigenous goats from Ethiopia. This discrepancy may be due to the different goat breeds, which have specific morphological traits, as well as environmental conditions that can increase variability in many traits. The relationship between BW and testicular circumference (TC) was significant ($p < 0.05$) and highly positive ($r = 0.976$) only in the group of bucks aged between three and four years (L34). This significance aligns with the findings of Ahmed and Kawmani (2019) [64], who reported a significant positive correlation between BW and SC in male sheep and goats.

Age and other factors, such as season and breed, are often associated with sperm quality and fertility in domestic animals. Many studies have evaluated the relationship between male age and sperm parameters, but the effects of age and season have not been comprehensively assessed. Changes in sperm quality from young males to adults and older males have been identified in bulls, rams, stallions, boars, dogs, and stallions [64].

In the present research, the Pearson correlation coefficients obtained for various traits related to reproductive capacity and buck fertility with respect to age and season indicate both positive and negative correlations. Thus, in the spring season, it was found that the correlation between ejaculate volume (EV) and sperm concentration (SC) was significant in L14 ($p < 0.05$), while in L34 and L56 it was not significant ($p > 0.05$). In the summer season, EV was negatively correlated with sperm motility (SM) in L14 ($r = -0.658$) and L34 ($r = -0.038$), while the correlation was positive ($r = 0.110$) in bucks aged between five and six years (L34). In the autumn season, the correlation between EV and SC in young bucks (L14) was high but negative ($r = -0.889$) and significant ($p > 0.05$), with no statistical significance for L34 and L56. Additionally, in L14 and L56, the determined EV levels in the autumn season were negatively correlated with SM (r = −0.331 and r = −0.339) and positively correlated with sperm mass (MS) (r = 0.625 and r = 0.565). In the winter season, the correlation between EV and SC in samples collected from L34 was positive (r = 0.90) and statistically significant ($p < 0.05$), while in the other groups it was not significant. Based on the obtained data, it can be concluded that the main semen characteristics improve in certain seasons and in male sheep and goats of a specific age, and then they decline as the animals age. All of these results, as well as the main conclusion reached, are in full agreement with many specialized scientific publications [65–68].

## 5. Conclusions

Age significantly influences many traits that determine bucks' fertility. These factors cause changes in semen quality traits, with a notable decline outside the natural breeding period. Also, during non-breeding seasons, there is a decrease not only in body weight but also in testicular size, ejaculate volume, semen concentration, motility, and the percentage of viable sperm. Additionally, changes in blood testosterone levels during these periods suggest that both environmental factors and age have a substantial impact on reproductive behavior and sexual reflexes. The average values recorded for various semen traits, body development, testicular dimensions, sexual behavior, and sexual reflexes suggest that young bucks, aged between 14 and 23 months, are suitable for breeding. However, while bucks can reproduce year-round, their semen quality and mating capacity are typically lower during the winter, spring, and summer compared to the optimal breeding season at the start of autumn. This reinforces the seasonal variation in reproductive performance and the need for careful management to ensure optimal breeding outcomes.

**Author Contributions:** Conceptualization, C.P. (Constantin Pascal) and V.M.; methodology, C.P. (Constantin Pascal) and I.N.; software, R.-M.R.-R. and C.P. (Claudia Pânzaru); validation, C.P. (Claudia Pânzaru) and A.M.F.; formal analysis, I.N., V.M. and A.M.F.; investigation, C.P. (Constantin Pascal), I.N. and A.M.F.; data curation, C.P. (Constantin Pascal); writing—original draft preparation, C.P. (Constantin Pascal), C.P. (Claudia Pânzaru) and R.-M.R.-R.; writing—review and editing, C.P. (Constantin Pascal), C.P. (Claudia Pânzaru) and R.-M.R.-R.; visualization, C.P. (Constantin Pascal), I.N. and R.-M.R.-R.; supervision, C.P. (Constantin Pascal) and V.M. All authors have read and agreed to the published version of the manuscript.

**Funding:** This research received no external funding.

**Institutional Review Board Statement:** The experimental procedure was conducted in accordance with the protocols approved by the Institutional Animal Care and Use Committee of the Research and Development Station for Sheep and Goats Breeding Popăuți Botoșani (Registration number: 278, Date: 10 March 2023).

**Data Availability Statement:** The raw data supporting the conclusions of this article will be made available by the authors on request.

**Conflicts of Interest:** The authors declare no conflicts of interest.

## References

1. Pascal, C. *Treaty for Sheep and Goat Breeding*; Ion Ionescu de la Brad Publ. House: Iași, Romania, 2015.
2. Hahn, K.; Failing, K.; Wehrend, A. Effect of temperature and time after collection on buck sperm quality. *BMC Vet. Res.* **2019**, *15*, 355. [CrossRef] [PubMed]
3. Ishaku, B.; Ajanwachukwu, N.P.; Markum, H.J.; Otaka, A.F.; Tshara, G.T.; Yohana, D.; Mngusuur, R.K.; Abednego, C. Determinants of puberty and age of sexual maturity in Nigerian indigenous breeds of male goats. *Dutse J. Pure Appl. Sci.* **2023**, *9*, 95–108. [CrossRef]
4. Ridler, A.L.; Smith, S.L.; West, D.M. Ram and buck management. *Anim. Reprod. Sci.* **2012**, *130*, 180–183. [CrossRef]
5. Pascal, C.; Zaharia, N. Evaluation of the reproduction function of goats in Romania. *Sci. Pap.—Anim. Sci. Ser.* **2016**, *64*, 9–13.
6. Chasles, M.; Chesneau, D.; Moussu, C.; Abecia, J.A.; Delgadillo, J.A.; Chemineau, P.; Keller, M. Highly precocious activation of reproductive function in autumn-born goats (*Capra hircus*) by exposure to sexually active bucks. *Domest. Anim. Endocrinol.* **2019**, *68*, 100–105. [CrossRef] [PubMed]
7. Kadam, P.D.; Raut, M.R.; Sontakke, S.H.; Khadase, J.R. A Study of Age, Body Weight and Scrotal Circumference on Semen Production in Buck. *Int. J. Curr. Microbiol. Appl. Sci.* **2020**, *9*, 1284–1288. [CrossRef]

8. Abecia, J.; Forcada, F.; González-Bulnes, A. Hormonal control of reproduction in small ruminants. *Anim. Reprod. Sci.* **2012**, *130*, 173–179. [CrossRef]
9. Rosa, H.J.D.; Bryant, M.J. Seasonality of reproduction in sheep. *Small Rumin. Res.* **2003**, *48*, 155–171. [CrossRef]
10. Ford, D.; Okere, C.; Bolden, O. Libido test scores, Body confirmation and testicular traits in Boar and Kiko Goats Bucks. *J. Agric. Biol. Sci.* **2009**, *4*, 1–8.
11. Kridli, R.T.; Tabaa, M.J.; Sawalha, R.M.; Amashe, M.J. Comparative study of Scrotal circumference and semen characteristics of mountain Black Goat and its crossbred with Damascus Goat as affected by different factors. *Jordan J. Agric. Sci.* **2005**, *1*, 37–40.
12. Available online: https://insse.ro/cms/ro/tags/comunicat-efective-animale (accessed on 7 August 2024).
13. Zamiri, M.J.; Khalili, B.; Jafaroghli, M.; Farshad, A. Seasonal variation in seminal parameters, testicular size, and plasma testosterone concentration in Iranian Moghani rams. *Small Rumin. Res.* **2010**, *94*, 132–136. [CrossRef]
14. Aké-López, J.R.; Aké-Villanueva, N.Y.; Segura-Correa, J.C.; Aké-Villanueva, J.R.; Montes-Pérez, R.C. Effect of age and season on semen traits and serving capacity of Pelibuey rams under tropical conditions. *Livest. Res. Rural Dev.* **2016**, *28*, 166.
15. Barkawi, A.H.; Elsayed, E.H.; Ashour, G.; Shehata, E. Seasonal changes in semen characteristics, hormonal profiles and testicular activity in Zaraibi goats. *Small Rumin. Res.* **2006**, *66*, 209–213. [CrossRef]
16. Rachiți-Short Monography. Available online: https://zmbotosani.ro/membri/rachiti (accessed on 25 March 2024).
17. Păunescu, A.; Șadurschi, P.; Chirica, V. *Territorial Repertoire of Botoșani County*; Publishing house ASP Bucharest: Bucharest, Romania, 2020.
18. Available online: https://www.ansvsa.ro/legislatie/nutritie-animala-si-sncu/ (accessed on 25 July 2024).
19. Jha, P.K.; Paul, A.K.; Rahman, M.B.; Tanjim, M.; Bari, F.Y.; Golam, M. Improvement of preservation quality of chilled bull semen using α-tocopherol as an antioxidant. *J. Emb. Trans.* **2013**, *28*, 31–39. [CrossRef]
20. Goshme, S.; Banerjee, S.; Rekik, M.; Haile, H.; Yitagesu, E.; Getachew, T. Evaluation and characterization of semen quality in rams of Menz, Dorper and Awassi crosses in different seasons in Ethiopia. *Livest. Res. Rural Dev.* **2020**, *32*, 180. Available online: http://www.lrrd.org/lrrd32/11/shenk32180.html (accessed on 8 July 2024).
21. Available online: https://www.anarz.eu/AnarzAdministratorSite/CMSContent/Procedura%20operationala%20pt%20implement%20normelor%20tehnice%20privind%20organiz%20si%20desfas%20MN%20autorizate%202020.pdf (accessed on 15 February 2023).
22. Kozioł, K.; Koziorowski, M. Steroid hormones in peripheral blood plasma and androgen receptors in testis and epididymis of roe deer male (*Capreolus capreolus*) during the reproduction season. *Small Rumin. Res.* **2013**, *115*, 86–93. [CrossRef]
23. Kerketta, S.; Singh, M.; Patel, B.H.M.; Dutt, T.; Upadhyay, D.; Bharti, P.K.; Sahu, S.; Kamal, R. Relationships between age, body measurements, testicular measurements and total ejaculation of semen in local goat of Rohilkhand region. *Small Rumin. Res.* **2015**, *130*, 193–196. [CrossRef]
24. Chenoweth, P.J. Sexual Behavior of the Bull: A Review. *J. Dairy Sci.* **1983**, *66*, 173–179. [CrossRef]
25. Raji, A.O.; Igwebuike, J.U.; Aliyu, J. Testicular biometry and its relationship with body weight of indigenous goats in a semi-arid region of Nigeria. *J. Agric. Biol. Sci.* **2008**, *3*, 24.
26. Mahmud, M.A.; Shaba, P.; Zubairu, U.Y. Live Body Weight Estimation in Small Ruminants-A Review. *Glob. J. Anim. Sci. Res.* **2014**, *2*, 102–108.
27. Akounda, B.; Ouédraogo, D.; Soudré, A.; Burger, P.A.; Rosen, B.D.; Van Tassell, C.P.; Sölkner, J. Morphometric Characterization of Local Goat Breeds in Two Agroecological Zones of Burkina Faso, West Africa. *Animals* **2023**, *28*, 4725. [CrossRef]
28. Ataç, F.E.; Takma, C.; Gevrekci, Y.; Altınçekiç, S.O.; Aya¸san, T. Estimates of Genetic Parameters for Direct and Maternal Effects on Pre-Weaning Growth Traits in Turkish Saanen Kids. *Animals* **2023**, *13*, 940. [CrossRef] [PubMed]
29. Ouchene-Khelifi, N.A.; Ouchene, N. Relationships between age, body measurements, and testicular measurements in Arabia bucks in Algeria. *Trop. Anim. Health Prod.* **2021**, *53*, 91. [CrossRef] [PubMed]
30. Samir, I.; Fatma, A.; Mohamed, R.; Yousef, N.; Elsebaey, A.; Ibrahim, M.R.; Noseer, E.; Nour, S.; Hussein, H.A. Testicular Biometry, Spermigram, and Biochemical Parameters in Male Goats. *Egypt. J. Vet. Sci.* **2024**, *55*, 671–679. [CrossRef]
31. Mohammed, I.D.; Amin, J.D. Estimating body weight from morphometric measurements of Sahel (Borno White) goats. *Small Rumin. Res.* **1997**, *24*, 1–5. [CrossRef]
32. Agga, G.E.; Udala, U.; Regassa, F.; Wudie, A. Body measurements of bucks of three goat breeds in Ethiopia and their correlation to breed, age and testicular measurements. *Small Rumin. Res.* **2010**, *95*, 133–138. [CrossRef]
33. Toe, F.; Rege, J.; Mukasa-Mugerwa, E.; Tembely, S.; Anindo, D.; Baker, R.; Lahlou-Kassi, A. Reproductive characteristics of Ethiopian highland sheep. I. Genetic parameters of testicular measurements in ram lambs and relationship with age at puberty in ewe lambs. *Small Rumin. Res.* **2000**, *36*, 227–240. [CrossRef]
34. Rege, J.E.; Toe, F.; Mukasa-Mugerwa, E.; Tembely, S.; Anindo, D.; Baker, R.L.; Lahlou-Kassi, A. Reproductive characteristics of Ethiopian highland sheep. II. Genetic parameters of semen characteristics and their relationships with testicular measurements in ram lambs. *Small Rumin. Res.* **2000**, *37*, 173–187. [CrossRef]
35. Abba, Y.; Igbokwe, I.O. Testicular and Related Size Evaluations in Nigerian Sahel Goats with Optimal Cauda Epididymal Sperm Reserve. *Vet. Med. Int.* **2015**, *2015*, 357519. [CrossRef]
36. Patni, M.; Singh, S.K.; Singh, D.V.; Palod, J.; Kumar, A.; Singh, M.K.; Sathapathy, S. Studies on body weight, body measurements and scrotal morphology in local Pantja goats. *Indian J. Anim. Res.* **2016**, *50*, 105–111. [CrossRef]

37. Koyuncu, M.; Kara Uzun, S.; Ozis, S.; Duru, S. Development of testicular dimensions and size, and their relationship to age and body weight in growing Kivircik (Western Thrace) ram lambs. *Czech J. Anim. Sci.* **2005**, *50*, 243–248. [CrossRef]
38. Chentouf, M.; Arrebola, F.M.; Bister, J.L.; Abbadi, N. Evaluación mensual de los parámetros testiculares y espermáticos de los machos cabríos del norte de Marruecos. In Proceedings of the En Memorias del XXXIV Congreso Nacional de Ovinotecnia y Caprinotecnia, Barbastro, Spain, 16–19 September 2009; pp. 368–371.
39. Prado, V.; Orihuela, A.; Lozano, S.; Pérez-León, I. Effect on ejaculatory performance and semen parameters of sexually-satiated male goats (*Capra hircus*) after changing the stimulus female. *Theriogenology* **2003**, *60*, 261–267. [CrossRef] [PubMed]
40. Brito, L.F.; Althouse, G.C.; Aurich, C.; Chenoweth, P.J.; Eilts, B.E.; Love, C.C.; Luvoni, G.C.; Mitchell, J.R.; Peter, A.T.; Pugh, D.G.; et al. Andrology laboratory review: Evaluation of sperm concentration. *Theriogenology* **2016**, *85*, 1507–1527. [CrossRef] [PubMed]
41. Türk, G.; Gür, S.; Kandemir, F.M.; Sönmez, M. Relationship between seminal plasma arginase activity and semen quality in Saanen bucks. *Small Rumin. Res.* **2011**, *97*, 83–87. [CrossRef]
42. Nechifor, I.; Florea, M.A.; Radu-Rusu, R.-M.; Pascal, C. Influence of supplemental feeding on body condition score and reproductive performance dynamics in Botosani Karakul Sheep. *Agriculture* **2022**, *12*, 2006. [CrossRef]
43. Pascal, C.; Nechifor, I.; Florea, M.A.; Pânzaru, C.; Simeanu, D.; Mierliță, D. Diet influence on sperm quality, fertility, and reproductive behavior in Karakul of Botoșani rams. *Agriculture* **2023**, *13*, 2168. [CrossRef]
44. Simeanu, D.; Radu Rusu, M.R. Animal Nutrition and Productions. *Agriculture* **2023**, *13*, 943. [CrossRef]
45. Available online: https://www.angoras.co.za/article/angora-goat-fertility-testicle-traits (accessed on 1 October 2024).
46. Rachmawati, L.; Ismaya, I.; Panjono, P. Perbandingan kuantitas dan kualitas sperma kambing kacang, kejobong, dan peranakan Etawah. In *Prosiding Seminar Nasional Prospek dan Potensi Sumberdaya Ternak Lokal Dalam Menunjang Ketahanan Pangan Hewani*; Fakultas Peternakan Univ. Jenderal Sudirman: Purwokerto, Indonesia, 2011; pp. 509–518.
47. Ahmad, N.; Noakes, E.D. Seasonal variations in the semen quality of young British goats. *Br. Vet. J.* **1996**, *152*, 225–236. [CrossRef]
48. Freshman, J.L. Semen collection and evaluation. *Clin. Tech. Small Anim. Pract.* **2002**, *17*, 104–107. [CrossRef]
49. Al-Anazi, Y.; Al-Mutary, M.; Al-Ghadi, M.; Alfuraiji, M.; Al-Himaidi, A.; Ammari, A. Seasonal variations in scrotal circumference and semen characteristics of Naimi and Najdi rams in Saudi Arabia. *S. Afr. J. Anim. Sci.* **2017**, *47*, 454–459. [CrossRef]
50. Polat, U.; Yesilbag, D.; Eren, M. Serum Biochemical Profile of Broiler Chickens Fed Diets Containing Rosemary and Rosemary Volatile Oil. *J. Biol. Environ. Sci.* **2011**, *5*, 23–30.
51. Rachmawati, L.; Ismaya, I.; Astuti, P. Level of testosteron, libido, and sperm quality of bligon, Kejobong and Etawah cross-breed bucks. *Anim. Prod.* **2013**, *15*, 76–82.
52. Bintara, S.D.; Maharani, I.B.; Mujadidyyati, A.N. The correlation between scrotal circumference, scrotal volume, and semen quantity and quality on fat tailed rams. In Proceedings of the 7th International Seminar on Tropical Animal Production, Yogyakarta, Indonesia, 12–14 September 2017; pp. 719–723.
53. Rizal, M.; Riyadhi, M.; Suleiman, A. The Quality of Boer Goat Semen Preserved with Sugar Palm Juice. *Bul. Peternak.* **2018**, *42*, 97–102. [CrossRef]
54. Gororo, F.; Price, T.Z.; Fungayi, P.C.; Mhuca, C. Effects of different extenders and storage temperatures on longevity of small East African goat (Capra hircus) semen. *Small Rumin. Res.* **2019**, *175*, 83–88. [CrossRef]
55. Delgadillo, J.; Canedo, G.; Chemineau, P.; Guillaume, D.; Malpaux, B. Evidence for an annual reproductive rhythm independent of food availability in male creole goats in subtropical northern mexico. *Theriogenology* **1999**, *52*, 727–737. [CrossRef]
56. Ponce, J.; Velázquez, H.; Duarte, G.; Bedos, M.; Hernández, H.; Keller, M.; Chemineau, P.; Delgadillo, J. Reducing exposure to long days from 75 to 30 days of extra-light treatment does not decrease the capacity of male goats to stimulate ovulatory activity in seasonally anovulatory females. *Domest. Anim. Endocrinol.* **2014**, *48*, 119–125. [CrossRef]
57. Aké-Villanueva, J.R.; Aké-López, J.R.; Magaña-Monforte, J.G.; Segura-Correa, J.C. Reproductive behavior in hair sheep rams under tropical conditions. *Trop. Anim. Health Prod.* **2019**, *51*, 1627–1635. [CrossRef]
58. Godfrey, R.W.; Collins, J.R.; Gray, M.L. Evaluation of sexual behavior of hair sheep rams in a tropical environment. *J. Anim. Sci.* **1998**, *76*, 714–717. [CrossRef]
59. Ponce-Covarrubias, J.L.; García y González, E.C.; Pineda-Burgos, B.C.; Guevara-Arroyo, A.M.; Hernández-Ruiz, P.E.; Torres-Agatón, F.; Ruiz-Ortega, M.; Paredes-Alvarado, M.; Robles-Robles, M.J.; Rodríguez-Castillo, J.; et al. Annual Sexual Behavior in Boer Bucks Located in the Guerrero Tropics in Mexico. *Ruminants* **2023**, *3*, 149–157. [CrossRef]
60. Shende, V.H.; Sontakke, S.H.; Potdar, V.V.; Tejsjree, V.; Shirsath, H.; Khadse, J.R. Scrotal Circumference and its Relationship with Testicular Growth, Age, and Body Weight in Murrah Buffalo Bulls. *Int. J. Curr. Microbiol. Appl. Sci.* **2019**, *8*, 2270–2274. [CrossRef]
61. Waheed, A.; Khan, M.S.; Ahmed, A.N.; Tariq, M.M.; Rauf, M.; Eyduran, E. Relationships among Testicular Traits, Body Measurements and Body Weight in Beetal Male Goats in Pakistan. *J. Inst. Sci. Tech.* **2011**, *1*, 59–62.
62. Mathapo, M.C.; Tyasi, T.L. Relationship between Testicular Traits, Body Measurements and Body Weight in Boer Goat Bucks. *Pak. J. Zool.* **2022**, *54*, 2957–2960. [CrossRef]
63. Gemeda, A.E.; Workalemahu, K. Body Weight and Scrotal-Testicular Biometry in Three Indigenous Breeds of Bucks in Arid and Semiarid Agroecologies, Ethiopia. *J. Vet. Med.* **2017**, *2017*, 5276106. [CrossRef] [PubMed]
64. Ahmed, A.; Kawmani, A.L. Testicular Parameters and Scrotal Measurements in Relation to Age and Body Weight in Growing Naemi-Rams. *J. Appl. Sci.* **2019**, *19*, 605–611. [CrossRef]
65. Abah, K.O.; Fontbonne, A.; Partyca, A.; Nizanski, W. Effect of male age on semen quality in domestic animals: Potential for advanced functional and translational research? *Vet. Res. Commun.* **2023**, *47*, 1125–1137. [CrossRef] [PubMed]

66. Atara, V.B.; Chaudhari, C.F.; Ramani, U.V.; Chaudhary, M.M.; Patel, D.K.; Patel, Y.R.; Patel, N.G. Semen characteristics in young and adult Surti buck. *Indian J. Anim. Health* **2018**, *57*, 219–224. [CrossRef]
67. Pascal, C.; Ivancia, M.; Gherasim, N. The influence of some factors on the reproductive function of Romanian local sheep. *Reprod. Domest. Anim.* **2008**, *43*, 99.
68. Chella, L.; Kunene, N.; Lehloenya, K. A comparative study on the quality of semen from Zulu rams at various ages and duringdifferent seasons in KwaZulu-Natal, South Africa. *Small Rumin Res.* **2017**, *151*, 104–109. [CrossRef]

agriculture

Article

# The Impact of Feed Management Technologies on Mineral Oil Hydrocarbons (MOH) Contamination: A Comparative Farm Level Approach

Mădălina Matei, Silvia Ioana Petrescu, Bianca Maria Mădescu, Dragoș Mihai Lăpușneanu, Daniel Simeanu, Paul Corneliu Boișteanu and Ioan Mircea Pop *

Faculty of Food and Animal Sciences, "Ion Ionescu de la Brad" Iasi University of Life Sciences, 8 Mihail Sadoveanu Alley, 700489 Iasi, Romania; madalina.matei@iuls.ro (M.M.); silvia.petrescu@iuls.ro (S.I.P.); bianca.madescu@iuls.ro (B.M.M.); dragos.lapusneanu@iuls.ro (D.M.L.); daniel.simeanu@iuls.ro (D.S.); paul.boisteanu@iuls.ro (P.C.B.)

* Correspondence: mircea.pop@iuls.ro

**Abstract:** Legislative frameworks about contamination are often limited to foods and underestimate the role of animal nutrition for safe production. This study aims to assess mineral oil hydrocarbon (MOH) contamination in feed and identify the technological factors that are contributing to this issue, particularly focused on mechanised harvesting and processing. Three dairy farms, classified by contamination risk (low, medium, and high), were selected, and fifteen feed samples were analysed using the coupled liquid chromatography–gas chromatography–flame ionisation detection (LC–GC–FID) method, with a microwave-assisted saponification (MAS) step to determine mineral oil saturated (MOSH) and aromatic (MOAH) hydrocarbon levels. Important contamination levels were observed depending on the technological development of each farm. MOSH levels ranged from 11.4 mg/kg to 81.40 mg/kg, while MOAH levels ranged from 0.5 mg/kg to 4.6 mg/kg. MOAH accounted for 4.74% of the total MOH content. The results showed a connection between feed production technologies and MOH contamination levels. Factors such as the mechanisation, the machinery used, and the storage conditions were potentially contributors to contamination, while chemical treatments had no direct impact but some potential risks. The contamination levels varied across farms, indicating certain contamination sources beyond technological factors. Advanced technological measures and proper equipment maintenance are suggested to mitigate MOH contamination risks in feed.

**Keywords:** animal feed; contamination; food safety; toxicity; public health

## 1. Introduction

In the agrifood system, the study of the traceability of contaminants is not well addressed. Legislative frameworks are often limited to the final food sector and underestimate the major role of animal nutrition in ensuring safe production. Monitoring potential sources of contamination and implementing specific prevention measures could contribute positively to food safety [1,2]. However, these efforts are often hampered by legislative gaps and a lack of standardised practices at the farm level.

Approaches to contamination of animal products must include the entire production chain, from soil and crops to feed and livestock. The presence and accumulation of contaminants in any of these components can lead to the transfer and bioaccumulation of residues in the animals' bodies, affecting their health [3]. The negative repercussions of the presence of contaminants in the animal body are reflected both in terms of food security and animal productions' quality and safety [4]. The accumulation of foreign substances compromises the safety and innocuity of the products, generating considerable risks and negative effects on consumers' health [5,6].

Among the products of animal origin, milk holds a special place in the human diet due to the important supply of nutrients. As a result of the high fat content, milk has the unwanted ability to retain contaminants, making it vulnerable to their accumulation. Therefore, ensuring milk innocuity becomes essential to guaranteeing food safety.

In particular, the relationship between the consumption of contaminated feed and the safety of animal production, especially milk, has been highlighted in numerous studies [7–10]. These works show us that the massive presence of contaminants in animal feed can lead to their transfer during production, affecting milk safety and consumer health. Also, long-term exposure to contaminants can have serious consequences for both animals and humans, but also for the productivity and economic viability of farms, highlighting the importance of suitable feed management [11].

Mineral oil hydrocarbons (MOHs) are contaminants of petrogenic origin, complex mixtures of saturated (mineral oil saturated hydrocarbon (MOSH)) or aromatic (mineral oil aromatic hydrocarbon (MOAH)) hydrocarbons.

The presence of MOH is the result of environmental pollution or contamination that occurs throughout the production cycle [12–14].

In the food chain, the presence of MOH has been a persistent problem for more than a decade, raising concerns about its ecotoxicological health impact [15–17]. Recent findings, based on various contamination situations [18,19], have led to a deeper awareness of the problems related to the incidence of MOH in the environment and food [20–23].

In the agrifood system, MOHs mainly contaminate when they are released as residues, especially from using lubricating oils in various technological processes [14,24].

MOH contamination of feed can be greatly influenced by the technological factors carried out when obtaining them, especially through mechanised agricultural machinery. This can result from incorrect harvesting and processing practices of plant raw materials, from intentional use of mineral oils as lubricants or release agents, from accidental contamination, but also from contact materials containing mineral oils [25]. Moreover, feed storage and handling techniques are other technological factors that can influence their MOSH and MOAH contamination.

Using lubricating oils for the agricultural machinery and equipment operation presents a considerable risk of contamination, since the oils used (engine oils and hydraulic oils) can enter the environment in unrefined or partially refined form, bringing with them a wide range of secondary compounds. Grob et al. [26] underlined the alarming realities of these oils, because, in addition to mineral hydrocarbons, they also contain polyalphaolefins, high-molecular-weight compounds ($n–C_{25–30}$, and they can reach up to $n–C_{45}$), polyesters, and different proportions of additives with toxic potential. This contamination is also compounded by the presence of other chemicals, including cleaning solvents and thinners from machine maintenance processes, as well as paints or other protective substances.

Recent studies by Hoccheger et al. [24] and Van Heyst et al. [27] highlighted concerns related to MOH contamination, including the licensed use of these oils as ingredients in various products. According to Regulation (EC) 1107/2009 [28] and the Commission Implementing Regulation (EU) 540/2011 [29], paraffinic mineral oils are permitted as components in substances intended for agricultural crop protection, such as pesticides, insecticides, and acaricides. This authorised use can contribute to feed contamination and, by extension, food product contamination, raising serious concerns about food safety and consumer health. In addition, the presence of these compounds in the food chain can generate long-term risks; therefore, careful assessment of the risks associated with the use of mineral oils in agriculture is insisted upon.

MOHs' toxicity remains a distinct concern in specialist research because of the uncertainties related to their effects on living organisms. Recent studies indicated an increased capacity of MOSH to accumulate in human organs and tissues, which may generate health risks [30–32]. More, MOAHs are structurally similar to polycyclic aromatic hydrocarbons (PAHs), have carcinogenic potential, and can exert genotoxic effects, especially the variants with three or more aromatic rings [22,33].

The toxicity of mineral oils has been studied for about three decades, but the subject is still open due to the chemical complexity of these substances. MOHs are characterised by a diversity of constituents and structures, so the variable toxicological profile contributes to the uncertainty regarding their health effects. The risks associated with mineral oils are influenced by the molecular weight distribution of hydrocarbons and the presence of MOAH, which is considered the most toxic of the fractions [34].

Their high fat content and the absence of specific functional barriers make products of animal origin vulnerable to MOSH and MOAH contamination through the migration of these compounds from the animal tissues [21]. Over time, varying proportions of MOH have been identified in products of animal origin as well as other food products [35]. According to Bratinova and Hoekstra [36] and Bratinova et al. [37], Commission Recommendation (EU) 2017/84 [38] emphasises the importance of monitoring the presence of these contaminants, especially in fats, meat, dairy, fish, and derived products.

In relation to food safety, although their toxicological impact is well documented [22,30–34], up to the present, very few specific approaches regarding MOH contamination of feed have been developed. This gap highlights the need to investigate the sources and mechanisms by which these contaminants can affect the food chain, especially in the context of their negative effects on animal and consumer health.

Limited data and the harmful nature of these contaminants have led to the need to critically investigate the incidence of MOHs in animal production.

The aim of this research is to assess the mineral oil hydrocarbon (MOH) contamination of feed and to understand the impact on food safety and consumer health. The study aims to identify the sources and the risky technological factors that contribute to feed contamination, with a particular focus on mechanised feed harvesting and processing.

Through the analysis of MOH compounds and contamination mechanisms, this study aims to contribute to the awareness of feed safety issues in relation to food safety, providing a basis for the development of more efficient production practices in the agrifood system. The results will have the final objective of proposing measures and recommendations to reduce the risks associated with feed contaminants and their sustainable management at the farm level.

## 2. Materials and Methods

### 2.1. Samples and Sampling Sites

For this study, three dairy farms were selected by the levels of potential contamination, established on the basis of some preliminary assessments. These evaluations were carried out through the observation method, structured interviews, and an evaluation questionnaire designed to identify possible sources of contamination for each farm. For the objectivity of the results, the farms were monitored and analysed during the study according to the protocol developed by Matei and Pop [39].

The questionnaire focused on (1) the feed base and the diet specificity, (2) the traceability of agricultural practices, soil fertilisation, and phytosanitary treatments applied to crops, (3) the use of equipment and technical oils with contamination risk, (4) feed processing, storage, and handling practices, and (5) farms' proximity to urban or industrial areas.

Figure 1 shows the locations of the farms and some brief images illustrating the specific feed base. The feed base of the farms was provided mostly from internal production, obtaining feeds directly from the crop land using its own infrastructure. For some farms, the external purchase of feed from different sources was performed for supplementation when the internal production was not sufficient.

Our hypothesis was focused on the different levels of contamination from each farm, which may directly influence the contamination levels of the feed samples. Table 1 summarises some characteristics of the sampling sites (area, number of inhabitants, and traffic distance) and of the feed samples (origin and participation rate in the ration).

(**a**) (**b**)

**Figure 1.** Sampling sites: (**a**) Farms' locations within Romania, with codes indicating the contamination risk level: low (L), medium (M), and high (H). (**b**) The location of $L_1$ farm (mountain area—low contamination) in relation to the nearest major city in the region (Suceava County); the locations of $M_2$ farm (rural area—medium contamination) and $H_3$ farm (urban area—high contamination), highlighting the proximity to the nearest urban area (Iasi city). Symbols show the main sources of pollution in the vicinity of the farms. The colour indicates the estimated pollution risk level of the sources near the sampling sites: red (high pollution risk, intense air transport), yellow (moderate risk, small-scale industry), green (low pollution risk, sporadic pollutant emissions), and blue (urban pollution).

Farms were grouped into three categories of expected contamination risk, based on the responses and assessments from the evaluation questionnaire. Each farm was coded ($L_1$, $M_2$, and $H_3$) to indicate its contamination category and risk: low (L), medium (M), and high (H). Farm $L_1$ is located in the mountainous area and is represented by a low use of machinery in feed management, semi-intensive animal husbandry and agricultural activities, potentially improper feed handling and storage practices, and a large distance from sources of pollution and urban traffic. Farm $M_2$ is located in a rural area and is represented by a moderate use of agricultural machinery and potentially improper feed handling and storage practices. Farm $H_3$ is closer to the urban area and is characterised by intensive use of machinery, potentially improper feed handling and storage practices, and is positioned closer to industrial areas.

Information on some possible sources of pollution (road distances, traffic intensity, distance from urban areas, and number of inhabitants) is also provided. $H_3$ farm sites were located close to one of the biggest cities in Romania (Iasi), near an industrial area, with an airport positioned in the vicinity of the farm as well as other potential sources of contamination (municipal waste, construction activities, and traffic infrastructure). Samples from $M_2$ farm, such as alfalfa hay ($M_2$–AH), corn grains ($M_2$–C), and corn silage ($M_2$–CS), were also located ~5 km from the urban centre. There were no important sources of contamination for $L_1$ farm, even though it is located 6 km from one important urban centre of Suceava County, except the mountain pasture ($L_1$–MP), which is located in an ex-mining area (activity stopped in 2002).

For some samples, information about technological processes (crop-care treatments, the harvesting mode, equipment, and storage area) was also reported. Table 2 lists the main substances used in these treatments.

**Table 1.** Characteristics of the feed samples in relation to sampling sites.

| Farm and Category | Sample Code | Sample Name | % of the Ration | Origin | Crop Location | |
|---|---|---|---|---|---|---|
| | | | | | d—Traffic/Roads * | d—Inhabited Areas (No.) |
| $L_1$ Low risk | $L_1$–NH | Natural hay | 100 (summer) | Internal | 500 m/medium ~6 km/intensive (urban) | 500 m/medium ~6 km/intensive (urban) |
| | $L_1$–MP | Mountain pasture | 100 (winter) | Internal | ~5 km/medium | >7 km (269) |
| | | | | TOTAL ($L_1$): 2 | | |
| $M_2$ Medium risk | $M_2$–AH | Alfalfa hay | 22.4 | Internal | >10 km/intensive ($S_1$) ~2 km/intensive ($S_2$) | ~6 km (4.577)/rural ~4 km (271.692)/urban |
| | $M_2$–CS | Corn silage | 56 | Internal | ~6 km/intensive | ~6 km (271.692)/urban |
| | $M_2$–C | Corn grains | 11.20 | Internal | <15 km/intensive | ~4 km (4.577)/rural |
| | $M_2$–S | Soya | 8.40 | External purchase | - | - |
| | $M_2$–CF | Combined feed *** | 100 | Internal | AH, CS, C, S | |
| | | | | TOTAL ($M_2$): 5 | | |
| $H_3$ High risk | $H_3$–AH | Alfalfa hay | 5.45 | Internal | ~1 km/intensive ~1 km/medium ** | ~2 km (271.692) |
| | $H_3$–AS | Alfalfa Silage | 10.9 | Internal | ~1 km/intensive ~1 km/medium ** | ~2 km (271.692) |
| | $H_3$–CS | Corn silage | 45.45 | Internal | ~4 km/intensive ~1 km/medium ** | ~2 km (271.692) |
| | $H_3$–C | Corn grains | 6.35 | Internal | ~1 km/intensive ~1 km/medium ** | ~2 km (271.692) |
| | $H_3$–S | Soya | 7.1 | External purchase | - | - |
| | $H_3$–T | Triticale | 4.54 | Internal | ~5 km/medium | ~4 km (2.067) |
| | $H_3$–BSG | Brewer's grains | 18.18 | External purchase | - | - |
| | $H_3$–CF | Combined feed *** | 100 | Internal | AH, AS, CS, C, S, T, BSG | |
| | | | | TOTAL ($H_3$): 8 | | |

* Traffic conditions associated with urban areas (national roads). ** Air transport (25 landing–take-off sequences). *** Supplements are also included in the combined feed. d = distance; $S_{1,2}$ = parcelling area; $L_1$, $M_2$, $H_3$ = farm codes indicating contamination risk levels: low (L), medium (M), and high (H); AH, AS, BSG, C, CF, CS, MP, NH, S, T = abbreviations used for feed sample types derived from the sample name (column 3).

Different types of samples (green fodder, dry fodder, pickled fodder, concentrated fodder, and combined fodder) were collected for analysis from the feed base of the farms during the feed-obtaining campaign in 2021–2022. Some of the samples were taken directly from the crop area, and others were taken after harvesting steps, from their transport machine, or from the storage areas. A total of 15 feed samples were obtained for the laboratory (2 samples from the $L_1$ farm, 5 samples from the $M_2$ farm, and 8 samples from the $H_3$ farm) after dividing the elemental samples taken depending on the size of the sampling area and the specific ration from each farm. The samples were packaged, labelled, and transported to the laboratory for processing.

The sampling, the protocol, but also the reporting of the results followed the standards and performance criteria of the analytical approaches, but also different working protocols used by different authors [11].

**Table 2.** Market formula and active ingredients for crop-care treatments.

| Sample | Market Formula | Active Ingredient |
|---|---|---|
| $M_2$–AH | Pulsar 40 | 40 g/L Imazamox |
| $M_2$–CS | Sulfammo–25–APPM–1 | 25% N (18% ammoniacal N; 7% N nitric); 31% $SO_3$; 2% MgO |
| | Principal Plus | 9.2% Nicosulfuron; 55% Dicamba; 2.3% Rimsulfuron |
| $M_2$–C | DAP 18–46–0 | 18% $NH_4$; 46% $P_2O_5$ |
| | Sulfammo–25–APPM–1 | 25% N (18% ammoniacal N; 7% N nitric); 31% $SO_3$; 2% MgO |
| | Principal Plus | 9.2% Nicosulfuron; 55% Dicamba; 2.3% Rimsulfuron |
| $H_3$–CS and $H_3$–C | Urea | $CO(NH_2)_2$ |
| | NPK 20–20–0 Complex | 20% total N; 20% total $P_2O_5$; 60% $P_2O_5$ water soluble; 98% $P_2O_5$ soluble in citric acid 2%; max. 0.6% water |
| | Ammonium nitrate | 27% N; 7% CaO; 5% MgO |
| | Henik | 40 g/L Nicosulfuron |
| | Mustang | 6.25% Florasulfam; 30% Acid 2,4D EHE |
| | Adengo | 225 g/L Isoxaflutol; 90 g/L Thiencarbazone-methyl; 150 g/L Cyprosulfamides |
| $H_3$–AH and $H_3$–AS | 16–16–16 Complex | 16:16:16 N:P:K |
| | Corum | 480 g/L Bentazon; 22.4 g/L Imazamox |
| $H_3$–T | Urea | $CO(NH_2)_2$ |
| | Ammonium nitrate | 27% N; 7% CaO; 5% MgO |
| | Lebosol | 1.6% Cu—$Cu_2Cl(OH)_3$ 25 g/L; 11.5% Mn—$MnO_2$ 183 g/L; 4.9% Zn—ZnO 78 g/L |
| | Pixxaro Super | 12 g/L Halauxifen-methyl; 280 g/L Fluroxy-pyr meptyl; 12 g/L Cloquintocet-mexyl |
| | Orius | 250 g/L Tebuconazole |
| | Falcon Pro | 53 g/L Prothioconazole; 224 g/L Spiroxamine; 148 g/L Tebuconazole |
| | Mospilan | 20% Acetamiprid |

$M_2$, $H_3$ = farm codes indicating contamination risk levels: medium (M) and high (H). AH, AS, C, CS, T = abbreviations used for feed sample types derived from sample names: AH = alfalfa hay; AS = alfalfa silage; C = corn grain; CS = corn silage; t = triticale. APPM = activated poly-phenolic molecules; DAP = diammonium phosphate; NPK = nitrogen (N), phosphorus (P), and potassium (K) fertiliser; $SO_3$ = sulphur trioxide; MgO = magnesium oxide; $NH_4$ = ammonium; $P_2O_5$ = phosphorus pentoxide; CaO = calcium oxide; EHE = ethylhexyl ester; $Cu_2Cl(OH)_3$ = dicopper chloride trihydroxide; $MnO_2$ = manganese dioxide; ZnO = zinc oxide.

Feed samples were collected according to the SR EN ISO 6497:2005 [40] standard and Regulation (EC) 152/2009 Annex I [41]. Sample preparation was carried out according to the SR EN ISO 6498:2012 [42] and Regulation (EC) 152/2009 Annex II [41]. Samples were manually cut to 1–2 cm, dried to 8–12% moisture (60 °C, ESAC–100 model thermo-adjustable oven, Electronic April s.r.l., Cluj-Napoca, Romania), then finely ground using a Grindomix GM 200 (Verder GmbH, Vienna, Austria) laboratory mill. The samples were stored in polypropylene and aluminium packaging to prevent contamination until analysis.

### 2.2. Protocol, Reagents, and Standards

A method based on liquid chromatography–gas chromatography–flame ionisation detection (LC–GC–FID), including a microwave-assisted saponification (MAS) step, was applied to the feed samples for determining MOSH and MOAH fractions.

The protocol was adjusted to indicate the specificity of our samples, being similar to the methodology used by Bauwens et al. [21] for MOSH and MOAH analysis from fish feed. Due to the presence of natural n-alkanes and olefins, additional sample purification steps were required, such as epoxidation and passage through aluminium oxide (AlOx), according to the protocol developed by Nestola and Schmidt [43]. Method optimisation and protocol development were performed with adjustments based on the work of Biederman et al. [44] and Biedermann and Grob [45,46]. The method met the analytical performance criteria set out in the Joint Research Centre (JRC) Guide [36]. The protocol used for optimisation was proposed by Moret et al. [47] and applied with good results in other studies [25,48,49].

The following solvents and reagents were used for the preparation and analysis of feed samples: n-hexane (≥95%), methanol (≥99.9%), saturated KOH, metachloroperoxybenzoic acid (mCBPA; 70–75%, 200 mg/mL ethanol), anhydrous sodium thiosulphate, aluminium oxide, and sodium sulphate.

All reagents used were purchased from Merck Millipore (Burlington, MA, USA), Sigma-Aldrich (Saint Louis, MO, USA), or Supelco (Bellefonte, PA, USA). The mCBPA reagent was obtained from Acros Organics (Thermo Fisher Scientific, Waltham, MA, USA). Ultrapure water was obtained using a Milli-Q filtration system (Millipore, Bedford, MA, USA). To prevent contamination during sample preparation, all glassware was carefully cleaned and rinsed with pure solvents (acetone and n-hexane) before use.

A standard mixture purchased from Restek (Bellefonte, PA, USA) was the Internal Standard (IS) #31070 (150–600 μg/mL in toluene 99%) used for LC–GC performance evaluation, MOSH/MOAH separation, integration, and quantification of the results. This standard includes the following: n-Undecane (n–$C_{11}$; 99%; 0.3 mg/mL), cyclohexylcyclohexane (CyCy; 0.3 mg/mL), n-pentylbenzene (5B; 99%; 0.3 mg/mL), 1–methyl naphthalene (1–MN; 98%; 0.3 mg/mL), 2–methyl naphthalene (2–MN; 96%; 0.3 mg/mL), 1,3,5–tritertbutylbenzene (TBB; 99%; 0.3 mg/mL), n-tridecane (n–$C_{13}$; 99%; 0.15 mg/mL), 5-α-cholestane (Cho; 99%; 0.6 mg/mL), and perylene (Per; 99%; 0.6 mg/mL).

### 2.3. Sample Preparation

Organic phase separation was performed according to the protocol optimised by Moret et al. [47] for MOH extraction from cereal-based products. Saponification was performed using a microwave system (MARS 5, CEM Corporation, Bergamo, Italy), equipped with Teflon-lined cartridges. Each vial was filled with 5 g of feed sample, 10 mL of KOH (40%), 10 mL of n-hexane, and 20 μL of IS, then subjected to microwave extraction for 20 min at 120 °C. After extraction, the mixture was diluted with 40 mL of Milli-Q ultrapure water and 2 mL of methanol and set aside for phase separation. The extract was concentrated under vacuum up to 4 mL (Uniequip centrifuge, UNIVAPO–100H model, coupled with a V-700 vacuum pump and V-850 controller, Büchi AG, Flawil, Switzerland).

For a pure extract, a washing step with a mixture of methanol and water (2:1 *v*/*v*) was applied. The samples were vortexed and centrifuged, and the purified extract was concentrated to 700 μL. For alfalfa hay, corn silage, and compound feed samples, additional purification was required due to the presence of large amounts of natural n-alkanes. This step was performed according to the method described by Nestola and Schmidt [43]. The extract was epoxidised with mCBPA, to which sodium thiosulphate and ethanol were added, and 500 μL of this mixture was transferred for injection into the LC–GC–FID system. For the AlOx purification step, 40 μL of the epoxidised extract was diluted with n-hexane and passed through an AlOx and sodium sulphate cartridge. The extract obtained was concentrated to 250 μL, of which 75 μL was injected into the LC–GC–FID system.

### 2.4. LC–GC–FID Analysis and Instrument Conditions

MOSH and MOAH analysis was performed with a LC–GC 9000 Brechbuhler system (Zurich, Switzerland) composed of a Phoenix 9000 HPLC coupled to a Trace 1310 GC (Thermo Fisher Scientific, Waltham, MA, USA), configured with a dual channel for simultaneous analysis of the fractions. MOH fractions were transferred from LC to GC via a Y-interface using partial eluent evaporation, according to the method of Biedermann et al. [44].

For HPLC, a Lichrospher Si 60 column (25 cm × 2.1 mm, 5 µm) from DGB (Schlossboeckelheim, Germany) was used. The dual channel of the GC was composed of two PS-255 columns (15 m × 0.25 mm, 0.15 µm, Mega, Milan, Italy) connected to gap pre-columns and a solvent vapour removal system.

The HPLC elution program started with 100% n-hexane, followed by a switch to a 70/30 n-hexane/dichloromethane mixture at a flow rate of 300 µL/min. MOSH and MOAH fractions were transferred to GC between 2.1 and 3.6 min and 3.8 and 5.3 min, respectively. The carrier gas ($H_2$) was set at a constant pressure of 60 kPa, and the GC temperature was increased from 51 °C to 350 °C at a rate of 20 °C/min. The FID detector was heated to 350 °C with a collection rate of 10 Hz.

### 2.5. MOH Quantification and Method Validation

Data were processed using Chromeleon 7.3 software (Thermo Fisher Scientific, Waltham, MA, USA). The quantification was based on the internal standard CyCy for MOSH and average values of 5B, 1–MN, 2–MN, and TBB for MOAH. Methodologically, the total mass fractions for MOSH and MOAH were expressed in mg, related to the mass of the sample (expressed in kg), after the separation and removal of all possible interferences from the extract and the quantification and integration of the entire chromatographic signal between n–$C_{10}$ and n–$C_{50}$ retention times. Associated areas were integrated, and interferences were eliminated by running blanks for each batch of samples.

Analytical method performance was evaluated according to the JRC [36] and Eurachem [50] guidelines, using blind samples. These allowed for checking the possible contributions of the reagents used in the analytical process of the measurement signal, thus eliminating the possibility of external interferences. The limit of quantification (LOQ) for each n–C fraction, as well as for total MOSH and MOAH, was determined according to the recommendations in the SANTE/12682 guidelines [51]. Method performance met the JRC guideline criteria, with recovery values between 70 and 120%, as well as suitable intermediate repeatability for method validation.

## 3. Results

The feeds were analysed based on the technological development degree of each farm to assess the level of MOSH/MOAH contamination and identify the technological factors contributing to this contamination.

## 4. MOSH and MOAH in Animal Feed

MOSH and MOAH contamination for all feeds is indicated by the hydrocarbon range (n-C; 6 MOSH sub-fractions and 4 MOAH sub-fractions) and the total area (n–$C_{10–50}$). The data synthetically present the average results and summarise the overall situation of contamination levels with MOSH (Figure 2) and MOAH (Figure 3) of each type of feed. MOSH and MOAH concentrations are expressed in mg/kg. The quantification in the n-$C_{10–50}$ range was achieved by integrating the peaks, respecting the performance criteria described in the JRC guide [36], according to the European Commission regulations [52].

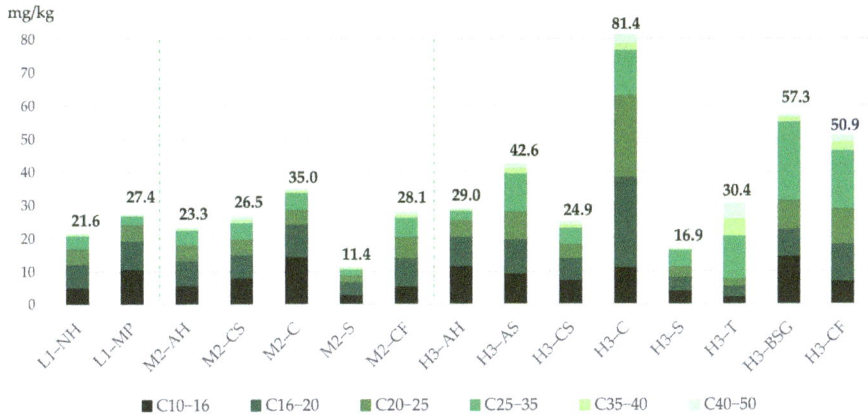


**Figure 2.** MOSH concentrations of feed sampled from each farm. $L_1$, $M_2$, $H_3$ = farm codes indicating contamination risk levels: low (L), medium (M), and high (H). AH = alfalfa hay; AS = alfalfa silage; BSG = brewer's grain; C = corn grain; CF = combined feed; CS = corn silage; MP = mountain pasture; NH = natural hay; S = soya; T = triticale. Values are the mean of two replicates per sample, shown as carbon fractions ($n–C_{10–50}$).

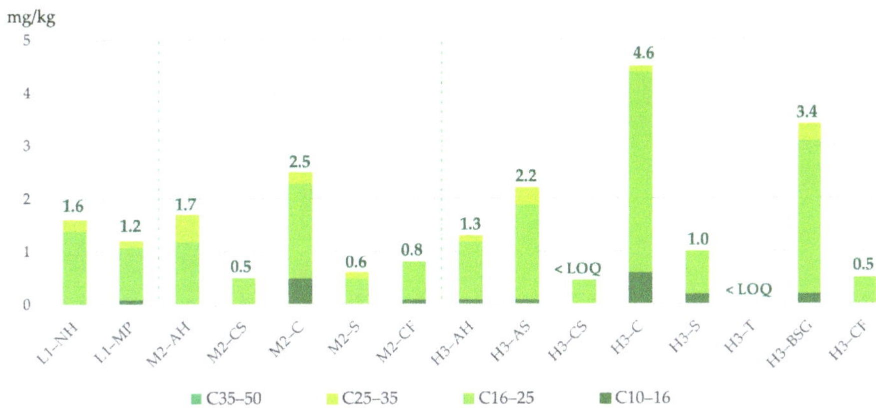


**Figure 3.** MOAH concentrations of feed sampled from each farm. $L_1$, $M_2$, $H_3$ = farm codes indicating contamination risk levels: low (L), medium (M), and high (H). AH = alfalfa hay; AS = alfalfa silage; BSG = brewer's grain; C = corn grain; CF = combined feed; CS = corn silage; MP = mountain pasture; NH = natural hay; S = soya; T = triticale. Absence of data labels indicates levels below the limit of quantification (LOQ; 0.5 mg/kg). Values are the mean of two replicates per sample, shown as carbon fractions ($n–C_{10–50}$).

The results following the analyses performed confirmed the MOH contamination of most of the feed samples. Important value differences were observed depending on the degree of technological development of each farm.

The $L_1$ farm, practicing semi-intensive agricultural and livestock activities, presented important levels of MOH contamination of feed, on average between 21.6 mg/kg and 27.4 mg/kg MOSH and 1.2 mg/kg and 1.6 mg/kg MOAH. These results show a serious potential contamination determined by applying some rudimentary agricultural and technological practices. The lack of effective contamination prevention measures also contributes.

In the feed from the medium-risk farm ($M_2$), MOSH levels ranged from 11.4 mg/kg to 35.0 mg/kg, while MOAH levels ranged from 0.5 mg/kg to 2.5 mg/kg. These results indicate a remarkable contamination linked to the higher degree of technological development, likely due to inadequate monitoring of technological processes.

Compared to the less advanced farms, the farm utilising intensive machinery and possibly inadequate feed handling or storage practices (high-risk farm, $H_3$) exhibited higher MOH contamination levels. This may suggest a greater exposure to contamination sources during cultivation, harvesting, transport, and storage. The average MOSH contamination ranged from 16.9 mg/kg to 81.4 mg/kg, and MOAH levels reached a maximum of 4.6 mg/kg.

Related to the technological specifics of each farm, the contamination levels were relatively proportional to the degree and intensity of exposure to various sources of contamination, as we found in the sampling sheets. Different numbers of samples were collected from the three farms (two samples from $L_1$ farm, five samples from $M_2$ farm, and eight samples from $H_3$ farm). While this variation in sample size could suggest differences between results, all samples were analysed according to the same methodological standards and validation criteria to ensure data comparability and consistency.

In analysing MOH contamination of feed, particular attention was especially focused on the proportion of the MOAH fraction, as it poses a risk factor to the safety of animal products. MOAH is considered more toxicologically concerning because it may include polycyclic aromatic compounds, some of which may have carcinogenic potential.

Figure 4 highlights the lack of uniformity among the 15 feed samples, with variability accentuated by the differing technological levels of the farms. The differences in MOSH and MOAH proportions can be attributed to specific feed-processing technologies, including the technical quality and maintenance of the machinery or processing equipment, the quality of mineral oils, but also the origin of raw materials, particularly those sourced from industrialised areas.

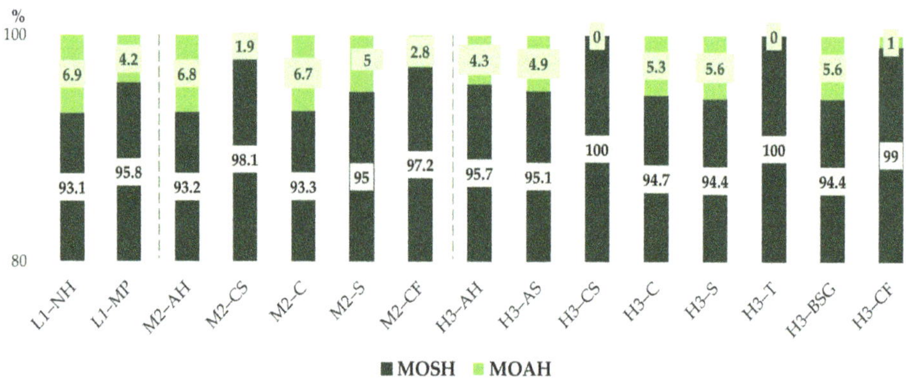


**Figure 4.** Proportion (%) of MOSH and MOAH in feed relative to total MOH content. $L_1$, $M_2$, $H_3$ = farm codes indicating contamination risk levels: low (L), medium (M), and high (H). AH = alfalfa hay; AS = alfalfa silage; BSG = brewer's grain; C = corn grain; CF = combined feed; CS = corn silage; MP = mountain pasture; NH = natural hay; S = soya; T = triticale.

MOAH concentrations showed some variability, with certain samples, such as $H_3$–AH (12.7%) and $H_3$–BSG (5.6%), presenting higher proportions. On average, in our samples, MOAH accounted for 4.74% of the total MOH content, which raises concerns because of the toxic nature of this fraction. Although this percentage may appear small, even the small amounts of MOAH in feed can be alarming for food safety if transferred into food products. Considering the lack of clearly defined European limits for MOAH in food or feed, it is recommended to minimise contamination as much as possible, ideally reaching an absence of MOAH in feed, to reduce the risk of transfer into the food chain.

A selection of chromatograms that confirmed MOSH and MOAH presence in various feed samples are shown in Figures 5 and 6. The contamination profiles were analysed for natural hay ($L_1$–NH), corn grain ($M_2$–C and $H_3$–C), alfalfa hay ($M_2$–AH), and corn silage ($M_2$–CS).

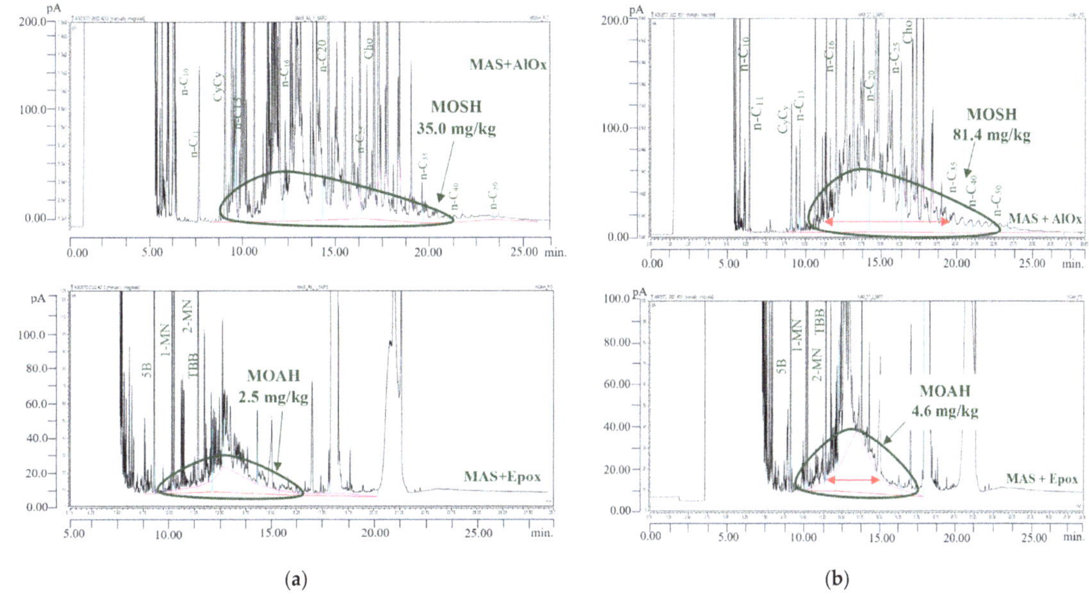


**Figure 5.** MOSH (**top**) and MOAH (**bottom**) HPLC–GC–FID chromatograms confirming contamination of corn grain samples $M_2$–C (**a**) and $H_3$–C (**b**) collected from the storage area. The green arrows and circles highlight the contamination concentrations and common profiles for MOSH and MOAH, particularly around specific molecular humps (n–$C_{10–35}$ MOSH and n–$C_{16–25}$ MOAH). The red arrow in (**b**) evidences the boundary distinguishing the MOSH and MOAH areas. Retention time (x): 0–28.5 min, and detector signal (y): 0_200 pA MOSH/0_100 pA MOAH.

Corn grain samples ($M_2$–C and $H_3$–C) indicated the highest levels of MOSH and MOAH, with a common contamination profile focused on the same molecular humps (n–$C_{10–35}$ MOSH and n–$C_{16–25}$ MOAH). This increased contamination can be attributed to factors such as mechanised harvesting, handling, and long-term storage, where mineral oils from equipment likely served as contamination sources. Additionally, the mineral oils used in post-harvest treatments to prevent infestation and preserve the feeds, along with the specific structure and chemical composition of corn, making it more prone to absorbing contaminants, also contributed to this contamination profile.

For the natural hay ($L_1$–NH), alfalfa hay ($M_2$–AH), and corn silage ($M_2$–CS) samples, a common contamination profile was observed, probably associated with the mineral oils used in agricultural machinery, as was also reported by Srbinovska et al. [49] in a study on certain plant products. Contamination can be further attributed to the specific technological processing methods. Mowing, baling, and compacting hay and silage involve direct contact with different equipment and materials (e.g., polythene sheeting), which can also be factors in MOSH and MOAH contamination (Figure 6).

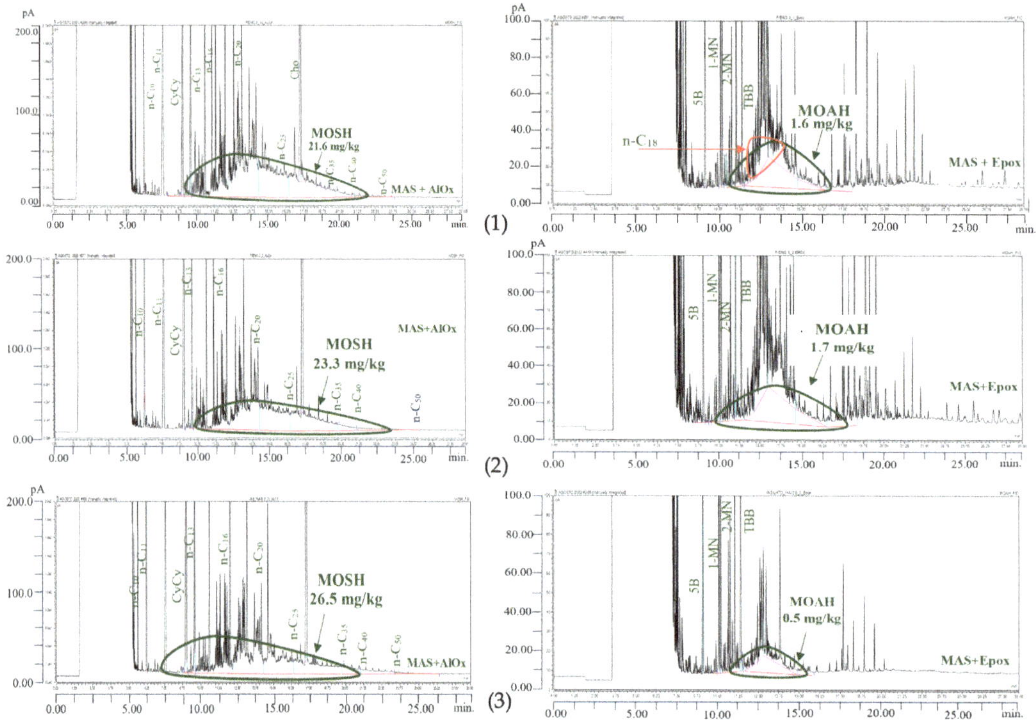


(**a**) (**b**)

**Figure 6.** MOSH (**a**) and MOAH (**b**) HPLC–GC–FID chromatograms confirming contamination of natural hay $L_1$–NH (**a1**,**b1**), alfalfa hay $M_2$–AH (**a2**,**b2**), and corn silage $M_2$–CS (**a3**,**c3**). The green arrows and circles highlight the contamination concentrations and the common profiles between samples for MOSH and MOAH. The red line in (**b1**) evidences the contamination profile centred around a specific molecular hump (n–$C_{16–35}$). Retention time (x): 0–28.5 min, and detector signal (y): 0_200 pA MOSH/0_100 pA MOAH.

## 5. Technological Operations

We considered it important to assess the potential feed contamination factors, some predictable factors, focusing on those typically associated with the specific technological processes and the level of mechanisation within the analysed farms. Table 3 summarises the main activities evaluated for their potential contribution to MOSH/MOAH contamination against the level of feed contamination.

The high contamination levels in the feed samples, as indicated by sampling sheets, suggest, to a certain extent, a direct connection between the MOH content in the feed and the diversity and complexity of the technological operations carried out on the farm.

An overview of the data from the sampling sheets reported that typical fertilisation or pest-control treatments were applied for crop protection on farms $M_2$ and $H_3$. The treatments applied did not seem to have a direct and express connection with the contamination levels detected. According to their labels, the substances used did not specifically contain MOH risk compounds. However, within a legislative framework that lacks strict regulation, the absence of specific mentions on the labels does not exclude the possible presence of MOH as co-formulants, since current regulations do not require their declaration. Moreover, the diversity of treatments applied (on average, 6–7 different substances) suggests that these practices cannot be entirely dismissed as potential sources of MOSH and MOAH contamination.

**Table 3.** Feed contamination factors linked to technological processes and mechanisation levels in analysed farms.

| Sample | MOSH | MOAH | Feeding | Technological Operations Applied to Crops | | | | Storage | |
|---|---|---|---|---|---|---|---|---|---|
| | | | | Phytosanitary Treatments (P) and Fertilisation (F) | | Harvesting/ Handling | Equipment | Area | Type/ Material |
| | mg/kg | | | Type | Formula */ Quantity (ha) | | | | |
| $L_1$–NH | 21.6 | 1.6 | Manually | Organic/Manually | - | Mechanised | (a) Mowing/harvesting machine (b) Transport vehicle | Half-open | Traditional wooden construction |
| $L_1$–MP | 27.4 | 1.2 | Manually | - | - | - | - | - | - |
| $M_2$–AH | 23.3 | 1.7 | Technological trailer | Organic and Chemical/Mechanised | P: Pulsar 40 (1.1 L/ha) | Mechanised | (a) Sprinkler pump (b) Mowing/harvesting machine (c) Transport vehicle (d) Baler | Open | Unwrapped bales |
| $M_2$–CS | 26.5 | 0.5 | Technological trailer | Organic and Chemical/Mechanised | F: Sulfammo–25–APPM–1 (170 kg/ha) P: Principal Plus (440 g/ha) | Mechanised | (a) Sprinkler pump (b) Harvesting machine (c) Transport vehicle (d) Crawler tractor | Open | Concrete cell covered with polyethylene film |
| $M_2$–C | 35.0 | 2.5 | Technological trailer | Organic and Chemical/Mechanised | F: DAP 18–46–0 (250 kg/ha); Sulfammo–25–APPM–1 (250 kg/ha) P: Principal Plus (440 g/ha) | Mechanised | (a) Sprinkler pump (b) Harvesting machine (c) Transport vehicle | Closed | Polypropylene bags |
| $M_2$–S | 11.4 | 0.6 | Technological trailer | - | - | Mechanised | Transport and unloading vehicle | Closed | Polypropylene bags |
| $M_2$–CF | 28.1 | 0.8 | Technological trailer | - | - | Mechanised | Technological trailer | - | - |
| $H_3$–AH | 29.0 | 1.3 | Technological trailer | Organic and Chemical/Mechanised | F: Complex 16–16–16 (250 kg/ha) P: Corum (1.2 L/ha) | Mechanised | (a) Sprinkler pump (b) Mowing/harvesting machine (c) Transport vehicle (d) Baler and foil press | Open | Polyethylene foiled bales |
| $H_3$–AS | 42.6 | 2.2 | Technological trailer | Organic and Chemical/Mechanised | F: Complex 16–16–16 (250 kg/ha) P: Corum (1.2 L/ha) | Mechanised | (a) Sprinkler pump (b) Harvesting machine (c) Transport vehicle (d) Crawler tractor | Open | Concrete cell covered with polyethylene film |
| $H_3$–CS | 24.9 | <LOQ ** | Technological trailer | Organic and Chemical/Mechanised | F: urea (100 kg/ha); NPK 20–20–0 (100 kg/ha); Ammonium nitrate (150 kg/ha) P: Henik (1.5 L/ha); Mustang (0.6L/ha); Adengo (0.4 L/ha) | Mechanised | (a) Sprinkler pump (b) Harvesting machine (c) Transport vehicle (d) Crawler tractor | Open | Concrete cell covered with polyethylene film |
| $H_3$–C | 81.4 | 4.6 | Technological trailer | Organic and Chemical/Mechanised | F: urea (100 kg/ha); NPK 20–20–0 (100 kg/ha); Ammonium nitrate (150 kg/ha) P: Henik (1.5 L/ha); Mustang (0.6L/ha); Adengo (0.4 L/ha) | Mechanised | (a) Sprinkler pump (b) Harvesting machine (c) Transport vehicle | Closed | Silo |
| $H_3$–S | 16.9 | 1.0 | Technological trailer | - | - | Mechanised | Transport and unloading vehicle | Closed | Silo |

**Table 3.** *Cont.*

| Sample | MOSH | MOAH | Feeding | Technological Operations Applied to Crops | | | | Storage | |
|---|---|---|---|---|---|---|---|---|---|
| | | | | Phytosanitary Treatments (P) and Fertilisation (F) | | Harvesting/ Handling | Equipment | Area | Type/ Material |
| | mg/kg | | | Type | Formula */ Quantity (ha) | | | | |
| $H_3$–T | 30.4 | <LOQ ** | Technological trailer | Organic and Chemical/Mechanised | F: urea (150 kg/ha); Ammonium nitrate (150 kg/ha); Lebosol (1.5 L/ha); P (I): Pixxaro Super (0.3 L/ha); P(II): Orius, Falcon Pro (0.5 L/ha) P(III): Mospilan (0.15 L/ha) | Mechanised | (a) Sprinkler pump (b) Harvesting machine (c) Transport vehicle | Closed | Silo |
| $H_3$–BSG | 57.3 | 3.4 | Technological trailer | - | - | Mechanised | Transport and unloading vehicle | Open | Concrete platform |
| $H_3$–CF | 50.9 | 0.5 | Technological trailer | - | - | Mechanised | Technological trailer | - | - |

* The treatments and active substances can be found in the previous sections. ** Absence of data labels indicates levels below the limit of quantification (LOQ; 0.5 mg/kg). $L_1$, $M_2$, $H_3$ = farm codes indicating contamination risk levels: low (L), medium (M), and high (H). AH, AS, BSG, C, CF, CS, MP, NH, S, T = abbreviations used for feed sample types derived from sample names: AH = alfalfa hay; AS = alfalfa silage; BSG = brewer's grain; C = corn grain; CF = combined feed; CS = corn silage; MP = mountain pasture; NH = natural hay; S = soya; T = triticale. APPM = activated poly-phenolic molecules; DAP = diammonium phosphate; NPK = nitrogen (N), phosphorus (P), and potassium (K) fertiliser.

Although no clear connection was established between the number or type of treatments applied and the level of MOH contamination, several noteworthy observations and trends emerged.

Samples $H_3$–C and $H_3$–BSG showed the highest contamination levels, with 81.4 mg/kg MOSH and 4.6 mg/kg MOAH, and 57.3 mg/kg MOSH and 3.4 mg/kg MOAH, respectively, despite undergoing relatively few chemical treatments. Even for crops with fewer treatments, such as alfalfa, similarly high contamination levels were reported (29.0 mg/kg to 42.6 mg/kg MOSH, and 1.3 mg/kg to 2.2 mg/kg MOAH) as for crops with more diversified treatments (24.9 mg/kg to 81.4 mg/kg MOSH, and 0.5 mg/kg to 4.6 mg/kg MOAH). This suggests that other factors, such as mechanised handling, equipment, and storage conditions, may play an important role in contamination.

Samples such as $L_1$–NH and $L_1$–MP, which were treated organically without using chemicals, showed lower contamination levels compared to those treated with both organic and chemical methods, such as $H_3$–C. While a direct causal connection could not be established, it can be inferred that organic treatments did not seem to have an important contribution to MOSH and MOAH contamination.

Some of the primary sources of MOSH and MOAH contamination are the residues, emissions, and technical oils from agricultural equipment and machinery used in feed processing. On the studied farms, the moderate to high technological level indicated an important contamination with MOSH and MOAH. Specifically, on farms $M_2$ and $H_3$, the extensive use of mechanised agricultural processes stood out as a major technological factor contributing to contamination. The variety of agricultural machinery used for harvesting and processing feed, such as harvesters, balers, or crawler tractors, presented a high risk of contamination due to the use of engine oils, lubricants, or hydraulic oils, which are known sources of MOH contamination. Moreover, for some feeds, such as hay or silage, the risk of contamination increased even more because of the specific processing methods involved, particularly direct contact of the feed with mechanised equipment and materials used in the process (e.g., polyethylene film).

Samples collected and transported using mechanised methods, such as $L_1$–NH, $M_2$–C, and $H_3$–C, showed high levels of MOH contamination. This confirmed that mechanised equipment can be an important source of contamination. Furthermore, the use of multiple complex machines for samples $M_2$–CS, $M_2$–C, $H_3$–AS, and $H_3$–C appeared to be associated with high contaminations (ranging from 26.5 mg/kg to 42.6 mg/kg MOSH and 0.5 mg/kg to 2.2 mg/kg MOAH). This reinforces the hypothesis that prolonged contact between feed and various mechanised components of equipment during harvesting and transport processes may introduce contaminants. Nevertheless, in the case of the $L_1$-MP sample, where animals grazed freely, no mechanised operations were involved that could have influenced the level of contamination; thus, the hypothesis of technological contamination was excluded in this case.

Regarding the causal connection between contamination levels and feed storage conditions, both the type of storage (closed or open) and the materials used (polyethylene, polypropylene, and concrete) seemed to have a certain influence on MOSH and MOAH contamination. Prolonged exposure to open environments and the use of synthetic materials appeared to be associated with higher contamination. Although various chemicals were used on some farms to sanitise animals or shelters, they did not seem to have a marked impact on the level of MOH contamination. No clear results could be directly linked to these substances, suggesting that other sources are likely more important to feed contamination.

## 6. Discussion

The results of our research reported varying levels of MOSH and MOAH contamination in the analysed feeds, influenced by technological factors on the selected farms.

Although the topic of this paper has gained attention in recent years and relevant studies have been conducted, it is possible that existing data are limited to food contamination or only to certain types of feed and technologies. However, similar data to those in the present research have been reported in other studies. Jaén et al. [53] obtained comparable conclusions regarding MOSH and MOAH contamination levels in food, assigning this to contaminant migration from packaging materials and direct contact with processing equipment. In the study of Srbinovska et al. [25], part of the contamination was also attributed to the technological equipment and materials used during production. High levels of MOH were detected in feed stored in polyethylene bags, similar to the $H_3$–C and $H_3$–S samples in this study, which were maintained under similar conditions. In contrast, $L_1$ samples were stored in different conditions, without polyethylene materials, which may account for their contamination levels.

Recently, Menegoz Ursol et al. [33] investigated the impact of certain agricultural technologies on MOH contamination. The risks associated with the use of mechanised equipment, such as a mechanised comb, straddle harvester, or pneumatic comb, were highlighted, and their findings seem to support our hypothesis regarding the role of advanced equipment in feed harvesting, especially the use of hydraulic oils and other technical oils as sources of contamination. The same studies mentioned the critical factors involved and mechanisms through which feed-processing technologies influence MOH contamination. Various conclusions suggested that hydrocarbon migration from contact materials into feed can occur through direct contact or infiltration of residues from the storage spaces, especially under conditions of material degradation or exposure to extreme conditions [53]. These data aligned with other studies investigating feed contamination from packaging and coating materials. In particular, research emphasised the risk of hydrocarbon infiltration from polymers during prolonged storage or in environments exposed to temperature and humidity fluctuations [54].

The confirmation of MOSH and MOAH presence in animal feed can have serious implications for the safety of animal products. Albendea et al. [55] showed that MOH-contaminated feed can be a direct source of contamination in animal productions. The authors further detailed how MOH from feed can be transferred into animal tissues, directly affecting the safety of meat. The presence of MOH in products such as dairy is even more concerning, as these are regularly consumed by vulnerable groups, including children, the elderly, or the ill. In addition to the toxic risks for consumers, high levels of contamination can lead to non-compliance with food safety standards established by international organisations, such as the European Food Safety Authority [23]. In the

European Union, strict limits are recommended for MOSH and MOAH contamination in food, and producers who exceed these limits face economic sanctions or even the withdrawal of their products from the market [2,20,23,52].

The results of this study, along with the importance of assessing technological risk in preventing MOSH and MOAH contamination, are supported by the literature. However, more extensive monitoring may be necessary to fully understand the implications for food safety and public health.

The contamination levels identified in the analysed feed emphasised the need for certain measures to minimise contamination risks, especially during the harvesting and processing stages. Adopting good working practices is an effective solution, as contamination can occur at all stages of the feed production chain. More awareness of the contamination risks caused by technical oils used in agricultural equipment is necessary.

To prevent MOSH/MOAH contamination, it is recommended to replace lubricants and oils with high-grade, refined, or food-grade products that do not contain MOH. For enhanced safety, completely MOH-free alternatives should be considered. Furthermore, proper maintenance of agricultural machinery and equipment is particularly important for reducing the risk of contamination. Regular sanitation and proper usage of equipment can help prevent the accumulation and transfer of contaminants to feed. Additionally, improvements in the design and development of agricultural equipment that minimise the impact on feed could provide long-term solutions, even though they may require substantial technological changes.

## 7. Conclusions

This study showed a clear connection between the technologies employed in feed production and the levels of MOSH and MOAH contamination. Specifically, the storage conditions, as seen in certain samples, appeared to be associated with elevated contamination levels, underscoring the importance of storage materials and methods in the technological process.

The comparative assessment of contamination levels generally indicated the presence of MOSH and MOAH across all types of feed at each studied farm. The proportional differences were attributed to the specific characteristics of each unit concerning their exposure to various contamination sources, including the degree of technologisation in farm operations, the location of agricultural crops, and the technological practices employed.

Factors such as the mechanisation of the agricultural process, the machinery used, and the storage conditions considerably influenced contamination levels, highlighting the essential requirement of implementing advanced technological measures to mitigate this issue. The data clearly showed that the use of mechanised equipment, such as combines, balers, crawler tractors, and transport vehicles, was associated with increased contamination, particularly when the equipment came into direct contact with feed. Storage conditions and the materials used further exacerbated contamination, indicating that other stages of the technological processes may also contribute to the transfer of contaminants. The chemical treatments applied did not appear to have a direct impact on MOH contamination; however, the presence of unclarified co-formulants and the mechanised application processes could represent indirect sources of contamination.

In the context of the findings concerning MOSH and MOAH contamination in feed, it is necessary to embrace optimal technological measures to mitigate these risks. Farmers and producers are encouraged to replace the lubricants and technical oils used in agricultural equipment with food-grade products or MOH-free alternatives. They should also ensure regular and appropriate maintenance of mechanised equipment to prevent leaks of technical oils and the accumulation of contaminants on surfaces. Moreover, using well-protected storage areas and avoiding materials that may contribute to contamination is important. Lastly, supporting improvements in the design of agricultural equipment to minimise direct contact with feed can help reduce contamination risks. As a forward-looking solution, transitioning towards the electrification of agricultural machinery by replacing combustion engines with electric motors could further reduce MOH contamination, enhancing overall safety in feed production.

The elevated levels of contamination in feed samples cannot be totally attributed to technological factors. The MOSH and MOAH concentrations suggested the presence of multiple sources of contamination that are likely more diverse than initially expected.

To further substantiate the research findings, future studies could explore in depth the impact of additional technological steps on contamination, including processing methods and the influence of atmospheric conditions, as well as exposure to urban pollution sources. A more comprehensive approach should also extend to analysing MOSH and MOAH contamination in various types of feed, but especially in animal products. Such an approach would provide a broader perspective on risks throughout the food chain and have a direct impact on food safety.

**Author Contributions:** Conceptualisation, M.M. and I.M.P.; methodology, M.M. and D.M.L.; software, M.M.; validation, D.S., P.C.B. and I.M.P.; formal analysis, M.M.; investigation, S.I.P., B.M.M. and D.M.L.; resources, M.M. and P.C.B.; data curation, D.S. and I.M.P.; writing—original draft preparation, M.M.; writing—review and editing, S.I.P. and B.M.M.; supervision, I.M.P.; project administration, P.C.B. All authors have read and agreed to the published version of the manuscript.

**Funding:** This research received no external funding.

**Institutional Review Board Statement:** Not applicable.

**Data Availability Statement:** Data supporting the reported results are available from the authors upon request.

**Acknowledgments:** This research is part of a broader initiative investigating feed contamination, the impact on animals, and the transfer of contaminants into animal products. The study was conducted through collaboration between the Iasi University of Life Sciences and the University of Udine, Department of Agricultural, Food, Environmental, and Animal Sciences. We are grateful to the Food Chemistry Laboratory team for providing the necessary facilities and resources, and we also extend our appreciation to them for their valuable support in data analysis and technical guidance throughout the study.

**Conflicts of Interest:** The authors declare no conflicts of interest.

## References

1. Lăpușneanu, D.M.; Radu-Rusu, C.G.; Matei, M.; Petrescu, S.I.; Pop, I.M. Aflatoxins occurrence and levels in maize from a Romanian feed mill. *Sci. Papers Ser. D. Anim. Sci.* **2024**, *LXVII*, 110–114.
2. Postolache, A.N.; Chelmu, S.S.; Ariton, A.M.; Ciorpac, M.; Pop, C.; Ciobanu, M.M.; Creangă, Ș. Analysis of RASFF notifications on contaminated dairy products from the last two decades: 2000–2020. *Rom. Biotechnol. Lett.* **2020**, *25*, 1396–1406. [CrossRef]
3. Boișteanu, P.C.; Flocea, E.-I.; Anchidin, B.G.; Mădescu, B.M.; Matei, M.; Murariu, O.C.; Frunză, G.; Postolache, A.N.; Ciobanu, M.M. Essential and toxic elements analysis of wild boar tissues from north-eastern Romania and health risk implications. *Front. Sustain. Food Syst.* **2024**, *8*, 1406579. [CrossRef]
4. Ciobanu, M.M.; Munteanu, M.; Postolache, A.N.; Boișteanu, P.C. Toxic heavy metals content in wild boar and venison meat: A brief review. *Sci. Pap. Ser. D Anim. Sci.* **2020**, *LXIII*, 435–441.
5. Kim, M.; Cho, B.H.; Lim, C.M.; Kim, D.G.; Yune, S.Y.; Shin, J.Y.; Bong, Y.H.; Kang, J.; Kim, M.A.; Son, S.W. Chemical residues and contaminants in foods of animal origin in Korea during the past decade. *J. Agric. Food Chem.* **2013**, *61*, 2293–2298. [CrossRef]
6. Mierliță, D.; Teușdea, A.C.; Matei, M.; Pascal, C.; Simeanu, D.; Pop, I.M. Effect of Dietary Incorporation of Hemp Seeds Alone or with Dried Fruit Pomace on Laying Hens' Performance and on Lipid Composition and Oxidation Status of Egg Yolks. *Animals* **2024**, *14*, 750. [CrossRef]
7. Lake, I.R.; Foxall, C.D.; Fernandes, A.; Lewis, M.; Rose, M.; White, O.; Dowding, A. Seasonal variations in the levels of PCDD/Fs, PCBs and PBDEs in cows' milk. *Chemosphere* **2013**, *90*, 72–79. [CrossRef]
8. Rusu, L.; Harja, M.; Suteu, D.; Dabija, A.; Favier, L. Pesticide Residues Contamination of Milk and Dairy Products. A case study: Bacău District Area, Romania. *J. Environ. Prot. Ecol.* **2016**, *17*, 1229–1241.
9. Piskorska–Pliszczynska, J.; Maszewski, S.; Mikolajczyk, S.; Pajurek, M.; Strucinski, P.; Olszowy, M. Elimination of dioxins in milk by dairy cows after the long–term intake of contaminated sugar beet pellets. *Food Addit. Contam. Part. A* **2017**, *34*, 842–852. [CrossRef]
10. Bedi, J.S.; Gill, J.P.S.; Kaur, P.; Aulakh, R.S. Pesticide residues in milk and their relationship with pesticide contamination of feedstuffs supplied to dairy cattle in Punjab (India). *J. Anim. Feed Sci.* **2018**, *27*, 18–25. [CrossRef]

11. Matei, M.; Pop, I.M. Mineral oil hydrocarbons (MOH) analysis in animal feed: A characterization based on modern pollution. *Sci. Pap. Ser. D Anim. Sci.* **2023**, *LXVI*, 113–122.
12. Grob, K. Update on recycled paperboard and its compliance for food contact. *J. Verbrauch. Lebensm.* **2014**, *9*, 213–219. [CrossRef]
13. Canavar, Ö.; Kappenstein, O.; Luch, A. The analysis of saturated and aromatic mineral oil hydrocarbons in dry foods and from recycled paperboard packages by online HPLC-GC-FID. *Food Addit.Contam. Part A* **2018**, *35*, 2471–2481. [CrossRef] [PubMed]
14. Matei, M.; Petrescu, S.-I.; Flocea, E.-I.; Lăpușneanu, D.M.; Simeanu, D.; Pop, I.M. Variation in mineral oil hydrocarbons content of milk during processing. *Sci. Pap. Ser. D. Anim. Sci.* **2024**, *LXVII*, 490–499.
15. Barp, L.; Kornauth, C.; Würger, T.; Rudas, M.; Biedermann, M.; Reiner, A.; Concin, N.; Grob, K. Mineral oil in human tissues, part I: Concentrations and molecular mass distributions. *Food Chem. Toxicol.* **2014**, *72*, 312–321. [CrossRef]
16. Barp, L.; Biedermann, M.; Grob, K.; Blas-Y-Estrada, F.; Nygaard, U.C.; Alexander, J.; Cravedi, J.P. Accumulation of mineral oil saturated hydrocarbons (MOSH) in female Fischer344 rats: Comparison with human data and consequences for risk assessment. *Sci. Total Environ.* **2017**, *575*, 1263–1278. [CrossRef]
17. Barp, L.; Biedermann, M.; Grob, K.; Blas-Y-Estrada, F.; Nygaard, U.C.; Alexander, J.; Cravedi, J.P. Mineral oil saturated hydrocarbons (MOSH) in female Fischer 344 rats; accumulation of wax components; implications for risk assessment. *Sci. Total Environ.* **2017**, *583*, 319–333. [CrossRef]
18. Grob, K. Could the Ukrainian sunflower oil contaminated with mineral oil wake up sleeping dogs? *Eur. J. Lipid Sci. Technol.* **2008**, *110*, 979–981.
19. Biedermann, M.; Grob, K. How "white" was the mineral oil in the contaminated Ukrainian sunflower oils? *Eur. J. Lipid Sci. Technol.* **2009**, *111*, 313–319. [CrossRef]
20. European Food Safety Authority (EFSA). Scientific opinion on mineral oil hydrocarbons in food. *EFSA J.* **2012**, *10*, 2704.
21. Bauwens, G.; Conchione, C.; Sdrigotti, N.; Moret, S.; Purcaro, G. Quantification and characterization of mineral oil in fish feed by liquid chromatography-gas chromatography-flame ionization detector and liquid chromatography-comprehensive multidimensional gas chromatography-time-of-flight mass spectrometer/flame ionization detector. *J. Chromatogr. A* **2022**, *1677*, 463208. [PubMed]
22. Nestola, M. Automated workflow utilizing saponification and improved epoxidation for the sensitive determination of mineral oil saturated and aromatic hydrocarbons in edible oils and fats. *J. Chromatogr. A* **2022**, *1682*, 463523. [CrossRef] [PubMed]
23. European Food Safety Authority (EFSA). Update of the risk assessment of mineral oil hydrocarbons in food. *EFSA J.* **2023**, *21*, e08215.
24. Hochegger, A.; Moret, S.; Geurt, L.; Gude, T.; Meitner, E.; Mertens, B.; O'Hagan, S.; Poças, F.; Simat, T.; Purcaro, G. Mineral oil risk assessment: Knowledge gaps and roadmap. Outcome of a multi-stakeholders workshop. *Trends Food Sci. Technol.* **2021**, *113*, 151–166. [CrossRef]
25. Srbinovska, A.; Conchione, C.; Celaj, F.; Menegoz Ursol, L.; Moret, S. High sensitivity determination of mineral oils and olefin oligomers in cocoa powder and related packaging: Method validation and market survey. *Food Chem.* **2022**, *396*, 133686.
26. Grob, K.; Vass, M.; Biedermann, M.; Neukom, H.-P. Contamination of animal feed and food from animal origin with mineral oil hydrocarbons. *Food Addit. Contam.* **2001**, *18*, 1–10. [CrossRef]
27. Van-Heyst, A.; Goscinny, S.; Bel, S.; Vandevijvere, S.; Mertens, B.; Elskens, M.; Van Hoeck, E. Dietary exposure of the Belgian population to mineral oil. *Food Addit. Contam. Part A* **2020**, *37*, 267–279. [CrossRef]
28. European Commission. Regulation (EC) No 1107/2009 of the European Parliament and of the Council of 21 October 2009 concerning the placing of plant protection products on the market and repealing Council Directives 79/117/EEC and 91/414/EEC. *Off. J. Eur. Union* **2009**, *309*, 1–50.
29. European Commission. Commission implementing regulation (EU) No 540/2011 of 25 may 2011 implementing regulation (EC) No 1107/2009 of the European Parliament and of the Council as regards the list of approved active substances. *Off. J. Eur. Union* **2011**, *153*, 1–186.
30. Grob, K. Toxicological Assessment of Mineral Hydrocarbons in Foods: State of Present Discussions. *J. Agric. Food Chem.* **2018**, *66*, 6968–6974. [CrossRef]
31. Brühl, L. Occurrence, determination, and assessment of mineral oils in oilseeds and vegetable oils. *Eur. J. Lipid Sci. Technol.* **2016**, *118*, 361–372. [CrossRef]
32. Hidalgo Ruiz, J.L.; Arrebola Liebanas, J.; Martinez Vidal, J.L.; Garrido Frenich, A.; Romero Gonzalez, R. Offline Solid-Phase Extraction and Separation of Mineral Oil Saturated Hydrocarbons and Mineral Oil Aromatic Hydrocarbons in Edible Oils, and Analysis via GC with a Flame Ionization Detector. *Foods* **2021**, *10*, 2026. [CrossRef] [PubMed]
33. Menegoz Ursol, L.; Conchione, C.; Peroni, D.; Carretta, A.; Moret, S. A study on the impact of harvesting operations on the mineral oil contamination of olive oils. *Food Chem.* **2023**, *406*, 135032. [CrossRef]
34. Mertens, B.; Van Heyst, A.; Demaegdt, H.; Boonen, I.; Van Den Houwe, K.; Goscinny, S.; Elskens, M.; Van Hoeck, E. Assessment of hazards and risks associated with dietary exposure to mineral oil for the Belgian population. *Food Chem. Toxicol.* **2021**, *149*, 112034. [CrossRef]
35. Foodwatch. International Tests of Various Food Products for Their Contamination by Mineral Oil Hydrocarbons (MOSH/MOAH), Project-Report. 2021. Available online: https://www.foodwatch.org/fileadmin/-INT/mineral_oil/documents/2021-12-03_technical__minoil__project_report.pdf (accessed on 2 April 2023).

36. Bratinova, S.; Hoekstra, S. *Guidance on Sampling, Analysis and Data Reporting for the Monitoring of Mineral Oil Hydrocarbons in Food and Food Contact Materials*, 1st ed.; Publications Office of the European Union: Luxembourg, 2019.
37. Bratinova, S.; Robouch, P.; Hoekstra, E.; Bratinova, S. *Guidance on Sampling, Analysis and Data Reporting for the Monitoring of Mineral Oil Hydrocarbons in Food and Food Contact Materials*, 2nd ed.; Publications Office of the European Union: Luxembourg, 2023.
38. European Union. Commission Recommendation (EU) 2017/84 of 16 January 2017 on the monitoring of mineral oil hydrocarbons in food and in materials and articles intended to come into contact with food. *Off. J. Eur. Union* **2017**, *12*, 95–96.
39. Matei, M.; Pop, I.M. Monitoring of dairy farms to assess the potential level of pollution of animal feed and animal production. *Sci. Pap. Ser. D. Anim. Sci.* **2022**, *LXV*, 129–136.
40. *SR EN ISO 6497:2005*; Animal Feeding Stuffs. Sampling. International Organization for Standardization: Geneva, Switzerland, 2005.
41. European Commission. Commission Regulation (EC) No 152/2009 of 27 January 2009 laying down the methods of sampling and analysis for the official control of feed. *Off. J. Eur. Union* **2009**, *52*, 1–130.
42. *SR EN ISO 6498:2012*; Animal Feeding Stuffs. Guidelines for Sample Preparation. International Organization for Standardization: Geneva, Switzerland, 2012.
43. Nestola, M.; Schmidt, T.C. Determination of mineral oil aromatic hydrocarbons in edible oils and fats by online liquid chromatography–gas chromatography–flame ionization detection—Evaluation of automated removal strategies for biogenic olefins. *J. Chromatogr. A* **2017**, *1505*, 69–76. [CrossRef]
44. Biedermann, M.; Fiselier, K.; Grob, K. Aromatic hydrocarbons of mineral oil origin in foods: Method for determining the total concentration and first result. *J. Agric. Food Chem.* **2009**, *57*, 8711–8721. [CrossRef]
45. Biedermann, M.; Grob, K. On-line coupled high performance liquid chromatography- gas chromatography for the analysis of contamination by mineral oil. Part 1: Method of analysis. *J. Chromatogr. A* **2012**, *1255*, 56–75. [CrossRef]
46. Biedermann, M.; Grob, K. On-line coupled high performance liquid chromatography- gas chromatography for the analysis of contamination by mineral oil. Part 2: Migration from paperboard into dry foods: Interpretation of chromatograms. *J. Chromatogr. A* **2012**, *1255*, 76–99. [CrossRef] [PubMed]
47. Moret, S.; Scolaro, M.; Barp, L.; Purcaro, G.; Conte, L.S. Microwave assisted saponification (MAS) followed by on-line liquid chromatography (LC)-gas chromatography (GC) for high-throughput and high-sensitivity determination of mineral oil in different cereal-based foodstuffs. *Food Chem.* **2016**, *196*, 50–57. [CrossRef] [PubMed]
48. Menegoz Ursol, L.; Conchione, C.; Srbinovska, A.; Moret, S. Optimization and validation of microwave assisted saponification (MAS) followed by epoxidation for high-sensitivity determination of mineral oil aromatic hydrocarbons (MOAH) in extra virgin olive oil. *Food Chem.* **2022**, *370*, 130966. [CrossRef]
49. Srbinovska, A.; Gasparotto, L.; Conchione, C.; Menegoz Ursol, L.; Lambertini, F.; Suman, M.; Moret, S. Mineral oil contamination in basil pesto from the Italian market: Ingredient contribution and market survey. *J. Food Compos. Anal.* **2023**, *115*, 104914. [CrossRef]
50. Magnusson, B.; Ornemark, U. *Eurachem Guide: The Fitness for Purpose of Analytical Methods–A Laboratory Guide to Method Validation and Related Topics*, 2nd ed.; Eurachem: Middlesex, UK, 2014. Available online: https://www.eurachem.org/images/stories/Guides/pdf/MV_guide_2nd_ed_EN.pdf (accessed on 2 April 2023).
51. SANTE/12682/2019. Guidance Document on Analytical Quality Control and Method Validation Procedures for Pesticides Residues Analysis in Food and Feed. Available online: https://www.eurl-pesticides.eu/userfiles/file/EurlALL/AqcGuidance_SANTE_2019_12682.pdf (accessed on 5 May 2023).
52. European Commission. Summary Report. Standing Committee on Plants, Animals, Food and Feed. Section Novel Food and Toxicological Safety of the Food Chain, 21 April 2022. Available online: https://food.ec.europa.eu/system/files/2022-07/reg-com_toxic_20220421_sum.pdf (accessed on 31 March 2023).
53. Jaén, J.; Domeño, C.; Úbeda, S.; Aznar, M.; Nerín, C. Migration of mineral oil aromatic hydrocarbons (MOAH) from cardboard containers to dry food and prediction tool. *Food Control* **2022**, *138*, 109016. [CrossRef]
54. Buijtenhuijs, D.; van de Ven, B.M. *Mineral Oils in Food; a Review of Occurrence and Sources, RIVM Letter Report 2019-0048;* National Institute for Public Health and the Environment: Bilthoven, The Netherlands, 2019; pp. 1–59.
55. Albendea, P.; Conchione, C.; Menegoz Ursol, L.; Moret, S. A Study on Mineral Oil Hydrocarbons (MOH) Contamination in Pig Diets and Its Transfer to Back Fat and Loin Tissues. *Animals* **2024**, *14*, 1450. [CrossRef]

agriculture

MDPI

Article

# On the Digestibility of Mulberry Leaf Fed to *Bombyx mori* Larvae

**Marius Gheorghe Doliș [1], Claudia Pânzaru [1,*], Marius Giorgi Usturoi [1], Alexandru Usturoi [2], Cristina-Gabriela Radu-Rusu [2] and Mădălina Alexandra Davidescu [2]**

1 Department of Animal Resources and Technologies, Faculty of Food and Animal Sciences, "Ion Ionescu de la Brad" Iasi University of Life Sciences, 8 Mihail Sadoveanu Alley, 700489 Iasi, Romania; marius.dolis@iuls.ro (M.G.D.); marius.usturoi@iuls.ro (M.G.U.)

2 Department of Control, Expertise and Services, Faculty of Food and Animal Sciences, "Ion Ionescu de la Brad" Iasi University of Life Sciences, 8 Mihail Sadoveanu Alley, 700489 Iasi, Romania; alexandru.usturoi@iuls.ro (A.U.); cristina.radurusu@iuls.ro (C.-G.R.-R.); madalina.davidescu@iuls.ro (M.A.D.)

* Correspondence: claudia.panzaru@iuls.ro

**Abstract:** Considering that sericulture is an important branch of animal husbandry, not only for the production of silk but also as a valuable source of protein, it is necessary to constantly study the possibilities for its improvement as a branch of this domain. Therefore, the purpose of this paper is to assess the nutritional value and digestibility of mulberry leaves from the Kokuso 21 and Eforie varieties, as consumed by silkworms (*Bombyx mori* L., *Bombicidae* Family, *Lepidoptera* Order) during a summer study, in 2021. The Japanese variety (Kokuso 21) and the Romanian variety (Eforie) were used as food sources for the Triumf hybrid, developed in Romania; the larvae were divided in two batches of 300 larvae and each set was subdivided into six groups of 50 larvae, which were raised in paper trays based on their age and size. The research indicated that mulberry leaves have an average digestibility value of 54.46%; the aging process of the leaves altered their chemical composition, with most nutrients showing a decreasing trend in digestibility throughout the larval growth period, except for crude fiber, which remained unchanged in the early larval stages and increased to 26.78% towards the end of the experiment. Overall, the Kokuso 21 variety demonstrated superior nutrient digestibility compared to Eforie. An important finding from this study is the need for future research to determine the degree of nutrient metabolism and conversion into silk.

**Keywords:** silkworm; sericulture; mulberry; leaves; nutrition; silk

## 1. Introduction

These days, sericulture extends beyond merely producing silk; it now serves as a significant protein source for a diverse range of animal species, including fish, birds, mammals, and even humans [1–5]. Longvah et al., 2011, highlight the rich nutritional content provided by silkworm species, boasting 16% protein, 8% fat, and essential minerals, with an impressive high protein digestibility score (amino acid score of 86—PDCAAS) [2]. Consequently, *B. mori* L. (*Bombicidae* Family, *Lepidoptera* Order), among other insects, is frequently considered a valuable nutrient source for humanity [6–14]. Moreover, insect farming for this purpose carries a significantly lower environmental impact compared to activities like cattle ranching [9].

Furthermore, the nutritional quality of mulberry leaves has led to their increased use in feeding various farm animals, such as sheep, goats, cattle, pigs, rabbits, chicken, quail, and fish. These leaves can reduce dependence on other food sources, such as soybeans for laying hens. For instance, in an experiment involving Leghorn breed hens, the incorporation of silkworm meal showcased beneficial effects on their health and pro-duction, establishing it as a viable alternative protein source to soybean meal [10]. The age of mulberry trees has been observed to impact digestibility, as evidenced in goat farming [11–13]. Additionally, mulberry leaves have been shown to positively influence protein intake for swine when constituting up to 0.3% of a sow's diet [14,15]. Similarly, research on sheep has indicated favorable limits in mulberry digestibility [16], as they exhibit a preference for these leaves due to their palatability [17].

The Kokuso 21 mulberry (*Morus alba* L.) is a cultivar known for its highly appreciated characteristics such as the leaf yield, the nutritional content, the resistance to harsh environmental conditions and diseases, the growth rate, and palatability for the leaves. Kokuso 21 is renowned for its high leaf yield, which provides ample food for silkworms, rich in nutrients, particularly protein and essential amino acids, which are vital for the growth and health of silkworms, as well as for silk production. This variety is known for its resilience in arid and semi-arid conditions, making it suitable for cultivation in regions with limited water availability, being also highly resistant to common mulberry diseases (it helps ensure a very good quality and stable leaf supply. It also has a fast growth rate, allowing for multiple harvests of leaves within a growing season (which is beneficial for continuous sericulture operations)). Furthermore, Kokuso 21 is preferred by the silkworms because of its high palatability (it encourages better feeding and growth rates, ultimately leading to higher silk yields). The listed characteristics make this variety a popular choice among sericulturists, particularly in regions where environmental conditions can be challenging [14–16].

The Eforie mulberry is a specific variety known for certain distinct characteristics, particularly in silkworm farming. It is appreciated for the high leaf yield (it is known for producing a significant amount of leaves, which are rich in nutrients, especially in protein content, crucial for the health and development of *B. mori*, L., *Bombicidae* Family, *Lepidoptera* Order) and high resilience to cold and diseases. Eforie mulberry trees are noted for their resistance to cold weather, making them suitable for cultivation in regions with cooler climates, and they tend to have good resistance to common mulberry diseases, helping to maintain a healthy crop and ensuring a consistent supply of quality leaves. This variety also has a great growth rate (it typically exhibits a rapid growth rate, which is beneficial for producing leaves throughout the growing season) and a high adaptability (it is adaptable to different soil types and climatic conditions, which makes it a versatile choice for various geographic regions). All these traits make Eforie preferred in certain sericulture practices, especially in areas where climate and disease resilience are critical factors [14,15]. Irrespective of the intended purpose of sericultural activities, whether for silk production or other advantages, the primary focus of research in this field revolves around assessing the nutritional value of mulberry leaves, the factors influencing this value, and how *B. mori* larvae utilize these nutrients. Additionally, it is noteworthy to mention the use of various foliar additives aimed at enhancing the nutritional value of mulberry leaves [18–20].

Sericulture specialists are highly focused on producing premium biological materials, developing superior mulberry varieties or hybrids and silkworm new breeds, bloodlines, and hybrids. Also, the breeding of biological material, as well as microclimate conditions (temperature, humidity, brightness), hygiene, nutrition, and the use of specific equipment, are part of the mulberry hybrids' enhancement. Over the years, different types of mulberry varieties and hybrids have been developed to produce high-quality leaf crops adapted with resilience against environmental factors such as drought, frost, diseases, and pests. Larvae of *B. mori* originating from these varieties exhibit enhanced productivity and an increased ability to withstand environmental factors and diseases, effectively utilizing the nutrients provided by mulberry leaves.

To optimize the peak productivity, especially in larvae used within intensive silk production systems, it is crucial to fine-tune every aspect of the growth process, where nutrition plays a pivotal role. Both the quantity and quality of mulberry leaves fed to the larvae substantially impact their growth rate, overall health, vitality, and silk production. Leaf quality is affected by diverse factors such as soil conditions, climate, seasonal changes, mulberry variety, harvesting methods, and storage [15–18,20,21].

Mulberry leaves furnish silkworm larvae with essential proteins, carbohydrates, fats, minerals, vitamins, and water necessary for growth and silk yield. The quality of leaves is determined by their chemical composition, which is influenced by various factors like leaf maturity, their position on the branch, harvesting time and method, mulberry type, cli-mate, use of fertilizers, and maintenance techniques applied to the mulberry plantations [16,17].

Young leaves, abundant in protein, possess lower energy due to fewer carbohydrates, leading to increased protein consumption. Leaves at branch tips contain more water and protein, while those at the base contain higher cellulose and minerals. Typically, feeding silkworms with leaves harvested 30–40 days from the middle of branches yields consistent results. However, feeding leaves harvested 90–100 days from the base prolongs the larval period, raises mortality rates, and diminishes cocoon quality by reducing silk content [20].

Leaves collected in the morning exhibit different chemical compositions compared to those collected in the evening, primarily due to the influence of direct sunlight synthesizing organic compounds [20,21]. Fertilizer application post-pruning for leaf production compensates for substantial organic matter loss. Enhancing soil fertility via proper fertilizer application remains crucial in increasing leaf yield. Mulberry leaf nutritional value can be enhanced by applying organic and mineral fertilizers, resulting in leaves with higher protein and ascorbic acid content [17–19,21].

Mulberry variety also affects leaf quality, showcasing variations in yield and nutritional value among different varieties. High temperatures and low humidity can impact leaf quality, potentially causing drought during summer heat, compromising leaf harvests to varying degrees [21].

Fresh leaves yield the best results when fed to the larvae. Chopped leaves are more efficiently consumed during the early larval stages as they provide more edges for consumption. In the final larval stages, whole leaves or leaves on branches are typically administered [22].

Hence, this study aims to contribute to the understanding of mulberry leaf quality from Romanian varieties and their utilization by *B. mori* larvae. Romania historically played a significant role in European sericulture, developing valuable mulberry varieties and hybrids, along with *B. mori* breeds and hybrids through extensive research.

This paper further aims to unveil the value of the Romanian Eforie mulberry variety, particularly its digestibility compared to the Japanese Kokuso 21 variety, in a summer experiment. These findings hold significant importance in the field of sericulture due to the uniqueness of the selected mulberry variety and the digestibility-related results obtained.

## 2. Materials and Methods

### 2.1. The Animal Material

The animal material involved two sets, designated as L1 and L2, each consisting of 300 *B. mori* larvae of the Romanian hybrid called Triumf. This hybrid demonstrates consistent productive traits and a significant degree of heterosis. To enable tracking, each set was subdivided into six groups of 50 larvae and raised in paper trays based on their age and size. Additionally, a backup group was established, comprising larvae raised separately but under identical conditions, ready to substitute any deceased individuals in the main experimental group. These larvae were reared throughout August in a controlled environment, meticulously monitoring all microclimate factors within an air-conditioned room.

For each subgroup, the quantities of administered leaves, unconsumed residues, and excreta were recorded separately for each of the five larval stages. Samples were collected from these and subsequently subjected to chemical analysis. Thus, each subgroup was given the same amounts of mulberry leaves, from which samples were collected in advance for chemical analysis. Daily, and at the same time, the uneaten mulberry leaf residues and excreta were collected, weighed, and recorded from the sublots. Additionally, samples of residues and excreta were collected from each sublot and then subjected to chemical analysis. To separate the larvae from the residues and excreta, nets with mesh sizes appropriate for the larvae's age were used. These were placed above the larvae, with fresh leaves scattered over them. To reach the fresh leaves above, the larvae had to cross the net. The uneaten leaf residues and excreta remained below the net, where they were collected, separated, weighed, and recorded. By knowing the quantities of administered mulberry leaves, uneaten residues, and excreta, as well as their chemical composition, the digestibility coefficients of the mulberry leaves could be calculated for each variety and period.

After the donuts were formed, those harvested from each batch were sectioned and emptied of content (such as chrysalis and the leftovers from the last shedding of the larvae). This is how the silk wrap was separated and then the dry matter content was determined. In the following calculation regarding the food conversion, the average content of dry matter of the silk wrap, determined in the harvested donuts of each subgroup, was utilized.

The efficiency of the use of nutrients from leaf by the larvae was expressed by the amount of ingested/digested dry matter required for increasing 1 g of body mass/weight (silk wrap), respectively, by the efficiency of conversion of ingested substances (ECIS%)/digested (ECDS%) in weight [22–25].

The efficiency of nutrient utilization from mulberry leaves [26–28] was expressed by the following:

- grams of dry matter (DM) ingested necessary to produce one gram of growth (grams of dry matter of ingested leaf/grams of dry matter of silk wrap); this ratio provides insight into the feed efficiency of the larvae, indicating how much of the leaf's dry matter is needed to produce a specific amount of silk (a lower ratio indicates higher efficiency, as less feed is required per unit of silk produced).
- grams of dry matter (DM) digested necessary to produce one gram of growth (grams of dry matter of digested leaf/grams of dry matter of silk wrap); this ratio reflects the efficiency of the larvae's digestive system, highlighting how much of the ingested dry matter is actually utilized (absorbed and metabolized) for silk production (a more efficient digestive system results in a lower DM digested value).
- efficiency of conversion of ingested substances (ECIS) into silk wrap (ECIS = dry matter of silk wrap/dry matter of ingested leaf × 100); a higher ECIS value indicates better utilization of the ingested nutrients for silk production, demonstrating the larvae's ability to convert the feed into the desired product effectively.
- efficiency of conversion of digested substances (ECDS) into silk wrap (ECDS = silk wrap/digested leaf × 100); this ratio provides an understanding of the efficiency with which the digested nutrients are converted into silk, accounting for the absorption and metabolic processes (a higher ECDS value suggests that a larger proportion of the digested nutrients contribute to silk synthesis).

The analysis of the nutritional content and digestibility of the mulberry leaves is crucial for interpreting these efficiency metrics. This study involved assessing the protein, carbohydrate, lipid, and fiber content of the leaves, as these components are vital for the growth and silk production of the larvae. Additionally, understanding the digestibility of these nutrients helps in determining how much of the ingested food can be metabolically converted into silk.

By integrating these detailed metrics and analyses, this study provides a thorough evaluation of the silkworms' efficiency in converting mulberry leaf nutrients into silk, which is essential for optimizing feeding strategies and improving silk yield.

### 2.2. The Vegetal Material

The vegetal material comprised Kokuso 21 mulberry leaves (adapted well to Romania's conditions, obtained through a crossbreeding of the Naganua, Gariin, and Shiso varieties) provided to the L1 set. The Eforie variety (a high-yielding Romanian strain) was given to the L2 group. Each subgroup within the larger sets received identical amounts of leaves (15.5 g at stage I, 26 g at stage II, 77 g at stage III, 242 g at stage IV, and ultimately reaching 1000 g at stage V, accumulating a total weight of 1254.5 g over the entire growth period). Samples were extracted beforehand from these leaves and subjected to chemical analyses. The mulberry leaf leftovers and excreta were collected daily and recorded at the same time. The unconsumed portions were weighed and recorded for each subgroup.

Samples were also collected and subjected to chemical analysis (the excreta for age I was 0.147 g in Kokuso 21 and 0.155 g in Eforie; for age II, 1.019 g in Kokuso 21 and 0.622 g in Eforie; for age III, 4.125 g in Kokuso 21 and 3.629 g in Eforie; for age IV, 19.858 g in Kokuso 21 and 22.204 g in Eforie; and for age V, 122.39 g in Kokuso 21 and 126.96 g in Eforie).

### 2.3. The Analytical Techniques

The analytical techniques were aimed at evaluating the nutritional content of mulberry leaves and relied on examining their chemical composition and digestibility of individual elements.

The chemical composition was determined using the "proximate analysis" method, which involved samples that had been previously dried at 65 °C and subsequently ground [29].

To ascertain the moisture content, collected samples underwent drying in a Matest oven at 105 °C for 4–5 h (Treviolo, Italy) [30]. The samples have to be prepared as follows: initially, samples of mulberry leaves were collected and subjected to drying at a temperature of 65 °C; this drying step is crucial to remove any free moisture content and stabilize the samples, preventing microbial growth or enzymatic activity that could alter the composition. After drying, the samples were ground into a fine powder; grinding increases the surface area and homogenizes the sample, ensuring that subsequent analyses accurately reflect the overall composition. Each sample was weighed and the results recorded (the samples of leaf leftovers consisted in the following: for larval age 1, Kokuso 21 had 5.007 g and Eforie 5.295 g; for age II, Kokuso 21 had 8.221 g and Eforie 9.039 g; for age III, Kokuso 21 had 23.146 g and Eforie 24.090 g; for age IV, Kokuso 21 had 67.940 g and Eforie 65.503 g; for age V, Kokuso 21 had 269.055 g and Eforie 269.837 g); these samples were then dried in a controlled temperature and airflow environment. The temperature in the first two larval stages ranged between 26 and 27 °C, in the third stage between 24 and 25 °C, and in the last two stages between 23 and 24 °C. Humidity levels were maintained at 85–90% during the first stage, 80–85% during the second and third stages, 70–75% during the fourth stage, and 65–70% during the fifth stage.

To determine the moisture content, a specific portion of the ground sample was weighed and then placed in a Matest oven. The oven temperature was set to 105 °C, and the samples were dried for 4–5 h. This temperature is chosen because it is sufficient to evaporate all free water without decomposing other components of the sample. After drying, the samples were reweighed. The difference in weight before and after drying represents the amount of moisture lost from the sample. The moisture content was then calculated using the following formula: (initial weight − final weight/initial weight) × 100. This percentage indicates the proportion of the sample that was water. The higher the moisture content, the more water was present in the original sample. The dry matter content, which represents the portion of the material excluding water, was calculated by subtracting the moisture content percentage from 100%. This step is essential because dry matter includes all the nutritional components of the leaves, such as proteins, fats, carbohydrates, and fiber: dry matter content (%) = 100% − moisture content (%). The DM value is crucial for subsequent nutrient analysis as it allows for the quantification of nutrients relative to the dry weight of the material, rather than the fresh weight, which can vary significantly due to moisture fluctuations. The determination of Crude Ash (CA) was carried out by incinerating leaf samples in a 40-Segment Muffle Furnace (Bayswater, NSW, Australia) [30]. The samples were weighed, incinerated at 550 °C to eliminate organic components, leaving behind inorganic mineral content (ash). The ash content was calculated as a percentage of the initial wet weight using a specific formula.

The Kjeldahl method (Cole-Parmer equipment, Vernon Hills, IL, USA) was employed to measure total nitrogen content and subsequently determine crude protein (CP) [30]. This process involved heating the samples with concentrated sulfuric acid, along with catalysts, to break down organic compounds containing nitrogen into ammonium sulfate. Subsequent steps included neutralizing the acidic solution, collecting liberated ammonia gas through distillation, and then titrating it to quantify nitrogen content.

The determination of crude fat (CFa) employed the direct Randall method (Dolphin Labware 6 Test apparatus, Mumbai, India). This method relies on the property of fats to dissolve in organic solvents (petroleum ether) [30]. The process involved solvent extraction through the sample, vaporization, condensation, and refluxing, resulting in the concentration and quantification of extracted lipids based on the difference in initial and final flask masses.

Crude fiber (CFi) determination involved sample preparation and initial weighing (samples of dried and ground mulberry leaves were accurately weighed to record their initial dry weight. This initial weight serves as the baseline for calculating the crude fiber content), acid hydrolysis (the first step in the hydrolysis process involved treating the samples with a sulfuric acid solution. This step is critical for breaking down complex carbohydrates like cellulose and lignin into simpler monosaccharides and lignin residues. During the acid hydrolysis, cellulose is partially broken down, while lignin remains largely unaffected. The acid treatment dissolves other polysaccharides, such as hemicelluloses, leaving behind a solid residue primarily composed of lignin and any unhydrolyzed cellulose. The mixture was then filtered to separate the solid residues from the liquid phase. The solid residues, containing the indigestible fibers, were collected on the filter paper), known as alkaline hydrolysis. The solid residues obtained from the acid hydrolysis were then subjected to alkaline hydrolysis using a sodium hydroxide solution. This step further breaks down any remaining cellulose into simpler sugars, which are soluble in the alkaline solution. The purpose of this step is to ensure that any remaining digestible carbohydrates are removed, leaving behind only the lignin and any unhydrolyzed cellulose. After alkaline hydrolysis, the mixture was again filtered to separate the liquid phase, containing the dissolved sugars, from the solid residues. The solid residues after this step were primarily composed of lignin and any remaining unhydrolyzed cellulose, which are considered indigestible, then the drying and weighing of residues took place (the solid residues from both the acid and alkaline hydrolyses were subjected to a drying process. This drying step is crucial for removing any remaining moisture, ensuring that the weight recorded corresponds to the dry matter content of the residues. The dry weights of these residues were carefully recorded. These weights are critical for calculating the crude fiber content), and then the calculation of crude fiber content was conducted. The crude fiber content was calculated as the difference between the initial dry weight of the sample and the combined dry weight of the solid residues after both hydrolyses. This calculation excludes other processed components such as sugars and soluble fibers that were removed during the hydrolysis steps. This value represents the amount of indigestible fiber present in the sample, which consists primarily of lignin and unhydrolyzed cellulose. The hydrolysable portion was removed, leaving crude fiber and mineral salts on the filter. Crude fiber was calculated by calcinating the mineral substances, starting with accurate weighing of samples and recording their dry weights. Acid hydrolysis with sulfuric acid broke down cellulose and lignin into constituent monosaccharides and lignin residues. Filtration separated solid residues, consisting of unhydrolyzed cellulose and lignin. Alkaline hydrolysis with sodium hydroxide (Missouri, USA) further broke down cellulose into sugars. Filtration separated the liquid phase (sugars) from solid residues (lignin). Solid residues from both acid and alkaline hydrolyses underwent a drying step, and their dry weights were recorded. Crude fiber content was calculated as the difference between the initial dry weight and the combined dry weight of solid residues after both hydrolyses, excluding other processed components [31].

The nitrogen-free extract (NFE) was calculated by subtracting the percentages of water, crude protein, crude fat, crude fiber, and crude ash from 100% [32].

The digestibility of mulberry leaf nutrients was measured using the "in vivo" method, specifically simple digestibility, with a single control period. Digestibility coefficients (DC%) were determined by considering the amounts of fed leaves, leftovers, excreta, and chemical analyses data. Ingesta (difference between fed and unconsumed leftovers) and digesta (difference between intake and excretion) were calculated. Digestibility coefficients, expressed as a percentage, indicated what portion of substances in mulberry leaves were digested by the larvae [32]. To calculate the digestibility coefficients for each nutritional substance, we analyzed the quantities of each nutrient intake, as well as those that were excreted and digested. The data were compiled separately for each mulberry variety, based on the larvae's age. In the Kokuso 21 variety, both the age of the larvae and the degree of leaf maturity significantly influenced the digestibility of organic matter; a gradual decrease from age I to age V was identified ($p < 0.001$). For the Eforie variety, the digestibility coefficients of dry matter (DM) significantly decreased ($p < 0.001$) as the larvae aged or the mulberry leaves also aged (age I vs. III, IV, and V, as well as ages II vs. IV and V, and ages III to V). A similar pattern was observed for other organic nutrients. All analytical assessments and computations were conducted in 18 repetitions (the total number of larvae was divided into 2 groups, each consisting of 300 larvae (Triumf hybrid); each group was further subdivided into 6 subgroups, with 50 larvae in each subgroup, resulting in 6 repetitions).

### 2.4. Statistical Data Processing

Statistical data processing involved calculating the main descriptors, such as arithmetic mean, variance, standard deviation, standard error of the mean, and coefficient of variation [33]. Graph Pad Prism 9.4.1 software was used to treat the database, in order to achieve (a) main descriptors (mean, standard deviation, standard error of mean, variation coefficient); (b) level of significance for one to one comparisons, using the unpaired Student ($t$) test, with Welch's correction between the means of the two varieties, within the same age period; (c) ANOVA single factor with Tukey post-hoc testing to run multiple comparisons between means (within the same variety, between more than two ages); (d) Pearson correlation coefficients when crude fiber values and the digestibility coefficients of the other organic matters were associated.

## 3. Results

### 3.1. The Proximate Composition of the Mulberry Leaves

Table 1 compiles the chemical composition data of mulberry leaves from both studied varieties based on larval age. Significant modifications in dry matter (DM) content were noted for Kokuso 21 across larval age groups (I vs. III, IV, V; II vs. IV, V; III vs. V), demonstrating notable differences ($p < 0.001$). Regarding crude protein (CP) content, substantial differences emerged between age groups: I vs. III, V, and II vs. III, IV, V ($p < 0.001$). Furthermore, significant variations were observed for crude fat (CFa) and crude fiber (CFi) contents between younger age groups I vs. II and older age groups IV vs. V for Kokuso 21.

**Table 1.** Proximate composition of mulberry leaves (mean ± standard deviation), based on the degree of maturity in relation to larvae age.

| Larval Age | Mulberry Variety | Dry Matter % (Mean ± Standard Deviation) | Crude Protein % (Mean ± Standard Deviation) | Crude Fat % (Mean ± Standard Deviation) | Crude Fiber % (Mean ± Standard Deviation) | Nitrogen-Free Extract % (Mean ± Standard Deviation) | Ash % (Mean ± Standard Deviation) |
|---|---|---|---|---|---|---|---|
| I | Kokuso 21 | 27.91 ± 0.25 | 6.31± 0.09 | 0.79 ± 0.09 | 4.74 ± 0.14 | 12.33 ± 0.18 | 3.75 ± 0.45 |
| | Eforie | 28.14 ± 0.90 | 6.23 ± 0.40 | 0.85 ± 0.20 | 4.79 ± 0.42 | 12.43 ± 0.50 | 3.84 ± 0.88 |
| II | Kokuso 21 | 28.34 ± 0.54 | 6.28 ± 0.18 | 0.88 ± 0.13 | 4.88 ± 0.17 | 12.34 ± 0.22 | 3.96 ± 0.67 |
| | Eforie | 28.03 ± 0.81 | 6.21 ± 0.33 | 0.88 ± 0.20 | 4.76 ± 0.49 | 12.24 ± 0.63 | 3.94 ± 1.27 |

**Table 1.** *Cont.*

| Larval Age | Mulberry Variety | Dry Matter % (Mean ± Standard Deviation) | Crude Protein % (Mean ± Standard Deviation) | Crude Fat % (Mean ± Standard Deviation) | Crude Fiber % (Mean ± Standard Deviation) | Nitrogen-Free Extract % (Mean ± Standard Deviation) | Ash % (Mean ± Standard Deviation) |
|---|---|---|---|---|---|---|---|
| III | Kokuso 21 | 29.70 ± 0.41 | 6.23 ± 0.15 | 1.14 ± 0.14 | 5.31 ± 0.62 | 12.64 ± 0.50 | 4.38 ± 1.22 |
| | Eforie | 29.32 ± 0.53 | 6.41 ± 0.39 | 1.17 ± 0.25 | 5.26 ± 0.31 | 12.30 ± 0.40 | 4.18 ± 0.94 |
| IV | Kokuso 21 | 29.87 ± 0.45 | 6.04 ± 0.10 | 1.16 ± 0.14 | 5.44 ± 0.42 | 13.09 ± 0.67 | 4.14 ± 1.28 |
| | Eforie | 30.47 ± 0.80 | 6.00 ± 0.39 | 1.22 ± 0.22 | 5.58 ± 0.50 | 13.37 ± 0.39 | 4.30 ± 1.13 |
| V | Kokuso 21 | 31.14 ± 0.91 | 6.15 ± 0.41 | 1.25 ± 0.33 | 5.93 ± 0.75 | 13.41 ± 0.61 | 4.41 ± 0.71 |
| | Eforie | 31.85 ± 0.69 | 6.06 ± 0.34 | 1.38 ± 0.36 | 6.15 ± 0.28 | 13.58 ± 0.44 | 4.69 ± 0.45 |
| I–V | Kokuso 21 | 29.39 ± 0.14 | 6.20 ± 0.10 | 1.04 ± 0.09 | 5.26 ± 0.27 | 12.76 ± 0.25 | 4.13 ± 0.50 |
| | Eforie | 29.56 ± 0.16 | 6.18 ± 0.08 | 1.10 ± 0.10 | 5.31 ± 0.23 | 12.78 ± 0.30 | 4.19 ± 0.34 |

Ash = inorganic mineral content.

In the case of the Eforie variety, the dynamics of DM content exhibited several distinctions when comparing age groups: I vs. IV ($p < 0.05$), I vs. V ($p < 0.001$), II vs. IV ($p < 0.05$), II vs. V ($p < 0.001$), and III vs. V $p < 0.05$).

Comparisons between varieties revealed no significant differences ($p > 0.05$), regardless of the analyzed nutrients or age.

### 3.2. Mulberry Leaves Digestibility

The data detailing the quantities of fed leaves, residues, excreta, and their respective chemical compositions, crucial for calculating ingesta and digesta, have been separately organized for each group in Tables 2–4 which present the chemical composition of the excreta, vital for computing digestibility coefficients.

**Table 2.** Digesta calculation for Kokuso 21 and Eforie leaves.

| Larval Age | Variety | F (g) | R (g) | I = F − R (g) | E (g) | D = I − E (g) |
|---|---|---|---|---|---|---|
| I | Kokuso 21 | 15.5 | 5.007 | 10.493 | 0.147 | 24.01 |
| | Eforie | 15.5 | 5.295 | 10.205 | 0.155 | 10.050 |
| II | Kokuso 21 | 26 | 8.221 | 17.779 | 1.019 | 16.760 |
| | Eforie | 26 | 9.039 | 16.961 | 0.622 | 16.339 |
| III | Kokuso 21 | 77 | 23.146 | 53.885 | 4.125 | 49.730 |
| | Eforie | 77 | 24.090 | 52.910 | 3.629 | 49.282 |
| IV | Kokuso 21 | 242 | 67.940 | 174.143 | 19.858 | 154.286 |
| | Eforie | 242 | 65.503 | 176.497 | 22.204 | 154.294 |
| V | Kokuso 21 | 1000 | 269.055 | 730.945 | 122.390 | 608.555 |
| | Eforie | 1000 | 269.837 | 730.163 | 126.960 | 603.203 |
| I–V | Kokuso 21 | 1365.5 | 373.368 | 987.215 | 147.539 | 839.676 |
| | Eforie | 1365.5 | 373.763 | 986.737 | 153.570 | 833.167 |

F = fed leaves quantity; R = residues; I = ingesta; E = excreta; D = digesta.

**Table 3.** Proximate composition of residues of Kokuso 21 and Eforie mulberry leaves (mean ± standard deviation), based on larvae age.

| Larval Age | Mulberry Variety | Dry Matter % (Mean ± Standard Deviation) | Crude Protein % (Mean ± Standard Deviation) | Crude Fat % (Mean ± Standard Deviation) | Crude Fiber % (Mean ± Standard Deviation) | Nitrogen-Free Extract % (Mean ± Standard Deviation) | Ash % (Mean ± Standard Deviation) |
|---|---|---|---|---|---|---|---|
| I | Kokuso 21 | 63.82 ± 0.41 | 14.04 ± 0.26 | 2.01 ± 0.37 | 14.59 ± 0.37 | 25.97 ± 0.44 | 7.23 ± 1.43 |
| | Eforie | 63.61 ± 0.41 | 15.01 ± 0.44 | 1.68 ± 0.43 | 13.92 ± 0.41 | 24.01 ± 0.42 | 9.00 ± 1.09 |
| II | Kokuso 21 | 60.03 ± 0.36 | 12.01 ± 0.81 | 2.11 ± 0.39 | 14.88 ± 0.43 | 21.01 ± 0.67 | 10.02 ± 1.24 |
| | Eforie | 59.16 ± 0.43 | 13.01 ± 0.32 | 2.01 ± 0.44 | 13.51 ± 0.45 | 22.61 ± 0.39 | 8.03 ± 1.13 |
| III | Kokuso 21 | 58.39 ± 0.33 | 10.78 ± 0.19 | 2.31 ± 0.25 | 15.99 ± 0.38 | 22.12 ± 0.31 | 7.20 ± 0.43 |
| | Eforie | 58.82 ± 0.63 | 12.33 ± 0.47 | 2.36 ± 0.32 | 15.77 ± 0.45 | 24.02 ± 0.42 | 4.33 ± 1.01 |
| IV | Kokuso 21 | 59.53 ± 0.40 | 12.62 ± 0.34 | 2.45 ± 0.42 | 14.49 ± 0.35 | 24.72 ± 0.54 | 5.27 ± 1.01 |
| | Eforie | 58.06 ± 0.36 | 12.06 ± 0.48 | 1.72 ± 0.45 | 15.71 ± 0.38 | 24.66 ± 0.41 | 3.91 ± 0.79 |
| V | Kokuso 21 | 56.73 ± 0.46 | 11.46 ± 0.41 | 1.88 ± 0.37 | 13.66 ± 0.39 | 23.97 ± 0.55 | 5.75± 1.12 |
| | Eforie | 58.56 ± 0.45 | 11.06 ± 0.55 | 2.61 ± 0.40 | 11.99 ± 0.72 | 24.88 ± 0.63 | 8.03± 1.88 |
| I–V | Kokuso 21 | 59.70 ± 0.17 | 12.18 ± 0.17 | 2.15 ± 0.11 | 14.72 ± 0.27 | 23.56 ± 0.24 | 7.09 ± 0.76 |
| | Eforie | 59.64 ± 0.12 | 12.69 ± 0.20 | 2.08 ± 0.28 | 14.18 ± 0.26 | 24.04 ± 0.32 | 6.66 ± 0.98 |

**Table 4.** Chemical composition of excreta from the Kokuso 21 and Eforie varieties (mean ± standard deviation), in relation to larvae age.

| Larval Age | Mulberry Variety | Dry Matter % (Mean ± Standard Deviation) | Crude Protein % (Mean ± Standard Deviation) | Crude Fat % (Mean ± Standard Deviation) | Crude Fiber % (Mean ± Standard Deviation) | Nitrogen-Free Extract % (Mean ± Standard Deviation) | Ash % (Mean ± Standard Deviation) |
|---|---|---|---|---|---|---|---|
| I | Kokuso 21 | 66.15 ± 0.41 | 21.87 ± 0.42 | 5.67 ± 0.41 | 2.33 ± 0.38 | 27.21 ± 0.48 | 9.07 ± 0.1.38 |
| | Eforie | 69.82 ± 0.42 | 14.33 ± 0.44 | 15.02 ± 0.44 | 3.35 ± 0.51 | 27.09 ± 0.70 | 9.13 ± 2.06 |
| *p* value | | 0.9999 | 0.9999 | >0.9999 | >0.9999 | 0.9999 | 0.9999 |
| II | Kokuso 21 | 64.69 ± 0.53 | 14.93 ± 0.62 | 3.64 ± 0.39 | 4.31 ± 0.35 | 27.85± 0.70 | 13.96 ± 1.87 |
| | Eforie | 63.22 ± 0.62 | 11.01 ± 0.59 | 4.01 ± 0.52 | 2.43 ± 0.43 | 29.78± 0.55 | 15.99 ± 1.91 |
| *p* value | | 0.9999 | >0.9999 | >0.9999 | 0.9999 | 0.9999 | >0.9999 |
| III | Kokuso 21 | 65.15 ± 0.44 | 14.99 ± 0.40 | 2.63 ± 0.43 | 7.99 ± 0.38 | 25.53 ± 0.53 | 14.00 ± 1.04 |
| | Eforie | 64.53 ± 0.64 | 16.29 ± 0.43 | 2.01± 0.20 | 6.09 ± 0.70 | 28.02 ± 0.52 | 12.12 ± 1.82 |
| *p* value | | 0.9999 | 0.9999 | >0.9999 | >0.9999 | 0.9999 | 0.9999 |
| IV | Kokuso 21 | 63.52 ± 0.40 | 11.01 ± 0.47 | 2.02 ± 0.41 | 14.09 ± 0.29 | 24.99 ± 0.48 | 11.41 ± 1.35 |
| | Eforie | 64.06 ± 0.46 | 11.01 ± 0.50 | 2.66 ± 0.34 | 12.04 ± 0.44 | 27.29 ± 0.42 | 11.06 ± 1.78 |
| *p* value | | 0.9985 | >0.9999 | >0.9999 | 0.9999 | 0.9999 | 0.9999 |
| V | Kokuso 21 | 61.03 ± 0.49 | 9.98 ± 0.70 | 3.22 ± 0.24 | 13.99 ± 0.45 | 25.01 ± 0.54 | 8.83 ± 1.53 |
| | Eforie | 62.58 ± 0.42 | 10.38 ± 0.43 | 3.01 ± 0.48 | 16.06 ± 0.23 | 24.02 ± 0.32 | 9.11 ± 1.15 |
| *p* value | | 0.9973 | >0.9999 | 0.9999 | 0.9999 | 0.9999 | 0.9999 |
| I–V | Kokuso 21 | 64.11 ± 0.25 | 14.56 ± 0.31 | 3.43 ± 0.26 | 8.54 ± 0.20 | 26.12 ± 0.44 | 11.46 ± 1.31 |
| | Eforie | 64.84 ± 0.26 | 12.60 ± 0.30 | 5.34 ± 0.18 | 8.00 ± 0.23 | 27.42 ± 0.45 | 11.48 ± 1.33 |
| *p* value | | 0.9999 | >0.9999 | 0.9979 | >0.9999 | >0.9999 | 0.9999 |

The average values of the digestibility coefficients of the mulberry leaves were calculated for each variety and stage of growth; the results are presented in Table 5.

**Table 5.** Digestibility coefficients of mulberry leaves.

| Larval Age | Mulberry Variety | Dry Matter % (Mean ± Standard Deviation) | Crude Protein % (Mean ± Standard Deviation) | Crude Fat % (Mean ± Standard Deviation) | Crude Fiber % (Mean ± Standard Deviation) | Nitrogen-Free Extract % (Mean ± Standard Deviation) |
|---|---|---|---|---|---|---|
| I | Kokuso 21 | 91.38 ± 1.66 | 88.30 ± 2.25 | 61.30 ± 7.36 | 0.17 ± 0.08 | 93.45 ± 1.26 |
| | Eforie | 89.08 ± 1.42 | 87.02 ± 1.67 | 46.01 ± 7.07 | 0.53 ± 0.21 | 93.36 ± 0.87 |
| *p* value | | >0.9999 | >0.9999 | $6.1 \times 10^{-6}$ | >0.9999 | >0.9999 |
| II | Kokuso 21 | 72.91 ± 1.10 | 76.46 ± 0.97 | 31.80 ± 2.68 | 4.40 ± 1.92 | 80.85 ± 0.79 |
| | Eforie | 79.87 ± 10.18 | 84.51 ± 7.84 | 47.50 ± 26.33 | 3.62 ± 1.10 | 83.81 ± 8.25 |
| *p* value | | 0.8709 | 0.5441 | $2.5 \times 10^{-6}$ | >0.9999 | >0.9999 |
| III | Kokuso 21 | 71.26 ± 0.10 | 73.09 ± 0.10 | 68.15 ± 0.12 | 15.44 ± 0.56 | 77.17 ± 0.08 |
| | Eforie | 72.20 ± 3.88 | 70.02 ± 4.23 | 77.94 ± 3.07 | 10.21 ± 2.19 | 72.47 ± 3.87 |
| *p* value | | >0.9999 | 0.9999 | 0.1042 | 0.9992 | 0.9999 |
| IV | Kokuso 21 | 60.42 ± 0.08 | 63.86 ± 0.08 | 65.25 ± 0.08 | 15.66 ± 0.29 | 66.68 ± 0.07 |
| | Eforie | 60.16 ± 0.87 | 63.13 ± 0.77 | 67.58 ± 0.83 | 16.76 ± 1.88 | 62.61 ± 0.83 |
| *p* value | | >0.9999 | >0.9999 | >0.9999 | >0.9999 | 0.9999 |
| V | Kokuso 21 | 52.96 ± 0.02 | 60.10 ± 0.02 | 46.62 ± 0.02 | 24.15 ± 0.03 | 56.01 ± 0.02 |
| | Eforie | 50.49 ± 0.89 | 57.14 ± 0.77 | 43.51 ± 1.03 | 30.09 ± 1.30 | 55.55 ± 0.80 |
| *p* value | | >0.9999 | >0.9999 | 0.9999 | 0.9882 | >0.9999 |
| I–V | Kokuso 21 | 55.42 ± 0.02 | 61.88 ± 0.03 | 49.79 ± 0.03 | 22.91 ± 0.06 | 59.48 ± 0.02 |
| | Eforie | 53.49 ± 0.74 | 59.19 ± 0.65 | 49.67 ± 0.84 | 28.62 ± 1.26 | 58.14 ± 0.67 |
| *p* value | | 0.0001 | $6.5 \times 10^{-8}$ | 0.9999 | $7.1 \times 10^{-15}$ | 0.0122 |
| I–V | Average values per larval period | 54.45 ± 0.38 | 60.53 ± 0.34 | 49.73 ± 0.43 | 25.67 ± 0.66 | 58.81 ± 0.69 |

In Table 6, the correlations between the crude fiber content of leaves and the digestibility of the organic matter achieved by larvae, for the entire period (ages I–V), revealed negative values in both Kokuso 21 and Eforie hybrids, but yet not significant in both situations ($p > 0.05$) and with stronger correlations in Eforie samples.

**Table 6.** Correlations between CFi and the digestibility of organic matter compounds in larvae (overall ages I–V).

| Variety | Organic Matter Compounds and Digestibility | | | |
|---|---|---|---|---|
| | Dry Matter % | Crude Protein % | Crude Fat % | Nitrogen-Free Extract % |
| Crude fiber **in Kokuso 21** | r = −0.33 | r = −0.39 | r = −0.36 | r = −0.30 |
| ***p* value** | 0.5288 | 0.4422 | 0.4875 | 0.5100 |
| Crude fiber **in Eforie** | r = −0.66 | r = −0.65 | r = −0.63 | r = −0.65 |
| ***p* value** | 0.1502 | 0.1608 | 0.1785 | 0.1641 |

r = correlations between the digestibility of organic matter compounds and the dietary crude fiber content.

Based on the determinations made on the cocoons collected from each subgroup, it appears that the silk wrap obtained from a single larva weighed approximately 0.4048 g DM (20.24 ± 0.23 g DM/subgroup) in the case of the batch fed with leaves from the Kokuso 21 variety, and 0.4096 g DM (20.48 ± 0.14 g DM/subgroup) in the case of the batch fed with leaves from the Eforie variety. From this perspective, no statistically significant differences were observed between the two batches ($p$ value = 0.060736728). The data regarding the conversion of mulberry leaves into silk have been summarized in Table 7.

**Table 7.** The efficiency of utilizing mulberry leaves by *B. mori* (mean ± standard deviation).

| Mulberry Variety | Ingested Dry Matter/Silk Wrap Dry Matter (Mean ± Standard Deviation) | Digested Dry Matter/Silk Wrap Dry Matter (Mean ± Standard Deviation) | ECIS Silk Wrap (%) (Mean ± Standard Deviation) | ECDS Silk Wrap (%) (Mean ± Standard Deviation) |
|---|---|---|---|---|
| Kokuso 21 | 10.06 ± 0.11 | 5.57 ± 0.06 | 9.90 ± 0.11 | 17.94 ± 0.04 |
| Eforie | 10.14 ± 0.08 | 5.42 ± 0.10 | 9.87 ± 0.08 | 18.45 ± 0.35 |
| $p$ value | $p < 0.05$ | $p < 0.05$ | $p > 0.05$ | $p < 0.05$ |

ECIS = efficiency of conversion of ingested substances; ECDS = efficiency of conversion of digested substances.

Based on the data in Table 7, it can be observed that for each gram of silk dry matter (DM) in the cocoon, the larvae ingested an average of 10.06 g of leaf DM in the case of the Kokuso 21 variety and 10.14 g of leaf DM in the case of the Eforie variety. In other words, the efficiency of the conversion of ingested matter into silken cocoon (ECIS) was on average 9.9% for the Kokuso 21 variety and 9.87% for the Eforie variety, with the differences not being significant.

When the comparison was made based on digested matter, it was observed that for each gram of silk DM in the cocoon, an average of 5.57 g of digested leaf DM was necessary for the Kokuso 21 variety and 5.42 g of digested leaf DM for the Eforie variety. In this way, the efficiency of the conversion of digested matter into silken cocoon (ECDS) was on average 17.94% for the Kokuso 21 variety and 18.45% for the Eforie variety, with the differences being significant this time.

Thus, although there were no significant differences in food conversion into silk based on ingested matter between the two varieties, the complex phenomena occurring during digestion make these differences significant, with the Eforie variety showing better leaf utilization by the larvae.

Therefore, it can be stated that, at least for summer rearing, the Eforie variety performs at least as well as the Kokuso 21 variety, which is why it can be recommended for the specific pedoclimatic conditions of Romania.

Regarding the efficiency of the use of the mulberry leaf by the *B. mori* larvae, the data obtained from the experiment are similar to those presented in the literature [33–37].

## 4. Discussion

### 4.1. The Proximate Composition of Mulberry Leaves

The intake of mulberry leaves significantly impacts the growth of silkworm larvae, especially during their initial stages when they prefer tender and well-hydrated leaves. Chemical analyses conducted throughout the larvae's growth period on the mulberry leaves used as feed (Tables 1 and 2) revealed an average leaf moisture content of 70.53% and a total solids (dry matter) content of 29.47%. Over the larvae's growth period, the leaf moisture decreased by an average of 3.78%.

In the literature, relative humidity values for mulberry leaves typically range from 65% to 75% [16–20,22,23,31,32]. Compared to common mulberry leaves, which have a moisture content ranging from 69.80% to 73%, the studied varieties had higher water content. In the larvae's bodies, water is found in larger quantities in the digestive tract and hemolymph, and in smaller proportions in tissues and muscles. The water content in larvae varies by age, with younger larvae having a higher percentage of water compared to adults. The water balance is maintained through its consumption with mulberry leaves and its elimination through processes such as excretion, respiration, and transpiration. The intensity of water metabolism is influenced by factors such as the water content, temperature, and humidity.

In the studied variety, the water content varies from 71.50% in the initial two phases to less than 65% in the later phases. The best results in silkworm growth are achieved when feeding starts with the appearance of the third or fifth leaf. Initiating growth when the seventh leaf appears leads to an extended larval period, and starting with the ninth leaf results in lower outcomes in terms of cocoon weight, silk cocoon shell, and silk's technological properties. Feeding larvae with young leaves results in better consumption, digestion, and utilization of protein substances compared to feeding them with mature leaves. As leaves age, their composition changes, with a decrease in water content, protein, and easily soluble carbohydrates, while the quantity of cellulose and ash increases, making the leaves coarse. Leaf maturity should largely correspond to larval development; thus, first- and second-stage larvae feed exclusively on young leaves. Young leaves, rich in protein, have lower energy value and are less rich in carbohydrates, leading to higher protein consumption. If the leaves are old, their nutritional efficiency is lower due to poor water content, which slows down biological reactions in the digestive tract. The same occurs if the leaves begin to wither. Leaf age significantly influences silk composition—to obtain leaves of suitable quality throughout the mulberry's entire growing period, intensive cultivation plantations employ harvesting practices that promote the growth of new leaves for repeated silkworm rearing cycles [38].

Fresh mulberry leaves have a dry matter content of 24.86%, of which 84.33% is organic matter [36]. The dry matter content of mulberry leaves harvested during the same period can vary depending on the variety or hybrid, with values ranging from 23.61% to 27.56%. Mulberry leaf humidity is influenced by the season, ranging from 71.85% to 77.81% in spring, 68.42% to 75.64% in summer, and 64.10% to 73.64% in autumn [39].

The protein content of mulberry leaves significantly influences the growth, development, and cocoon formation of silkworm larvae, as well as the overall silk yield. Throughout the study period, the average crude protein content in the mulberry leaves was 6.19%. In terms of dry matter, crude protein values ranged from 22.38% in the first larval stage to 19.39% in the fifth stage.

In the literature, the crude protein content of mulberry leaves is reported as 6.16% in fresh leaves [21], 16.57% [26], 16.67–21.62% [36], 19.60% [37], 20.34% [38], 20.97% [39], and 29.80% in dry matter, with 24.36% in organic matter [40]. Crude protein values in leaves can vary based on the season, time of day, mulberry variety/hybrid, with reported values ranging from 32.40% in spring to 24.53% in autumn [21], 26.80% in the morning to 29.10% in the evening [26], and between 20.20% and 26.72% depending on the variety [26–28] (these values are expressed as percent of crude protein in leaves' dry matter).

Protein metabolism is crucial in the nutrition of silkworms. The concept of crude protein used in experimentation encompasses both proteinaceous and non-proteinaceous nitrogenous substances. During digestion, proteins are broken down into peptides and amino acids, which are absorbed by the larva's body. After the cleavage of proteins from mulberry leaves, the resulting amino acids are re-synthesized into specific proteins within the organism. Some amino acids are synthesized by the organism. An increase in leucine, proline, and tryptophan is observed, along with a decrease in glycine, tyrosine, and other amino acids, in the stages of pupa and butterfly, during which the insect no longer feeds. Adding tryptophan and tyrosine to the silkworm's diet increases larval weight and shortens the larval period. Tryptophan and cysteine influence larval development, while alanine and tyrosine impact silk formation. Globulins are synthesized throughout the entire larval period (with an intensification in the fourth and fifth stages), while albumin synthesis occurs only in the last two stages. During the larval stage, various nitrogenous substances from mulberry leaves stored in the silkworm's body, such as globulins, prolamins, glutenin, and albumins, are particularly active in metabolism. Amino acids, besides their structural role, also serve as energy sources, degrading into $CO_2$, $H_2O$, and $NH_3$. The resulting nitrogen is eliminated as uric acid, which requires energy consumption. Depending on the larva's health, uric acid can also form from the nitrogen resulting from tissue cell breakdown. Additionally, amino acids not contributing to silk formation are either eliminated or degraded and converted into uric acid. Alongside uric acid, urates, small quantities of urea, ammonia, and oxalates, are found. The nitrogen balance in the silkworm's body during the larval period is positive [41].

Lipids are the most concentrated form of energy storage in any organism, including *B. mori*. In silkworms, fats serve as an energy source and play a structural role in cells. Lipid metabolism, though under-studied, involves the breakdown of fats into fatty acids and glycerol by alkaline gastric juice. These are absorbed in the midgut, where neutral fats are resynthesized and stored in adipose tissue and other cells. Lipids can also pass directly into tissues through the hemolymph. Larvae synthesize lipids from carbohydrates, with lipid metabolism becoming crucial during metamorphosis, especially after feeding ceases. In larvae, energy is primarily derived from carbohydrates, while in pupal and adult stages, lipids become the main energy source. The synthesized lipids contain both saturated and unsaturated fatty acids, with 75% being unsaturated, including oleic, linoleic, and linolenic acids [42].

The quality of mulberry leaves, which are essential for silkworm nutrition, is influenced by their crude fiber content. Higher cellulose content indicates leaf aging, making them tougher and harder to consume. The average crude fiber content in the analyzed samples was 5.29% (17.9% in DM), increasing by 1.28% over the study period. Literature values for crude fiber range from 11.63% to 15.50% in DM [43], with common mulberry varieties showing 10.43% to 13.70% [44]. The nitrogen-free extract had an average proportion of 12.77% (43.38% in DM), increasing by 1.12% during the period, similar to literature findings [45].

Carbohydrates are the primary energy source, degraded enzymatically into sugars like glucose, which are directly used by the larvae's tissues. Carbohydrates can also be resynthesized into fats due to the mulberry leaves' low lipid content. Glycogen, stored in the adipose tissue, is mobilized as needed, with its degradation controlled by a hormone from the prothoracic glands. Glycogen breakdown in muscles, primarily anaerobic, forms lactic acid, which is partly oxidized for energy and partly resynthesized into glycogen. Carbohydrate metabolism varies with muscular activity, temperature, humidity, and leaf water content, also protecting proteins from excessive energy consumption [46].

The ash content (total minerals) had an overall average of 4.16%, with an increase of approximately 0.76% recorded during the study period. In terms of dry matter, ash represented 14.06%. Total mineral values are in line with those found in the literature, such as 6.04–8.56% [21], 8.7–13.15% [26], 9.13–17.38% [27], 10.00–12.30% [29], 11.52–12.80% [33], and 13.37% [39]. Calcium accounted for 2.10–2.94%, and phosphorus was at 0.20–0.23% [24,25].

Chemical composition differences among mulberry varieties were minimal, except at the 5th larval age. The Eforie variety, comparable to the globally recognized Kokuso 21, demonstrated valuable potential and significant contributions to sericulture, matching the standard set by Kokuso 21 in leaf chemical composition. This highlights the Eforie variety's relevance and importance in scientific research.

### 4.2. Mulberry Leaf Digestibility

Throughout the intricate process of digestion, nutrients undergo breakdown into simpler compounds, which are subsequently absorbed across various segments of the digestive tract epithelium in silkworms. These compounds are utilized by silkworms for their growth and cocoon formation during their metamorphic phases. The quantification of substances digested showcases the difference between the ingested substances from their feed and those present in excretions. However, as not all substances in excretions originate directly from the diet (some are endogenous), the outcome derived from this distinction is essentially termed as apparent digestibility. The term "apparent digestibility" becomes particularly relevant when considering the presence of excreted products in

*B. mori*'s feces, which complicates the accurate evaluation of nutrient digestibility in mulberry leaves [31,33,36,39,40].

Measuring the digestibility of nutrients in mulberry leaves by silkworms holds significance due to several reasons: enhancing silk production—assessing how efficiently silkworms process nutrients from mulberry leaves aids silk producers in optimizing rearing conditions, potentially improving cocoon quality and silk production yields; nutrient balance and growth—understanding which nutrients silkworms readily absorb enables the formulation of diets that provide essential nutrients, fostering healthier and faster-growing silkworms; economic impact—maximizing nutrient digestibility can boost silk production's economic viability by enhancing cocoon quality and silk yield, benefiting both producers and local economies; resource efficiency—higher nutrient extraction by silkworms means less leftover material, potentially valuable for other uses like organic fertilizers; ecosystem sustainability—assessing nutrient digestibility contributes to understanding silkworm–mulberry tree interactions within ecosystems, impacting nutrient flow dynamics; research and innovation—insights into silkworm nutrition and digestibility can drive innovations in sericulture practices, guiding the development of specialized diets or enhanced digestibility in silkworm strains, significantly influencing the silk industry's future [39,40].

Our experiments revealed that, on average, the digestibility of mulberry leaf as a whole is 54.46%. This value represents the mean between the recorded digestibility coefficients for the two varieties during the larval period. The digestibility of dry matter decreased by an average of 38.54% throughout the larval growth period. As larvae develop, their enzymatic apparatus evolves, resulting in alterations in how they utilize nutrients from leaves. These alterations correlate with qualitative changes in the chemical composition of the leaves [21].

On average, mulberry leaf digestibility varies between 46.40% and 58.90% depending on the variety [21]. Some studies suggest that mulberry leaves fed to the 5th larval age exhibit an approximate digestibility ranging from 27.99% to 32.44% [26]. This parameter decreases from 71.07% in the first age to 39.99% for male larvae and 48.26% for female larvae in the 5th age [28]. The digestibility of mulberry leaves can reach a maximum of 70% [30].

The average digestibility of crude protein in mulberry leaves was around 60.54%. Throughout the larval growth period, digestibility coefficients showed an average decrease of 28.87%. The high digestibility in the 1st age may be attributed to the rich amide content present in young leaves, comprising simpler nitrogenous substances easier to digest compared to the complex protein-structured nitrogenous substances prevalent in older leaves. Reported digestibility coefficients of crude protein in the literature range from 69.21% to 78.92 [33], 60.06% to 74.69% [38], and 71.62% to 93.48% [41].

The average digestibility of crude fat in mulberry leaves was approximately 49.73% over the larval growth phase, displaying a sinusoidal pattern between 31.80% and 77.49%. It is crucial to note that interpreting crude fat digestibility test results from mulberry leaves can be inconclusive due to the possibility that many lipid compounds found in the excreta might originate from the larvae's digestive tract rather than solely from the leaves. This complexity underscores the importance of evaluating 'ether extract', which may contain significant pigment amounts. The observed fluctuations in crude fat digestibility during the study period might be attributed to these complexities.

In the larval digestive system, lipases break down fats from the leaves into glycerol and fatty acids. However, the presence of these lipases in gastric juice remains a subject of debate. Most authors agree that *B. mori* larvae digest and assimilate lipids, storing reserves for the chrysalis (cocoon) and butterfly stages. The mechanisms through which this occurs have sparked debates. In 1917, Hiratsuka observed a difference in the fat content of leaves and excrement, indicating larval utilization [42]. Some sources suggested that the lipase exists in the cells that make up the absorptive epithelium in the silkworm larval digestive tract [21,43,44]. Proteins, lipids, and carbohydrates (glycogen) are stored in the body tissues of *B. mori* larvae, especially in the fats [44]. Reported digestibility coefficient values for crude fat in the literature range between 63.28% and 74.19% [45].

The digestibility of crude fiber obtained from mulberry leaves showed a decline throughout the larval growth period, averaging 25.77% across both hybrids during the entire larval phase. It commenced at nearly negligible levels (almost zero) in young larvae, progressively increasing significantly towards the end (26.78%). This pattern in crude fiber digestibility aligned closely with the development of enzymatic equipment in the larval digestive tract. At the onset, enzymes responsible for cellulose digestion were barely present, gradually increasing and reaching their peak in the 5th age, coinciding with higher crude fiber content in mulberry leaves. According to certain sources, crude fiber was not digestible in the first two larval stages but reached 8% in the 3rd age and 21.13% in the 5th age [44].

The nitrogen-free extract (NFE) derived from mulberry leaves exhibited an average digestibility of 58.81% and encountered an average decrease of 37.62% throughout the studied period. These values align with existing literature, where NFE digestibility coefficients range between 63.40% and 94.97% depending on mulberry variety and larval age [46].

While no significant differences were noted in the proximate composition between the varieties, significant variations were observed within each variety across different larval ages ($p < 0.001$). When comparing the average digestibility coefficient values for leaf nutrients across the entire larval growth period, significant differences emerged between the varieties ($p < 0.001$ or $p < 0.05$), except for the digestibility coefficients of crude fat ($p > 0.05$). Conversely, comparing values for each larval age, except for the digestibility coefficients of crude fat in ages 1 and 2, the differences were insignificant, underscoring the consistent quality of Eforie variety leaves. Additionally, higher negative correlation coefficients in Eforie suggest a relatively lower capacity of silkworms to digest organic matter (as shown in Table 6), potentially due to the dietary crude fiber content in leaves, in contrast to Kokuso 21 leaves. This trend is similarly evident in the data from Table 5, observed over the entire period (ages I–V).

## 5. Conclusions

Throughout the developmental stages of the larvae, mulberry leaves undergo a dynamic shift in their components. This includes a decrease in certain nutrients alongside an increase in crude fiber content. This fluctuation in nutrients reflects an adaptive response by the larvae, which develop a specialized enzymatic complex, specifically geared toward breaking down fiber compounds. This adaptation enables them to more effectively utilize the cellulose richness in the leaves as they progress through their growth stages, contrasting with their initial developmental phases.

In general, most nutrients experience a decrease in digestibility as both the leaves and the larvae mature, with the notable exception of crude fiber, which displays an increase in digestibility.

In summary, we conclude that the nutritional value of the two mulberry varieties (Eforie and Kokuso 21) is comparable, making them suitable for utilization under similar conditions. The nutritional value of the leaves evolves throughout the larval growth period, indicating a reduction in the digestibility of all substances, except for crude fiber. Further research endeavors should involve the inclusion of additional silkworm hybrids from Romania, such as Băneasa Super, Zefir, Select, Miraj, or Record, as a follow-up. These hybrids showcase superior technological characteristics concerning cocoons and silk fibers. Additionally, it is crucial to investigate whether the nutrients in the leaves, whose digestibility we have studied, are effectively metabolized and transformed into the pupal silk wrap, both quantitatively and qualitatively.

On average, ECIS values for the two varieties were similar, with 9.90 ± 0.11% for Kokuso 21 and 9.87 ± 0.08% for Eforie. However, ECDS was significantly better for the Eforie variety (18.45 ± 0.35%) compared to the Kokuso 21 variety (17.94 ± 0.04%).

We consider the main novelty to be the comparison of the performance of the Romanian variety (Eforie) with that of a highly valuable and well-known variety (Kokuso 21), in a manner that has not been previously researched.

**Author Contributions:** Conceptualization, M.G.D.; methodology, M.G.D., M.G.U. and M.A.D.; software, M.G.D. and C.-G.R.-R.; validation, C.P., M.G.U. and A.U.; formal analysis, M.G.D. and C.-G.R.-R.; investigation, M.G.D.; data curation, M.G.D., C.-G.R.-R. and M.A.D.; writing—original draft preparation, M.G.D. and C.P.; writing—review and editing, M.G.D., C.P. and A.U.; visualization, M.G.U., C.-G.R.-R. and M.A.D.; supervision, M.G.D. All authors have read and agreed to the published version of the manuscript.

**Funding:** This research received no external funding.

**Institutional Review Board Statement:** This research received no external funding. Silkworm larvae were reared intensively for cocoon silk production and were not part of a research facility.

**Data Availability Statement:** Data supporting reported results available upon request from the authors.

**Conflicts of Interest:** The authors declare no conflicts of interest.

## References


1. Salem, M.; Khalafalla, M.M.E.; Saad, I.A.I.; El-Hais, A.M.A. Replacement of fish meal by silkworm *Bombyx mori pupae* meal in *Nile tilapia, Oreochromis niloticus* diets. *Egypt. J. Nutr. Feed.* **2008**, *11*, 611–624.
2. Longvah, T.; Mangthya, K.; Ramulu, P. Nutrient composition and protein quality evaluation of eri silkworm (*Samia ricinii*) prepupae and pupae. *Food Chem.* **2011**, *128*, 400–403. [CrossRef] [PubMed]
3. Bandlamori, S.V.; Mondal, M.; Singh, C.R.; Karkada, A.M. Evaluation of nutritional composition of hybrids of waste silkworm pupa *Bombyx mori* L. as a potential raw material for poultry feed-a sustainable technology for future. *Int. J. Eng. Res. Technol.* **2012**, *1*, 1–5.
4. Tran, G.; Heuzé, V.; Makkar, H.P.S. Insects in fish diets. *Anim. Front.* **2015**, *5*, 37–44.
5. Khan, S.H. Recent advances in role of insects as alternative protein source in poultry nutrition. *J. Appl. Anim. Res.* **2018**, *46*, 1144–1157. [CrossRef]
6. Dalle Zotte, A.; Singh, Y.; Squartini, A.; Stevanato, P.; Cappellozza, S.; Kovitvadhi, A.; Subaneg, S.; Bertelli, D.; Cullere, M. Effect of a dietary inclusion of full-fat or defatted silkworm pupa meal on the nutrient digestibility and faecal microbiome of fattening quails. *Animals* **2021**, *15*, 100112. [CrossRef] [PubMed]
7. Siriamornpun, S.; Thammapat, P. Insects as a delicacy and a nutritious food in Thailand. *Using Food Sci. Technol. Improv. Nutr. Promot. Natl. Dev.* **2008**, 16.
8. Mlček, J.; Rop, O.; Borkovcova, M.; Bednářová, M. A comprehensive look at the possibilities of edible insects as food in Europe-a review. *Pol. J. Food Nutr. Sci.* **2014**, *64*. [CrossRef]
9. Longvah, T.; Manghtya, K.; Qadri, S.S. Eri silkworm: A source of edible oil with a high content of α-linolenic acid and of significant nutritional value. *J. Sci. Food Agric.* **2012**, *92*, 1988–1993. [CrossRef]
10. Ullah, R.; Khan, S.; Khan, N.A.; Mobashar, M.; Sultan, A.; Ahmad, N.; Lohakare, J. Replacement of soybean meal with silkworm meal in the diets of white leghorn layers and effects on performance, apparent total tract digestibility, blood profile and egg quality. *Int. J. Vet. Health Sci. Res.* **2017**, *5*, 200–207.
11. Omar, S.S.; Shayo, C.M.; Uden, P. Voluntary intake and digestibility of mulberry (*Morus alba*) diets by growing goats. *Trop. Grassl.* **1999**, *33*, 177–181.
12. Azim, A.; Khan, A.G.; Ahmad, J.; Ayaz, M.; Mirza, I.H. Nutritional evaluation of fodder tree leaves with goats. *Asian-Australas. J. Anim. Sci.* **2002**, *15*, 34–37. [CrossRef]
13. Patra, A.K.; Sharma, K.; Dutta, N.; Pattanaik, A.K. Effect of partial replacement of dietary protein by a leaf meal mixture containing *Leucaena leucocephala, Morus alba* and *Azadirachta indica* on performance of goats. *Asian-Australas. J. Anim. Sci.* **2002**, *15*, 1732–1737. [CrossRef]
14. Tom, M.; Talaat, A.E.; Magid, D.A.; Eldeen, E.B. Investigation of the nutritional constituents of the wild silk moth *Epiphora Bauhiniae* (Guerin Meneville) Lepidoptera: Saturniidae in Sudan. *For. Prod. Ind.* **2013**, *2*, 69–74.
15. Leterme, P.; Botero, M.; Londoño, A.M.; Bindelle, J.; Buldgen, A. Nutritive value of tropical tree leaf meals in adult sows. *Anim. Sci.* **2006**, *82*, 175–182. [CrossRef]
16. Doran, M.P.; Laca, E.A.; Sainz, R.D. Total tract and rumen digestibility of mulberry foliage (*Morus alba*), alfalfa hay and oat hay in sheep. *Anim. Feed. Sci. Technol.* **2007**, *138*, 239–253. [CrossRef]
17. Todaro, M.; Sinacori, A.; Marinaro, G.; Alicata, M.L.; Giaccone, P. Palatability and in vivo digestibility of mulberry leaves (*Morus latifolia* CV. Kokusou 21) in sheep feeding. *J. Anim. Vet. Adv.* **2007**, *6*, 509–512.
18. Suchisree, J. Feeding preference of silkworm larvae depending on biochemical attributes related to mulberry genotypes. *Int. J. Pharm. Pharm. Sci* **2016**, *8*, 307–314.

19. Ruiz-Erazo, X.; Zambrano-González, G.; Castañeda-Vildozóla, Á.; Rodríguez-Ortega, L.T.; Pro-Martínez, A.; Hernández-Guzmán, F.J.; Rodríguez-Ortega, A. Silkworm (*Bombyx mori* L.) fed with mulberry (*Morus alba* L.) leaves and cow's milk. *Agro Product.* **2023**, *16*, 107–112. [CrossRef]
20. El-Banna, A.A.; Moustafa, M.N.; Mahmoud, S.M.; El-Shafei, A.M.; Moustafa, A.A. Effect of feeding different mulberry varieties on some biological characteristics of the silkworm, *Bombyx mori* L. *Egypt. J. Pure Appl. Sci.* **2013**, *51*, 55–60.
21. Lazar, S.; Vornicu, O.C. *Biological and Technical Foundation in Sericulture Production*; Ion Ionescu de la Brad Publ.: Iași, Romania, 2013; pp. 56–63.
22. Mărghitaş, L.A. *Silkworm Larvae Farming*; Ceres Publ.: Bucharest, Romania, 1995; pp. 84–89.
23. Doliș, M. *Sericulture*; Alfa Publ.: Iași, Romania, 2008; pp. 101–107.
24. Pătruică, S. *Beekeeping and Sericulture*; EUROBIT: Timișoara, Romania, 2013; pp. 23–27.
25. Pasca, I.; Marghitas, L.A.; Morar, R.; Cimpean, A.; Pusta, D. Testing correlation between biological traits of larvae and consumption and digestibility of the mulberry leaf in silkworm (*Bombyx mori* L.). *Anim. Sci. Biotechnol.* **2006**, *62*, 153–157.
26. Moise, A.R.; Marghitas, L.A.; Bobis, O.; Copaciu, F.M.; Dezmirean, D.S. *Morus* spp. Material conservation and characterization and its importance for romanian sericulture and GCEARS-PSP development—A review. *Bull. Univ. Agric. Sci. Vet. Med. Cluj-Napoca Anim. Sci. Biotechnol.* **2018**, *75*, 57–63. [CrossRef] [PubMed]
27. Pătruică, S. Successive growth of silkworms. *Agric. Banat.* 2008; 17, 23.
28. Pop, I.M.; Halga, P.; Avarvarei, T. *Animal Nutrition and Feeding*; Tipo Moldova: Iași, Romania, 2006; pp. 66–71.
29. *Regulation (EC) no. 152/2009 SR ISO 6496:2001*; Animal Feeding Stuffs—Determination of Moisture and Other Volatile Matter Content. International Organization for Standardization: Geneva, Switzerland, 2001.
30. *Regulation (EC) no. 152/2009 SR EN ISO 2171:2001*; Cereals, Pulses and by-Products—Determination of Ash Yield by Incineration. International Organization for Standardization: Geneva, Switzerland, 2001.
31. *Regulation (EC) no. 152/2009 SR ISO 6492:2001*; Animal Feeding Stuffs—Determination of Fat Content. International Organization for Standardization: Geneva, Switzerland, 2001.
32. *Regulation (EC) no. 152/2009 SR EN ISO 5983-2:2009 AOAC 2001.11*; Animal Feeding Stuffs—Determination of Nitrogen Content and Calculation of Crude Protein Content. International Organization for Standardization: Geneva, Switzerland, 2009.
33. *Regulation (EC) no. 152/2009 SR EN ISO 6865:2002*; Animal Feeding Stuffs—Determination of Crude Fibre Content—Method with Intermediate Filtration. International Organization for Standardization: Geneva, Switzerland, 2002.
34. Regulation (EU) No. 68/2013. Available online: https://eur-lex.europa.eu/legal-content/EN/ALL/?uri=CELEX:32013R0068 (accessed on 17 December 2023).
35. Ahanger, F.A.; Irfan, I.; Ramegowda, G.K.; Rao, R.J. Impact of road dust polluted mulberry leaves on the food ingestion, assimilation and conversion efficiency of silkworm, *Bombyx mori* L. in Kashmir Valley. Intern. *Int. J. Plant Anim. Environ. Sci.* **2015**, *5*, 86–94.
36. Diniță, G.; Tudorache, M.; Matei, A.; Antonescu, C.; Marmandiu, A. The inbreeding effect on food consumption and digestibility of the silkworms (*Bombyx mori* L.). In *Scientific Papers Series D. Animal Science*; Ion Ionescu de la Brad: Iași, Romania, 2004; Volume 47, pp. 684–687.
37. Review of Methods of Analysis, Chapter 7. Available online: https://www.fao.org/3/y4705e/y4705e12.htm (accessed on 17 December 2023).
38. Maciuc, V.; Creangă, Ș.; Maciuc, D.; Vidu, L. A New Programme for Data Management in Dairy Farms "ST 26733". In Proceedings of the International Conference "Agriculture for Life, Life for Agriculture", Bucharest, Romania, 4–6 June 2015; Volume 6, pp. 226–231.
39. Bura, M.; Acatincăi, S.; Pădeanu, I. *Silkworms, Biology and Growth*; Helicon Publ.: Timișoara, Romania, 1995; pp. 45–57.
40. Ifrim, S. *Natural Silk*; Ceres Publ.: Bucureşti, Romania, 1998; pp. 33–38.
41. Rath, S.S.; Narain, R.; Prasad, B.C.; Roy, G.C.; Sinha, B. Food allocation budgeting in tropical tasar, *Antheraea mylitta* fed on *Terminalia tomentosa*. *Sericologia* **2003**, *43*, 557–564.
42. De Marco, M.; Martínez, S.; Hernandez, F.; Madrid, J.; Gai, F.; Rotolo, L.; Belforti, M.; Bergero, D.; Katz, H.; Dabbou, S.; et al. Nutritional value of two insect larval meals (*Tenebrio molitor* and *Hermetia illucens*) for broiler chickens: Apparent nutrient digestibility, apparent ileal amino acid digestibility and apparent metabolizable energy. *Anim. Feed. Sci. Technol.* **2015**, *209*, 211–218. [CrossRef]
43. Rubia, B.; Kalpna, S. Analysis on the Effect of Different Mulberry Varieties on the Commercial Parameters of Mulberry Silkworm (*Bombyx mori*): A Review. *J. Exp. Agric. Int.* **2021**, *43*, 22–28.
44. Chorshanbi, B. Effect of feeding mulberry silkworms with leaves of different cultivars and hybrids on silk gland activity and yield of cultivated cocoons. *E3S Web Conf.* **2023**, *389*, 03088. [CrossRef]
45. Zhao, L.-C.; Hou, Y.-S.; Sima, Y.-H. Changes in glutathione redox cycle during diapause determination and termination in the bivoltine silkworm, *Bombyx mori*. *Insect Sci.* **2014**, *21*, 39–46. [CrossRef]
46. Ajage, D.A.; Tambe, A.S.; Pawar, S.S.; Khyade, V.B. Protein and carbohydrate digesting capability of syzigium seed powder in the tissue homogenate of mid gut in the fifth instar of silkworm, *Bombyx mori* (L.) race: Bivoltine cross breed [(CSR6 × CSR26) × CSR2 × CSR27)]. *Int. J. Curr. Microbiol. Appl. Sci.* **2019**, *8*, 2343–2354. [CrossRef]

*agriculture*

MDPI

*Review*

# Melatonin: An Overview on the Synthesis Processes and on Its Multiple Bioactive Roles Played in Animals and Humans

**Vasile-Cosmin Andronachi, Cristina Simeanu *, Mădălina Matei, Răzvan-Mihail Radu-Rusu and Daniel Simeanu**

"Ion Ionescu de la Brad" University of Life Sciences, 3 Mihail Sadoveanu Alley, 700489 Iasi, Romania; cosmin.andronachi@iuls.ro (V.-C.A.); madalina.matei@iuls.ro (M.M.); razvan.radu@iuls.ro (R.-M.R.-R.); daniel.simeanu@iuls.ro (D.S.)

* Correspondence: cristina.simeanu@iuls.ro

**Abstract:** Melatonin is a natural hormone synthesized mainly by the pineal gland of vertebrates, and, secondarily, by other tissues and organs as well. It is deemed a bioactive molecule due to the multiple roles and functions it performs in animals and humans. Research conducted up to 2024 has reported the presence of melatonin in a wide variety of plants and bacteria, as well. This review aims to collect some of the scientific data to identify and describe the main sources of melatonin, and to document the functions and roles it plays in animal organisms. It also includes a description of the main technological and nutritional factors that can positively or negatively influence the synthesis and secretion process of melatonin, which is subsequently transported from the animal body into some food products, such as milk. This paper also includes information on the interaction between melatonin and other bioactive compounds present in animal and human bodies, with the aim of identifying what other functions and roles this hormone performs, and whether it interacts with other substances present in the vertebrate organism.

**Keywords:** melatonin; pineal gland; bioactive molecule; nutritional factors; technological factors

## 1. Introduction

Melatonin, also known by its chemical name, N-acetyl-5-methoxytryptamine [1], is a natural hormone synthesized and secreted generally by the pineal gland of mammals [2], derived from the limiting essential amino acid tryptophan [3,4], which plays multiple roles and functions in both animal and human organisms. Melatonin is an endogenous idolamine [5] first discovered in 1958, in the pineal gland of cattle [6]; it was first isolated in 1960 [7], and is known as the hormone of darkness or the sleep hormone due to the fact that melatonin secretion occurs as an automatic response of organisms to the lack of light [8,9].

The pineal gland is an endocrine gland located on the third ventricle of the brain of all vertebrates, and its basic function is the production and secretion of melatonin [10–12].

The determination of circadian rhythms in vertebrates is carried out by an internal biological clock, which consists of the retina, hypothalamus, and pineal gland. Through the retinal, encephalic, and pineal photoreceptors, these three components (the retina, hypothalamus, and pineal gland) are synchronized with light cycles [13]. The pineal gland is an organ characteristic of vertebrates, being present in the animal body of all mammals, including in the human body [14], as well as in birds [15] and fish [16]. Through specialized studies, melatonin has been detected in other life forms that do not have a pineal gland, and its presence has also been reported in some species of microorganisms [17].

Melatonin is a bioactive molecule that acts in the regulation of sleep [18] and of the circadian rhythm [19], while also performing antioxidant [20], anti-inflammatory [21,22], immunomodulatory [23], anti-aging, anti-carcinogenic [24,25] and anti-apoptotic functions, being able to regulate apoptosis [26–28], and improving immune activity in organisms [29].

The term bioactive molecule (bioactive compound) is a notion whose definition is still being discussed and debated in 2024, because the opinions of different authors are divided into two spheres of classification of these substances [30]. Some researchers define bioactive molecules as substances with a positive or negative biologically active role, influenced by the nature of the bioactive compound, and by the quantity of molecules available in the sources of supply of these biologically active substances (the uptake of dietary bioactive molecules). Other authors strictly emphasize that a substance can be considered a bioactive molecule only if it fulfills an exclusively positive role on the organism in which it exerts its action [31,32].

Based on the literature survey, we have defined bioactive molecules as substances of dietary origin, with a biologically active role, fulfilling multiple roles and functions in the animal and human body, such as regulating the circadian rhythm, the normal and harmonious development of the physiological processes of growth and development of the body, maintaining the health of vertebrates, etc. These substances with bioactive role are widely distributed in nature, found in a multitude of sources, including raw materials and finished food products. They are classified from a chemical point of view as compounds different from the nutrients found in food. From a biological point of view, these substances fulfill an exclusively positive role in the animal and/or human body, acting either individually or in a synergistic relationship with other molecules present in the vertebrate organism [30–32].

## 2. Melatonin Synthesis and Secretion

Melatonin, with the chemical formula $C_{13}H_{16}N_2O_2$, can be synthesized and secreted in vertebrates by hydroxylation, decarboxylation, acetylation, and methoxylation processes [4,5]. Physically, melatonin in its pure state is a colorless powder with a whitish hue. The density of melatonin is 1.175 g/cm$^3$, with a molar mass of 232.28 g/mol and a boiling point reached at a temperature of +512.8 °C. The melting point of melatonin is within the limits of the thermal range of +116.5–+118 °C [7].

From a physiological point of view, melatonin can be secreted in the animal and human body via the pineal route (via the pineal gland) and via the extrapineal route in other organs and tissues of the body. In the vertebrate organism a much higher amount of melatonin is secreted at the extrapineal level, compared to the level of melatonin secreted exclusively by the pineal gland. However, it was observed that the quantitative level of melatonin secreted extrapineally cannot compensate for or replace the role played by melatonin secreted by the pineal gland, in terms of regulating the circadian rhythm and improving sleep quality. Compared to pineal melatonin, which has a well-established circadian rhythm, the melatonin secreted at the extrapineal level is not released into the blood, acting only locally, at the level of the tissues and organs that produce it, and does not have a circadian secretion pattern [1]. With the exception of intrapineal and extrapineal sources of melatonin secretion, the vertebrates' also have two other sources of this hormone, namely: gut microbiota and the dietary uptake [4]. Thus, by consuming food products rich in melatonin obtained through synthesis and natural secretion, the level of circulating melatonin in the body can be increased, leading to the improvement of people's health by manifesting the multiple roles and functions that melatonin can perform in the human body.

Pineal melatonin and extrapineal melatonin have the same chemical structure and perform similar roles in the animal and human body, but have different sources of origin

(Figure 1). In the animal and human body, both pineal melatonin and extrapineal melatonin have an antioxidant role by eliminating free radicals [33], also performing functions in modulating inflammatory responses at the intestinal level [34].

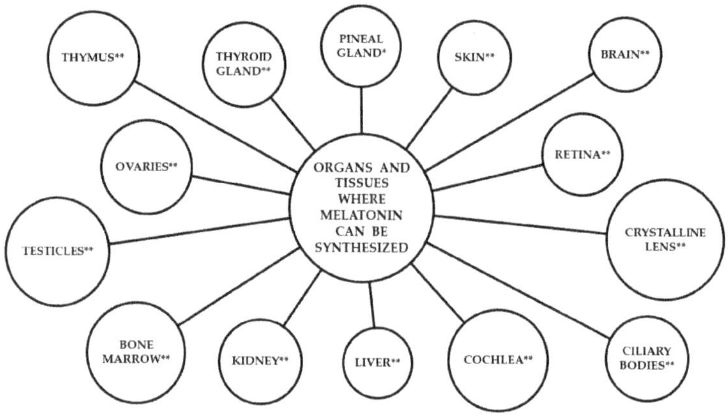


**Figure 1.** Organs and tissues where melatonin can be synthesized in animal and human bodies. * pineal-originated melatonin; ** extrapineal-originated melatonin.

Acuña-Castroviejo et al. (2014) [33] suggested that the absence of photoperiod-induced variations in the synthesis and secretion process of extrapineal melatonin is due to the existence of different synthetic pathways for this hormone at the extrapineal level, compared to melatonin synthesized and secreted exclusively by the pineal gland. These differences observed in the synthesis process of extrapineal melatonin may be due to states of adaptability in the organism, involved in cell survival. An example suggested by Acuña-Castroviejo et al. (2014) [33] is the antioxidant effect of melatonin, achieved through the neutralization of reactive oxygen species (ROS) and reactive nitrogen species (RNS). The production of ROS and RNS mainly occurs during phases of metabolic, motor, and neural activity, when oxygen consumption is at its maximum in both animal and human organisms [33]. Thus, in the case of diurnal animal species, which consume the largest amount of oxygen during the day, the organism's synchronization with environmental conditions and with physiological processes that are more intense during the day, leading to the production of large amounts of ROS and RNS, could represent an evolutionary factor that determined the adaptation of the organism through the establishment of an extrapineal melatonin synthesis process that is not dependent on day–night variations, and which provides a strong protective mechanism for cellular survival.

Melatonin has a very low toxicity, and in relatively high doses, due to its optimal dimensions, it is able to easily cross physiological barriers. N-acetyl-5-methoxytryptamine (melatonin) has an amphiphilic character, being partially soluble in water and highly soluble in lipids [35].

### 2.1. *Pineal-Originated Melatonin*

The pineal synthesis and secretion of melatonin begins with the help of noradrenaline, which is the main neurotransmitter involved in the activation of the pineal enzyme group, especially N-acetyltransferase. This activation of the pineal enzyme group is due to the cAMP (cyclic adenosine monophosphate) and cGMP (cyclic guanosine monophosphate) pathways that contributes to the activation of alpha1 ($\alpha$1), alpha2 ($\alpha$2) and beta1 ($\beta$1) receptors, located on the pinealocyte membrane [36].

Melatonin synthesis (Figure 2) is a sequential process, consisting of four phases/steps [37], and begins in the first phase with the transformation of tryptophan into 5-hydroxytryptophan, a transformation that occurs due to the action of the enzyme tryptophan hydroxylase (TPH). Subsequently, 5-hydroxytryptophan is transformed into serotonin (5-HT), and after transformation, serotonin undergoes the process of N-acetylation as a result of the action of arylalkylamine N-acetyltransformase (AANAT), forming NAS (N-acetylserotonin). In the last step, N-acetylserotonin is transformed into melatonin, and this process of transformation of N-acetylserotonin into melatonin is facilitated by the enzyme hydroxyindole-O-methyltransferase (HIOMT) [38].

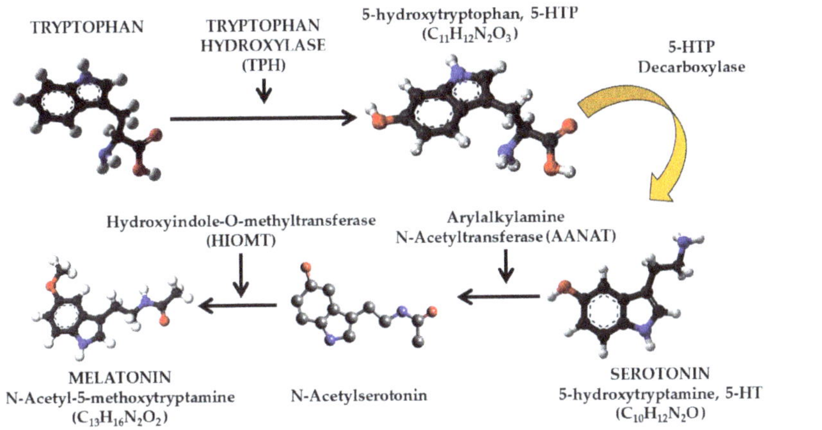


**Figure 2.** Synthesis of melatonin from limiting essential amino acid tryptophan. Original processing content. 3D models of Tryptophan, 5-Hydroxytryptophan, Serotonin, N-Acetylserotonin and Melatonin molecules are taken from online and copyrighted with permission to use them. Graphics: Tryptophan, 5-Hydroxytryptophan, Serotonin and N-Acetylserotonin—Copyright © "Creative Commons CC0 1.0 Universal Public Domain Dedication". Graphics: Melatonin—All rights reserved by the Free Software Foundation under the "GNU Free Documentation License".

Due to the fact that the enzyme arylalkylamine N-acetyltransformase (AANAT), which plays the role of catalyzing the conversion of serotonin into N-acetylserotonin, has minimal activity in the organism during the day, the melatonin production process is limited, thus favoring the accumulation of serotonin in pinealocytes, while melatonin synthesis at night is due to the increased activity of the AANAT as a result of the onset of darkness [39].

Melatonin synthesis in the pineal gland is rhythmically regulated by the body's "master biological clock" located in the hypothalamic suprachiasmatic nucleus (SCN). Melatonin synthesis begins with the conversion of the essential amino acid tryptophan (which is of dietary origin) into serotonin [40]. Furthermore, some studies have reported that mitochondria contain high levels of melatonin [41].

The process of capturing and transmitting light information from the retina to the pineal gland begins with photoreceptor cells in the retina, which receive light signals (the retina is responsible for transforming optical signals into biological signals), and transmit the information to the hypothalamic suprachiasmatic nucleus (SCN) via the retinohypothalamic tract. The main cells on the surface of the retina that receive light information are the photosensitive ganglion cells (ipRGC), which are highly sensitive to blue light. The suprachiasmatic hypothalamic nucleus (SCN) is the body's main biological clock, and through the information it receives from the retina, it determines whether it is day or night outside, depending on the intensity of light that has penetrated the surface of the retina. In turn, the suprachiasmatic hypothalamic nucleus sends the information received through the sympathetic preganglionic neurons that are located in the brainstem (especially in the modular area or in the lower part of the brainstem) to the superior cervical ganglia (SCG) [42].

In humans, melatonin synthesis is initiated immediately after sunset, reaching a peak in secretion in the middle of the night, and gradually decreasing in the second half of the night. About 80% of the melatonin present in the human body over a 24 h period is synthesized during the night. Serum melatonin levels during the night in humans reach values between 80 pg/mL and 120 pg/mL, respectively, and during the day the amount of melatonin decreases drastically to values of about 10–20 pg/mL [42,43].

### 2.2. Extra-Pineal-Originated Melatonin

Melatonin can also be synthesized and secreted in the animal and human body, and in other tissues and organs such as the skin [44–46], the retina [47,48], certain brain regions [49], the liver, kidneys, and female reproductive organs [33], the thyroid gland, the lens and bone marrow [50], the ciliary bodies [51], the thymus [52], and the cochlea [53], and is also present in the gastrointestinal tract [54,55].

It has been found that melatonin synthesized and secreted in other tissues and organs fulfills other roles in living organisms, compared to the ones played by the pineal melatonin, depending on the tissue or organ in which it is synthesized and secreted. Another characteristic of extrapineal melatonin is that it is not transported in the body through the blood and acts only locally, in the area where it was secreted. For instance, it was discovered that extrapineal melatonin synthesized in the skin protects the tissue against reactive oxygen species (ROS) and reactive nitrogen species (RNS) that are induced as a result of skin exposure to chemical toxins or ultraviolet radiation [4,56,57].

Extrapineal melatonin secreted in the skin is not able to directly neutralize ROS and RNS without assistance [1]. Thus, it has been demonstrated that when melatonin donates an electron in order to inactivate a radical species, it is transformed into another radical scavenger, called 3-hydroxymelatonin. This phenomenon of melatonin transformation into other derivatives with a role in the capture of reactive species is generally called the melatonin antioxidant cascade, or the free radical elimination cascade [4,55]. This transformation of melatonin into different radical scavengers includes the following forms: N-acetylserotonin (NAS), 5-methoxytryptamine (5-MT), cyclic 3-hydroxylated melatonin (c3OHM), N1-acetyl-N2-formyl-5-methoxykynuramine (AFKM), N1-acetyl-5-methoxykynuramine (AMK), 6-Hydroxymelatonin (6-OHM), 4-Hydroxymelatonin (4-OHM), and 2-Hydroxymelatonin (2-OHM) [58].

## 3. The Role of Melatonin

Multiple studies conducted in the field of knowledge of the mode of action of melatonin on vertebrate organisms have highlighted a series of roles and functions that this hormone performs in the animal and human body. Thus, depending on the nature of melatonin (pineal or extrapineal), and depending on the tissue or organ in which it is synthesized and secreted, this hormone performs different roles and functions at the level of each individual organism.

### 3.1. The Role of Melatonin in the Animal Body

In the animal body, melatonin plays a role in regulating the circadian rhythm [19], improves sleep quality by regulating the sleep–wake rhythm [59], increases the quality of milk in mammals, mediates seasonal reproductive changes [60,61], modulates energy metabolism [62], regulates cellular redox homeostasis [63], and plays a role in domestic species' reproduction [19].

Melatonin is rhythmically secreted in the animal body, as a result of the photostimulatory action caused by darkness, or by suppressing synthesis by exposing the eyeball to natural and/or artificial light [7]. After synthesis, pineal melatonin is released into the cerebrospinal fluid of the third ventricle, after which it is distributed to different regions of the brain in order to signal photoperiodic changes in the environment [64,65].

A study conducted on rats has shown that melatonin has the ability to inhibit the growth and development of certain cancerous tumors. Two groups of tumor-bearing rats were analyzed, infused with blood from human donors. The rats in the first group were infused with blood that had melatonin concentrations specific to the night period, and the rats in the second group were infused with blood that had melatonin levels specific to the day period. The results showed that in the case of rats infused with melatonin-rich blood, tumor growth was inhibited, while perfusion with melatonin-deficient blood led to tumor growth in the studied rats [66].

The involvement of melatonin in the reproductive process of animals is a well-known and documented function of this hormone. Melatonin primarily acts through the MT1 and MT2 receptors, whose presence has been reported in several specialized studies, in numerous brain and peripheral tissues, including the testes and ovaries. Specialized studies have reported that the melatonin receptor MT1 is more widely distributed in brain regions and endocrine tissues, compared to the MT2 receptor, which appears to be generally absent in the pituitary gland and hypothalamus. The brain regions and endocrine tissues are the main response areas for the circadian and physiological effects induced by N-acetyl-5-methoxytryptamine (melatonin), and the presence of the MT1 receptor in these organs indicates that this receptor (MT1) plays a major role in the physiological reproductive processes of mammals that are modulated by melatonin [67]. Melatonin is a hormone involved in the modulation of the hypothalamo–pituitary–gonadal (HPG) axis, which serves as a regulatory center for the reproductive process in both seasonally reproducing animals and non-seasonal reproducers (including humans) [68].

The influence of melatonin-rich diets on reproductive performance in rams was studied by Peña-Delgado et al. (2023) [69], who conducted a study in Spain on a group of 16 rams of the Aragonesa breed. In that study, the animals were divided into two groups, each consisting of 8 rams: a control group that received 500 g of a commercial diet, and an experimental group that received a modified diet, also administered at a rate of 500 g per day, but consisting of 20% agro-food by-products rich in phytomelatonin (plant melatonin), with the remaining 80% being the same commercial diet given to the control group. The study was conducted over a period of 5 months, from February to July, representing the non-reproductive season for the rams, and the animals were fed ad libitum with straw throughout the entire duration of the research. The phytomelatonin-rich by-products introduced into the diet of the experimental group consisted of pomegranate pomace with a melatonin content of $35.81 \pm 0.4$ ng/g, tomato pomace with a melatonin content of $23.76 \pm 1.37$ ng/g, and grape pulp with a melatonin content of $45.94 \pm 4.19$ ng/g. The plant components that made up the modified ration for the experimental group were mixed in equal proportions, and the melatonin levels in the plant by-products were determined by the authors using the HPLC-ESI-MS/MS technique (High Performance Liquid Chromatography–Electrospray Ionization–Tandem Mass Spectrometry). The authors of that study reported in their findings that a diet rich in phytomelatonin increased melatonin levels in seminal plasma and improved sperm viability and morphology. In the context of the same research, the authors highlighted that the introduction of plant by-products rich in phytomelatonin into the diets of farm animals offers economic benefits for both the agro-food industry and animal husbandry, as the reuse of plant materials, derived from certain production processes, reduces the waste levels resulting from the food and beverage processing processes, while the presence of a higher amount of melatonin in the animal's body exhibits protective effects against oxidative damage to sperm cells by reducing intracellular levels of reactive oxygen species [69].

### 3.2. The Role of Melatonin in the Human Body

Melatonin, a hormone generally synthesized by the pineal gland under the influence of light intensity, and whose synthesis and secretion are carried out in the highest quantities in the time interval 01:00–04:00 in the morning, as specified by some authors [70], fulfills multiple roles in the human body, some of which are similar to the roles performed in the animal body. The most studied roles and functions that melatonin fulfills in the human body are represented by the regulation of the circadian rhythm and the improvement of sleep quality [18], the reduction of oxidative stress at the level of the whole body [71,72], the intervention in the regulation of the immune system [73], the regulation of the functions of the cardiovascular system [74] and the nervous system [75], the participation in the regulation of the biological rhythm [76], the stimulation of immune cells, and the regulation of cytokine production [77].

Under optimal conditions of human activity, performed within 24 h, the share allocated to rest through quality sleep should represent one third of the duration of a circadian cycle [78]. Sleep-related disorders are situations encountered in all age groups among the human population, and various studies have demonstrated that a low quality of sleep achieved during a 24 h day of activity has multiple negative effects on the general health of the human body, such as fatigue, low performance in carrying out daily activities, and others [79,80].

According to some clinical research, it has been found that in the first months of a newborn's life, the pineal gland is not able to start a natural and individual synthesis and secretion of melatonin, so that infants must benefit exclusively from melatonin from external sources, such as breast milk [81,82]. The secretion and synthesis of melatonin, as well as the development of circadian sleep–wake rhythms, begin to manifest in infants only after the second to sixth month of life of newborns, or even after the sixth month in certain situations [81]. Due to these observations, the need for a more focused study of the way melatonin is secreted in cow's milk is justified, in order to allow the production of milk and dairy products derived from it, which can support the population suffering from sleep disorders due to melatonin deficiency.

According to other studies, melatonin secretion in human milk also exhibits a circadian rhythm [83], with a maximum melatonin hormonal content of $46.9 \pm 4.2$ pg/mL determined in milk collected at midnight, and very low levels (undetectably low) in milk collected during the day, when the retina was exposed to the influence of natural and/or artificial light [84].

An advantageous aspect of melatonin is that this hormone does not present toxicity in the human body when administered in high doses. Thus, it has been demonstrated through various clinical studies that daily oral administration of melatonin, even in high doses (between 1 and 300 g or between 4.3 and 1291.5 μmol), did not produce negative effects on the health of human patients who benefited from melatonin treatment [35], an aspect that may encourage the consumption of foods with a high content of melatonin, especially due to the simple fact that in milk and other food products, melatonin is found in small quantities (expressed in pg) compared to the levels of melatonin administered to human patients in clinical studies.

Some researchers have reported through their studies that melatonin has an inhibitory effect on intrinsic apoptotic pathways in neurodegenerative diseases, especially in

Alzheimer's disease, Huntington's disease, Parkinson's disease, vascular accidents, and amyotrophic lateral sclerosis [85].

Melatonin is synthesized and secreted in the human body depending on the age of each individual. A circadian rhythm is an oscillation between the light period and the dark period, carried out within approximately 24 h. In the vertebrate body, the suprachiasmatic nucleus (SCN) determines the synchronization of circadian rhythms by regulating body temperature, through various hormonal signals and by regulating neuronal activity. According to research conducted up to 2024, aging of people leads to various changes in sleep patterns, such as daily synchronization of rest hours, sleep duration, high latency of sleep onset, greater susceptibility to awakenings, sleep fragmentation manifested by periodic awakenings during a rest cycle, reduction in the quality of deeper sleep, amplification of periods spent in lighter stages of sleep, etc. Thus, it has been found that starting at the age of 50, the amount of melatonin secreted in the human body begins to decrease, and after the age of 70, the process of synthesis and secretion of melatonin naturally seems to be almost completely absent [86–88].

Milagres et al. (2013) [89] reported that the administration of cow's milk collected at 02:00 AM, which was rich in melatonin obtained through both synthetic production and natural secretion, increased plasma melatonin levels by 26.5% in adult Wistar rats, compared to plasma melatonin levels detected in adult Wistar rats that consumed daytime milk (cow's milk collected at 15:00 PM). In the same study, it was determined that the addition of tryptophan to nighttime milk increased plasma melatonin levels by 35.6% in adult Wistar rats that consumed this type of milk enriched with natural melatonin and added tryptophan [89]. Similarly, beneficial effects of consuming melatonin-rich milk have been reported in humans, through the improvement of sleep quality, which was manifested by greater satisfaction with the rest period achieved through sleep and by the improvement in the performance of daily activities [59].

## 4. Melatonin in Milk

Milk is a liquid mixture, formed of water and dry matter (the proportions of the two components being approximately 87.5% water and 12.5% dry matter, respectively). Milk is defined as a homogeneous and opalescent liquid, white in color and free of foreign bodies suspended in the liquid volume, being secreted by the mammary gland of female mammals [90].

The dry matter in milk is in turn represented by several nutritional components (proteins, fats, lipids, etc.) with a role in providing the body with the energy necessary to support healthy and harmonious growth and development [91], and by a series of molecules with a bioactive role in the animal and human body (vitamins, hormones, minerals). Due to the chemical composition of cow's milk, which is considered to be complex and complete, and due to the high nutritional value of this liquid, milk is considered to be one of the most important products of animal origin [92]. Naturally, raw milk has in its chemical composition several bioactive molecules (including free oligosaccharide structures, various hormones, peptides, lipids, etc.), which fulfill multiple active biological roles in the animal and human body with a different metabolic impact compared to the nutritional value of milk [93,94].

The natural secretion of melatonin is a process that occurs in the animal and human body during the night, so melatonin is also called the sleep hormone. Melatonin is naturally secreted first in the blood and in the cerebrospinal fluid of the third ventricle of the vertebrate brain [1], and in the case of female mammals, melatonin is subsequently released into the milk. For this reason, any variation in nutrients and bioactive molecules in the blood (including melatonin) will directly influence the chemical composition of milk, in terms of the milk's content in nutrients and bioactive molecules. Due to the nocturnal nature of melatonin secretion, numerous researchers have initiated the idea of harvesting animal milk at night, so as to obtain milk with a higher melatonin content, thus giving rise to the concept of daytime milk (milk harvested during the day) and nighttime milk (milk harvested at night).

Studies conducted in the zootechnical field on the factors that can influence the secretion of melatonin in cow's milk (but also in other species of animals of zootechnical interest), have demonstrated that the process of melatonin release in milk is a phenomenon influenced by a series of technological factors (breed and species of animals, animal health, conditions in shelters, photoperiods, the time when milk is harvested, milking frequency, environmental conditions in which animals are raised, etc.), and nutritional factors (animal nutrition and feeding).

According to the specialized literature and experiences in zootechnical research, it has been highlighted that the health status of animals (cattle, sheep, goats, birds, etc.), as well as the conditions of raising and care (generally exposure to stress factors) directly influence the quality and quantity of productions made [95].

Some authors reported in a study conducted in Switzerland, on a herd of 125 cows from eight farms with automatic milking systems, a correlation between the increase in the number of milkings performed in a night and the low content of salivary melatonin detected [96]. In this review, we will analyze and describe the influence of the main technological and nutritional factors that can significantly influence (positively or negatively) the natural secretion of melatonin in cow's milk, according to the information available in the specialized literature.

### 4.1. *Farming Technology Factors*

The most studied technological factors that can influence the melatonin content in cow's milk are represented by the species and breed of animals, environmental conditions, the daily milk production of the animals, the frequency of milking, the lighting conditions, and the intensity of artificial light in the animal shelters [5,97,98].

#### 4.1.1. Species and Breed

Animal species and breed are two factors that can directly influence the melatonin content in the milk of cows and other animals raised for dairy production. Studies on cattle and sheep have shown that there are major differences between the melatonin content released in milk collected during the day and the melatonin content released in milk collected at night, as well as between milk collected individually from the two animal species studied, according to the data presented in Table 1 [99,100]. These differences are mainly due to the circadian rhythm of melatonin synthesis, as well as to genetic differences between the two animal species.

**Table 1.** Melatonin content determined in daytime milk and nighttime milk collected from cattle and sheep.

| Species | Breed | Melatonin Content (pg/mL) | | | References |
|---|---|---|---|---|---|
| | | Daytime Milk | Nighttime Milk | Difference | |
| Cattle | Jersey | $2.924 \pm 0.216$ [a] | $6.954 \pm 0.567$ [a] | 4.03 | [99] |
| | Holstein Friesian | $2.912 \pm 0.266$ [a] | $11.314 \pm 1.1$ [a] | 8.402 | |
| Sheep | Awassi | $6.12 \pm 4.55$ [b] | $11.06 \pm 7.24$ [b] | 4.94 | [100] |

[a]—values are expressed as the mean of the samples $\overline{X} \pm S\overline{x}$; [b]—values are expressed as the mean of the samples $\overline{X} \pm$ s.e. (s.e.—standard error).

In sheep [100], it was demonstrated that keeping the animals for a period of 16 h in the dark and 8 h in the light did not influence the chemical composition of the milk in protein, fat, lactose, and salt, but led to the obtaining of milk with a higher melatonin content in the case of samples analyzed from milk collected at night (11.06 ± 7.24 pg/mL), compared to milk samples collected during the day (6.12 ± 4.55 pg/mL). The results presented in Table 1 highlight the nocturnal nature of melatonin secretion in bovine and ovine milk, due to the obtaining of higher melatonin values in nocturnal milk, compared to the results obtained in the case of daytime milk collected from both species.

#### 4.1.2. Environmental Conditions

Environmental conditions are other important factors in obtaining milk with a high melatonin content. Significant differences in the melatonin content of cow milk were also recorded in the milk of the same breed (Holstein) but located in different geographical regions, according to the data presented in Table 2.

**Table 2.** Melatonin content recorded in daytime milk and nighttime milk collected from Holstein cattle located in different geographical regions of the world.

| Melatonin Content (pg/mL) | | | Study Period | Duration of Photoperiod | N * | Geographical Area | References |
|---|---|---|---|---|---|---|---|
| Daytime Milk (Milking Time) | Nighttime Milk (Milking Time) | Difference | | | | | |
| 2.912 ± 0.266 (15:00–17:00) | 11.314 ± 1.1 (03:00–05:00) | 8.402 | January | 11 h of light 13 h of darkness | 27 | Konya, Turkey | [99] |
| 103.7 ± 6.61 [a] (07:00–16:00) | 163.13 ± 8.96 [a] (01:00) | 59.43 | Unspecified | Unspecified | 40 | Konya, Turkey | [101] |
| 4.03 [a,b] (15:00) | 39.43 [a,b] (02:00) | 35.4 | 2–16 June | 15 h of light 9 h of darkness | 10 | Viçosa, Brazil | [89] |
| 6.98 ± 3.05 (N.S. **) | 14.87 ± 7.69 (N.S. **) | 7.89 | S. *** | S. *** | 30 | Castro, Brazil | [102] |
| 5.36 ± 0.33 (12:30) | 30.7 ± 1.79 (04:30) | 25.34 | 1–15 November | 10.4 h of light 13.6 h of darkness | 28 | Israel | [103] |
| 3.3 ± 0.18 (12:30) | 17.81 ± 0.33 (04:30) | 14.51 | | | | | |
| 90.21 ± 7.21 [c] (15:00) | 120.07 ± 7.21 [c] (05:00) | 29.86 | August | 13 h of light 11 h of darkness | 10 | China | [104] |

* N—number of studied animals; ** N.S.—not specified; *** S (Specification)—data presented in that row of the table represent average melatonin content determined from milk samples collected in two seasons (winter and summer). All values are expressed as mean of samples ±SD. [a]—amount of that value is expressed in pg/mL$^{-1}$; [b]—variation index not stated; [c]—mean value ± SEM.

The significant differences between the values of melatonin levels in the milk of Holstein cows are mainly due to the conditions in which the experiments were carried out. Boztepe et al. [99] carried out their study in January (with a photoperiod of 11 h of natural light and 13 h of darkness) and with an artificial light intensity measured at eye level during the night of 150 lx. In that study, milk was collected in two different time intervals (between 15:00 and 17:00 for daytime milk, and between 03:00 and 05:00 in the morning for nighttime milk). In another study, Şahin et al. [101] collected milk from cows three times a day at 07:00 in the morning, 16:00 in the afternoon, and 01:00 in the morning. Milagres et al. [89] conducted a study on the differences in melatonin detected in the milk of Holstein cows during the summer season (between 2 and 16 June) for a period of 15 days, collecting milk at 02:00 in the morning for nighttime milk and 15:00 in the afternoon for daytime milk.

In the context of the study conducted by Romanini et al. [102], nighttime milk was collected between 05:00 and 06:00 in the morning, and daytime milk was collected during the day, when the animals were milked according to the milk collection schedule applied in the farm where the research was carried out.

The research conducted by Asher et al. [103] was carried out during the period of 1–15 November of the year, with a photoperiod duration of 10.4 h of natural light and 13.6 h of darkness. The study involved the formation of two experimental group that were subjected to different artificial lighting conditions during the night. In the case of cows in the Dark-Night batch, the lighting conditions applied in the animal shelters during the night were 648 ± 5.12 nm and 5.08 ± 0.04 lx, and for cows in the Night-Illumination group, lighting conditions of 462 ± 5.12 nm and 105 ± 3.91 lx were applied.

Teng et al. [104] conducted a study on the melatonin content of cow's milk at the end of August, and the milk was collected at 03:00 PM, representing daytime milk, and at 05:00 AM, representing nighttime milk.

The correlation of the data presented in Table 2, in relation to all the experimental conditions that were applied in the studies presented in this article, shows the high degree of variability of the melatonin content found in bovine milk. The experimental data presented in Table 2 indicate different values of the melatonin content range determined in both daytime and nighttime milk of cows. The presence of a higher level of melatonin in nighttime milk demonstrates the nocturnal nature of biosynthesis and secretion of this pineal hormone in the bovine body. The existence of a higher concentration of melatonin in nighttime milk, compared to daytime milk, indicates that a higher synthesis and secretion of melatonin achieved in the cow's body will lead to a subsequent release of this hormone in a larger quantity in milk, due to the nature of melatonin as a circulating molecule.

The large fluctuations in the minimum and maximum limits of melatonin found in both daytime milk (2.912–103.7 pg/mL) and nighttime milk of cattle (11.314–163.13 pg/mL), as well as the significant differences in melatonin content determined by different researchers (data presented in Table 2), are most likely due to the different experimental conditions (length of photoperiods, intensity of artificial light in animal shelters, and times at which milk samples were collected) in which the research was conducted [89,99,101–104]. Among the environmental conditions that represent technological factors that can influence the synthesis and secretion of melatonin, and the subsequent release of this hormone in cow's milk, are photoperiods (in the southern parts of the globe, the days are longer compared to the northern parts, which increases the period of exposure of the eyeball to natural light, thus inhibiting the synthesis and secretion of melatonin), and ambient temperature (the heat stress of animals is a factor that can influence the productivity of cattle).

Heat stress in dairy cows is a phenomenon that can lead to a decrease in the milk yield and to a decrease in the quality of the product obtained. It is well documented that the vertebrate organism must receive certain specific and distinct signals (the absence of light and the relatively low temperature of the environment) to initiate the process of entering rest. Cattle are animals tolerant to low temperatures (cows can also withstand temperatures in the thermal range of 0 ÷ +5 °C), and begin to show behavioral and physiological changes when they are exposed for a longer period of time to the influence of too high or even too low temperatures. The occurrence of heat stress in cattle is a phenomenon that can be regulated to certain extent through the animal's nutrition. In situations of heat stress, cattle will use more energy for thermoregulation, an aspect that will depreciate milk production from a qualitative and quantitative point of view, as a result of the manifestation of an energy unavailability at the animal's body level. This situation can also be accentuated by the low dry matter intake in forage, as a result of the unavailability of nutrients that should be assimilated by the animal's body [105].

It has been demonstrated (according to the data presented in Table 2) that milk has in its composition, different amounts of melatonin, depending on the environmental conditions to which the animals are exposed. Thus, for Holstein cows, in Konya, Turkey, melatonin amounts were recorded that varied in daytime milk between the limits of 2.912 pg/mL [99] and 103.7 pg/mL [101], and in nighttime milk, between the limits of 11.314 pg/mL [99] and 163.13 pg/mL [101]. In other studies conducted in different geographical regions of Brazil (Viçosa and Castro), the degree of variability of the melatonin content secreted in the milk of Holstein cattle was highlighted [89,102], with cattle from the Viçosa area recording a melatonin content of 4.03 pg/mL in daytime milk and 39.43 pg/mL in nighttime milk [89], while in the study conducted in the Castro region, the melatonin content level was 6.98 ± 3.05 pg/mL in daytime milk and 14.87 ± 7.69 pg/mL in nighttime milk [102].

In other studies conducted in Israel [103] and China [104], variations in melatonin content were also reported both between daytime and nighttime milk of cattle (highlighting once again the nocturnal nature of melatonin synthesis and secretion), and between milk collected from the experimental groups analyzed in the two works (highlighting the impact of environmental conditions to which the animals are exposed on the melatonin content in milk).

In the study conducted in Israel [103], milk samples were collected from two groups of animals kept under different lighting conditions throughout the rest period (night period). One group of cows was kept in dark conditions throughout the rest period, and the other group was kept under lighting conditions throughout the night. Differences were recorded both in terms of individual comparison (comparison made at group/lot level) and in terms of the common comparison made between the milk collected from the two groups of cattle studied. According to the results obtained by Asher and his collaborators, a higher melatonin content was recorded for both experimental groups studied, in the case of nighttime milk (30.7 ± 1.79 pg/mL for the group kept in dark conditions, and 17.81 ± 0.33 pg/mL for the group kept in light conditions), compared to the melatonin values detected in the daytime milk (5.36 ± 0.33 pg/mL for the group kept in dark conditions, and 3.3 ± 0.18 pg/mL for the group kept in light conditions).

The common comparison made on the research conducted by Asher et al. [103] was established between the results of the melatonin content determined in the daytime milk and in the nighttime milk, which was collected from the two groups/lots of animals studied. Thus, through the data presented in Table 2, it is observed that the melatonin level determined in the milk of cows was higher in the case of animals kept in the dark during the rest period, in the case of both types of milk (day and night). In the case of milk collected from cattle in the dark group, the melatonin level was 5.36 ± 0.33 pg/mL in daytime milk, and 30.7 ± 1.79 pg/mL in nighttime milk, compared to the melatonin level determined in milk collected from animals kept under light conditions (3.3 ± 0.18 pg/mL of melatonin determined in daytime milk, and 17.81 ± 0.33 pg/mL of melatonin determined in nighttime milk). These results demonstrate the nocturnal nature of melatonin synthesis and secretion, as well as the influence of artificial light as an inhibitory factor in the pineal gland action process. Due to the fact that differences were also recorded between the melatonin content in the daytime milk collected from the two groups of animals studied (5.36 ± 0.33 pg/mL of melatonin determined in the daytime milk of the animals in the group kept in dark conditions, and 3.3 ± 0.18 pg/mL of melatonin determined in the daytime milk of the animals in the group kept in light conditions), it can be deduced that the exposure of cattle to lighting conditions during the night (which should be intended for rest through sleep), can inhibit the synthesis and secretion of melatonin during the day, which will lead to circadian rhythm disorders, decreased sleep quality, and possible occurrence of health conditions in the animal body.

Data found in specialized literature indicate that the levels of melatonin that can be found in milk show a very high degree of variation, which can be explained by the management of the farms where the studied animals were raised and maintained.

#### 4.1.3. Animals' Productivity

Many authors have reported in their studies that they have recorded a much higher melatonin content in milk collected at night, and during winter periods. This phenomenon can be explained by the longer duration of winter nights, compared to other seasons, a factor that reduces the duration of exposure of animals to natural light conditions. At the same time, the shorter duration of winter days is also correlated with the lower milk production that animals have during the cold season. Thus, by extending the duration of the animals' exposure to darkness, a greater amount of melatonin is obtained in milk, and by obtaining a small amount of milk, the ratio of melatonin dissolved in the total volume of liquid increases. For these reasons, the amount of milk that the animal produces in a day is another factor that must be taken into account when it comes to obtaining milk rich in melatonin, because in the case of a high milk production, the amount of melatonin secreted will be diluted in a larger volume of liquid, compared to the situation in which a smaller amount of milk is obtained, in which case the melatonin content will be diluted in a smaller volume of liquid [102].

#### 4.1.4. Frequency of Milking

The frequency of milking is another important factor that influences the hormonal level of melatonin secreted in cow's milk. Some studies have shown that milk milked in the morning (recommended at 04:30) contains the highest amount of melatonin, and other authors have reported that the highest melatonin secretion in the animal organism reaches its maximum values at 00:00 [103]. Thus, by performing a milking process in the morning, a milk rich in melatonin can be obtained, as a result of the hormonal accumulation of N-acetyl-5-methoxytryptamine (melatonin) in the mammary gland fluid throughout the whole night.

From the point of view of productivity, the number of milkings performed in a day must also be taken into account. Under optimal management conditions of a dairy farm, two milkings per day are usually applied, although there are also situations in which this number can vary in an increasing or decreasing direction. Thus, some authors have reported that performing a single milking per day can reduce part of the farm's operating expenses, but it also reduces milk production in terms of quantity [106], and by increasing the number of milkings, from two to four milkings per day, milk production can be increased by modifying the expression of genes in the mammary gland [107], but it still remains debatable whether the milk loses its quality or not if a greater number of milkings per day is performed.

Harvesting milk twice a day (once during the day before sunset, and once in the early morning hours), and storing the harvested milk in different storage tanks, is a practice that can facilitate the obtaining and marketing of milk with a high content of melatonin that has been eliminated in the milk naturally.

#### 4.1.5. Lighting Conditions

Organizing a nighttime milk collection program is a process that would facilitate obtaining milk with a high melatonin content, but the involvement of the stress factor of animals subjected to numerous sleep interruption processes, as a result of the need to collect milk, must also be taken into account.

An inhibitory factor of the synthesis and secretion process of melatonin is represented by the intensity of light in the milk collection space. In order to carry out the milking stage, the existence of light in the collection room is necessary in order to facilitate the milking process in good conditions. This aspect could depreciate the quality of the next volume of milk collected, as a result of the inhibition of melatonin secretion, inhibition resulting from the penetration of light into the retina of the animals. For this reason, both the type of light used and the intensity of artificial light in the resting areas of the cows and in the milk collection spaces must be taken into account.

Numerous studies have demonstrated that the highest amount of melatonin in cow and sheep milk is secreted at night, in dark conditions, when the intensity of natural light is very low [5,108].

#### 4.1.6. Type and Intensity of Artificial Lighting

The intensity of artificial light that enters the retina is a very important factor for stimulating or inhibiting the synthesis and secretion of melatonin in the animal body. The high intensity of natural light throughout the day inhibits the secretion of melatonin and stimulates the secretion of serotonin, and the absence of light at night stimulates the secretion of melatonin and the release of this hormone into the blood and subsequently into the milk. Some studies have shown that the highest amount of melatonin can be obtained from cow's milk during the night, if the influence of the light intensity in the animal's shelters is suppressed to a minimum. At the same time, the duration of photoperiods directly influences the natural secretion of melatonin in milk, so that the highest quality milk, in terms of melatonin content, can be obtained from cows during the winter, as a result of the animals being exposed to darkness for a longer period of time. This longer exposure of animals to darkness is due to longer winter nights compared to the rest of the seasons, and as a result of obtaining a smaller amount of milk, which leads to the dilution of the melatonin content in a smaller volume of liquid [5,102].

Studies conducted to determine the impact of artificial light in cow sheds on the quality of raw milk have shown that exposing animals for a longer period of time to artificial light can increase milk production, but decreases the melatonin content in the body and in the milk. Thus, an effective method to stimulate the synthesis and secretion of melatonin in the body of cows, which will subsequently lead to the release of this hormone in the milk of cattle, is represented by increasing the photoperiod of darkness, and by using low-intensity light sources in animal sheds [5,108,109].

From a qualitative point of view, it has been demonstrated that the use of artificial light in cattle housing does not influence the chemical composition of milk in terms of dry matter, protein, fat, and lactose [104,108], but excessive use of artificial light in animal housing inhibits melatonin secretion and implicitly the release of this hormone in milk [5]. Due to the fact that the implementation of a nighttime milk collection program also requires the use of artificial light sources, it is recommended to use a certain type of artificial light and with a certain light intensity, which does not greatly inhibit melatonin synthesis and secretion. Research has been carried out and focused on studying the influence of the intensity and color of LED light on the melatonin content found in the milk of cows that were exposed during rest periods to artificial lighting conditions, with the application of different types of lights of different intensities and colors. It seems that LED light is the most useful artificial source of light propagation when it comes to the effects it has on the animal body, because LEDs have the ability to disperse light evenly over the entire area of action and to imitate natural light. The type and intensity of artificial light used in animal resting areas are two key factors in stimulating or inhibiting melatonin secretion, with the use of blue artificial light being found to produce milk with lower melatonin content than red or yellow light [110,111]. Other studies have also shown that keeping animals in natural darkness throughout the resting period stimulates melatonin secretion and reduces the number of somatic cells in milk, and by reducing the number of somatic cells, qualitative milk is obtained, the risk of mastitis is reduced, the health of the animals is improved, and stress in the cows is reduced by inducing a state of well-being [5].

Some studies have shown that LED lights with different color shades and wavelengths can inhibit to a greater or lesser extent the synthesis and secretion of melatonin in the animal body. The use of short wavelength (465–485 nm) blue light applied over a long period of time can quantitatively increase milk production but inhibits melatonin synthesis and secretion [110,111].

According to research conducted up to 2024, it has been demonstrated that in order to inhibit melatonin secretion from cow's milk to the baseline melatonin values recorded in daytime milk, it is necessary for cattle to be kept in shelters with artificial white light and a light intensity of at least 400 lx applied to both eyes of the animals, and in the case of using blue light, but applied to only one eye, to inhibit melatonin secretion, it is necessary to apply a light intensity of only 225 lx [109,110]. However, more concrete studies are needed in which blue light penetrates both eyes of the animals to determine how different shades of artificial light, applied under the same conditions, influence the level of melatonin secreted in the animal's body and subsequently released into the milk.

### 4.2. Nutritional Factors

Nutritional factors directly influence the chemical composition of milk secreted by the mammary gland of cows, through the intake of feed nutrients assimilated by the animal body. Any nutritional deficiency in the feed rations administered to animals has direct negative effects on the health of the animal body and on the quality of the production obtained, as a result of the non-assimilation of some chemical components with an energetic and bioactive role, necessary for the proper functioning of physiological and production processes [112].

Nutritional factors that can influence the melatonin content found in cow's milk include animal nutrition and feeding. Some studies have highlighted the fact that one of the sources of melatonin procurement by the body is represented by each individual's own diet [4], so that, also related to the process of synthesis and natural secretion of melatonin in the vertebrate body, the level of circulating melatonin can be increased, which can subsequently be eliminated in cow's milk by administering feeds rich in melatonin and/or tryptophan, as well as by administering rumen-protected tryptophan in the animal's rations.

#### 4.2.1. Feeding Melatonin Rich Feedstuffs

Some studies have reported the existence of melatonin in plant organisms [113], thus there is the possibility that by administering feeds containing plants with a high level of melatonin, a higher quality milk can be obtained, in terms of the content in some bioactive molecules, brought into the animal body through the diet. Melatonin from plants has been identified in the highest quantities in roots, stems, flowers, and leaves [114,115].

An important aspect that must be taken into account in the process of designing feed recipes is represented by the fact that animal rations must be calculated in such a way that they can cover the daily nutrient requirements that ensure the growth and development of the body in optimal conditions [112].

In mammals, it has been reported that the gastrointestinal tract contains higher levels of melatonin than the pineal gland, with melatonin in the rumen originating from the food consumed by the animals, from ruminal microorganisms, and from the ruminal wall [116].

According to a study conducted in the zootechnical field, regarding the influence of feed rations administered to cows on the hormonal secretion of melatonin in milk, it was found that supplementing a ration with ruminally protected B complex vitamins (Dpantothenic acid, pyridoxine, biotin, folic acid, cyanocobalamin) and ruminally unprotected vitamins A, D3, E, and B3, has no significant influence on the melatonin content in daytime milk, but negatively influences melatonin secretion in nighttime milk, with a melatonin content approximately 40.55% lower in nighttime milk collected from cows that consumed vitamin-supplemented feed, compared to the optimal melatonin values determined in the milk of cows that did not consume a vitamin supplement [117].

Table 3 presents the melatonin contents determined in various raw materials of plant origin, materials that can represent a feed base in the design of rations for dairy cows.

**Table 3.** Melatonin content in different cereals.

| Feedstuff | Melatonin Content | Samples | Assessment Method | References |
|---|---|---|---|---|
| Corn (whole, yellow) | 1.3 ± 0.28 ng/g | 5 | HPLC ** | [29] |
| Corn, germs floor | 1.0 ± 0.1 ng/g | | | |
| Wheat (*Triticum aestivum* L.) | 124.7 ± 14.9 ng/g FW * | 3 | HPLC-ECD *** | [118] |
| Barley (*Hordeum vulgare* L.) | 82.3 ± 6.0 ng/g FW * | | | |
| Oat (*Avena sativa* L.) | 90.6 ± 7.7 ng/g FW * | | | |

All values are expressed as mean ± SD; * FW—fresh weight; ** HPLC—high precision liquid chromatography; *** HPLC-ECD—high-precision liquid chromatography with electrochemical detection.

#### 4.2.2. Feeding Ruminally Protected L-Tryptophan

Tryptophan, with the chemical formula $C_{11}H_{12}N_2O_2$, and a molar mass of 204.22 g/mol, is an essential amino acid, introduced into the animal and human body through the diet. Tryptophan, along with phenylalanine and tyrosine, are amino acids that contain at least one six-membered benzene ring in their side chain [119].

Tryptophan is an amino acid that must be supplemented in the body through food, being the first precursor in the synthesis and secretion process of melatonin. From a biochemical point of view, the synthesis and secretion of melatonin begins with the transformation of the essential amino acid tryptophan into 5-hydroxytryptophan [38].

Supplementing dairy cow rations with different feeds containing high levels of tryptophan may be a useful method for stimulating melatonin secretion and synthesis. Supplementing feed rations with a higher intake of tryptophan in the animal body, which is absorbed and transported to the brain, can lead to the synthesis and secretion of a greater amount of serotonin, which subsequently, by ensuring optimal dark conditions, is transformed into melatonin, thus providing the body with much higher circulating melatonin hormonal amounts, compared to the situation in which the animals would not benefit from a tryptophan supplementation of the rations.

Various plant products rich in tryptophan (Table 4), which can be administered in the rations of cows intended for milk production, are represented by soybeans, soybean cake, alfalfa hay, oat hay, and wheat bran.

**Table 4.** Protein and tryptophan content of various feedstuffs.

| Feedstuff | Protein Content (g per 100 g) | Tryptophan (g per 100 g) | References |
|---|---|---|---|
| Soybean | 36.49 | 0.59 | [120] |
| Soybean meal | 47.46 | 0.53 | [121] |
| Alfalfa hay | 19.61 | 0.24 | |
| Oat hay | 8.88 | 0.08 | |
| Wheat bran | 20.15 | 0.26 | |

Some studies that had as their main purpose the stimulation of the hormonal secretion of melatonin in cow's milk, or the increase of the protein content of milk, by the administration of L-tryptophan, have highlighted that the supplementation of feed rations with the amino acid L-tryptophan can influence the melatonin content in the body only under certain conditions [105,122,123].

Studies conducted up to the year 2024 have highlighted that the administration of ruminally protected L-tryptophan in different amounts (20, 30, 50, 100, and 125 g) had different positive effects on animal productivity. The administration of 20 g of ruminally protected L-tryptophan had the effect of decreasing the amount of food consumed by each animal, increasing milk production, decreasing the plasma cortisol concentration by reducing the thermal stress to which the animals are subjected, and increasing the amount of melatonin in milk [122].

Supplementation of Holstein cows with 30 g of rumen-protected L-tryptophan increased milk production, altered the ratio of basic milk chemical composition (increased dry matter and reduced water content), and increased milk protein content [105]. Liu et al. [123] observed that supplementation of the rations with 50 g and 100 g of rumen-protected L-tryptophan, respectively, did not appear to influence the melatonin and tryptophan content of the milk or the tryptophan content of the blood of Holstein cows, but did appear to increase the circulating melatonin level in the blood of the animals. In the study conducted by Liu et al. (2024) [123], it was reported that supplementing the diet with rumen-protected L-tryptophan in Holstein cows during the preparation period had positive effects on reproductive performance and postpartum lactation, as a result of the increased serum concentrations of FSH with 100 g of rumen-protected L-tryptophan supplementation, and the increased serum LH content with 50 g of rumen-protected L-tryptophan supplementation, compared to the control group. FSH (follicle-stimulating hormone) and LH (luteinizing hormone) are two hormones involved in the regulation of reproductive functions in both males and females [123].

A comparative study conducted in Germany on a herd of 12 non-pregnant Brown Swiss heifers weighing 536 $\pm$ 13 kg and aged 22 $\pm$ 3 months and on a herd of 12 adult cows (also Brown Swiss) aimed to determine the effects of rumen-protected tryptophan supplementation at a dose of 125 g/day on plasma tryptophan content and hormonal levels in the heifers. The study showed that plasma levels of tryptophan (in heifers and cows) and melatonin (in heifers only) increased in response to dietary tryptophan supplementation [124]. An increase in the content of melatonin in the blood of cows was also observed in a study conducted in Viçosa, Brazil on an experimental batch of Holstein cows whose diet was supplemented with tryptophan [89].

The correlation of technological and nutritional factors that can influence the content of melatonin in the milk secreted by the mammary gland of cows is most likely a main reason why different authors have recorded different levels of melatonin in the milk of cows located in different geographical regions, both in daytime milk (between 2.912–103.7 pg/mL) and in nighttime milk (between 11.314–163.13 pg/mL), according to the results presented in Table 2 [99,101].

The management of livestock farms is the main starting point that determines the quality of the productions obtained, so that poor management of livestock units will have a negative impact on the welfare of the animals, as well as the quantity and quality of the resulting productions. Melatonin is a molecule that has been detected in cow's milk, in a very wide range of variation, mainly due to the lack of correlation of factors that can influence the synthesis and secretion of this hormone. The studies available in the specialized literature up to 2024, which focused both on determining the melatonin content in cow's milk and on the factors that can influence the quantitative variations of this hormone present in the mammary gland fluid, were carried out under particular and different conditions, so that we identified the following two situations: (1) different researchers collected milk samples from farms only to determine the melatonin content of that milk, without intervening in the management system of the livestock units from which they took the milk samples, and (2) other authors applied certain conditions for raising and maintaining the animals, to determine the degree of inhibition of melatonin synthesis and secretion, related to the evaluation of one or more factors, a situation in which it was not taken into account whether the hormonal secretion of melatonin can be stimulated or inhibited by correlating technological and nutritional factors. Thus, conducting a larger number of researches focused on the interrelationships that may arise between the application of different technological and nutritional factors on the processes of melatonin synthesis and secretion in the cow's body is a necessity for a deeper understanding of this subject.

## 5. Sources of Melatonin in Human Nutrition

Obtaining milk with a high content of naturally secreted melatonin is a complex process, influenced by a series of technological and nutritional factors, the correct management of which can lead to the desired result. However, in the milk processing process, in order to obtain derived food products with a high content of melatonin, two important factors intervene that must be taken into account and on which more studies must be carried out. These two factors are represented by the technological parameters applied during the raw milk processing process, and by the conditions for the development and obtaining of certain categories of dairy food products.

From a technological point of view, the possible influences that the work processes performed and the technological parameters used to obtain finished dairy products may have should be studied. Among these processes, greater attention should be paid to the stages of milk heat treatment (pasteurization and sterilization), homogenization, standardization/normalization, defatting, and concentration.

Heat treatment is a critical point in the technological flow of milk processing, because any variation in the technological parameters of pasteurization/sterilization can have irreversible repercussions on the finished product (for example, the appearance of the effect of over-pasteurization of milk). The impact that the temperature–time–thermal shock ratio can have on the melatonin concentration in the finished product should be studied. At the same time, the influences that the pressures used during the homogenization process, the milk centrifugation processes with the aim of separating a certain amount of fat from the liquid, and the milk filtration concentration processes can have on the melatonin content found in the obtained dairy products should also be studied.

The type of dairy product is another factor that should be taken into account when it is desired to process a milk rich in melatonin, in order to obtain finished dairy products with a similar melatonin content, compared to the melatonin content found in the raw material. Thus, greater attention should be paid to the influence that the acidic environment of fermented dairy products (yogurt, kefir, sana, acidophilus milk, buttermilk) can have on the melatonin content found in finished products.

Studies conducted up to 2024 on the amount of melatonin determined in different finished dairy products are few, which is why it would be useful to pay greater attention to this area. Table 5 presents the data available from specialized literature regarding the melatonin content found in some finished dairy products.

**Table 5.** Melatonin content of different dairy products.

| Product Name | Melatonin Content | Number of Samples Performed | Assessment Method | References |
|---|---|---|---|---|
| | | Fresh/processed milk | | |
| Whole cow milk | 14.45 ± 0.12 pg/mL$^{-1}$ | 6 | LC-MS/MS * | [125] |
| Skimmed cow milk | 18.41 ± 0.62 pg/mL$^{-1}$ | | | |
| UHT milk | 4.16 pg/mL | 16 | ELISA (RE54041; IBL) ** | [102] |
| | | Other dairy products | | |
| Colostrum, fresh | 0.06 ng/g | 5 | HPLC *** | [29] |
| Colostrum, powder | 0.6 ± 0.06 ng/g | | | |
| Yogurt | 0.13 ± 0.01 ng/g | 5 | LC-MS/MS * | [29] |
| Probiotic yogurt | 126.7 ± 9 pg/mL | 5 | LC-MS/MS * | [126] |
| Kefir | Not detected | | | |

All values are expressed as mean ± SD; * LC-MS/MS—Liquid chromatography with tandem mass spectrometry; ** ELISA (RE54041; IBL International, Hamburg, Germania); *** HPLC—high-precision liquid chromatography.

The lack of melatonin content in kefir [126] may be due to one of the following two causes (or even both of these related situations):

1. The use in the kefir production process of raw milk with a very low, almost non-existent, melatonin hormonal content, which would explain the absence of melatonin in the finished product;
2. The type of double fermentation (lactic + alcoholic) characteristic to kefir could be a factor in the depreciation of the melatonin content in the finished product.

A study conducted on determining the difference between the melatonin content in milk collected from temporary storage tanks, milk collected from cows individually, and milk heat-treated by UHT pasteurization process highlighted the fact that no major differences were recorded between the melatonin content determined in milk collected from storage tanks of different farms and milk heat-treated by UHT process. The study was conducted in Brazil (Castro), and the milk samples were collected from the same geographical area, as follows: milk samples collected from tanks were collected from 16 temporary raw material storage tanks, individual milk samples were collected from 30 Holstein cows, and UHT milk samples were collected commercially from 12 brands from different manufacturers. According to the results presented by the authors, there was no significant difference between milk collected from tanks and milk processed by the UHT heat treatment technique. Thus, for milk collected individually from Holstein cows, a melatonin content of 5.24 pg/mL was determined, for milk collected from raw material storage tanks, the melatonin content recorded was 4.08 pg/mL, and for UHT milk, a melatonin content of 4.16 pg/mL was recorded [102].

These results may suggest that the heat treatment by UHT pasteurization process does not negatively influence the melatonin content of milk. Romanini and his collaborators [102] collected milk from temporary storage tanks and from 30 Holstein cows in two different seasons, summer and winter. This aspect suggests that the collection of milk over a longer period of time, and from the same geographical area, correlated with the milk samples heat-treated by UHT process and purchased from the market, could strengthen the idea that the high temperatures of UHT pasteurization and the thermal shock to which the milk was subjected during sudden cooling, would not significantly influence the melatonin content of the finished product. However, it should also be taken into account that the samples analyzed did not follow a tradability flow, since the UHT milk samples collected from the market were not part of the same batch of milk with the samples collected from the tanks, or with the samples collected individually from each animal. Thus, there is a probability that to obtain the UHT milk analyzed in that study, a raw material milk with a higher melatonin content was used, and the heat treatment process negatively influenced the hormonal melatonin content of the finished product. This statement is also supported by the higher melatonin content found by the authors in the milk collected individually from Holstein cows (5.24 pg/mL), compared to the melatonin level found in the milk collected from the tanks (4.08 pg/mL). To explain this phenomenon, the authors proposed the idea that in the case of individually collected milk, there was a much better and more efficient farm management regarding the conditions of melatonin secretion in cow's milk, compared to the farms from which the samples were collected from the tanks.

Due to the fact that the respective research was not focused on studying the melatonin content in a volume of milk that would follow an optimal transability of the production flow (from farm to factory, and from factory to trade), we propose to carry out more precise research in which the level of melatonin found in raw milk, entirely obtained in livestock farms, would be compared with the finished milk/dairy products obtained from milk received from livestock farms and subjected to various technological processes (homogenization, centrifugal separation, concentration, thermal tartarization, seeding using production cultures, etc.). In this way, more precise results can be obtained that provide more concrete and realistic information.

Obtaining natural milk with a high melatonin content by managing technological and nutritional factors that can influence the process of melatonin synthesis and secretion in the animal body is a research topic the results of which would bring multiple benefits to farmers (by increasing the welfare of farm animals and by obtaining higher quality milk, with the same production costs, but with a higher selling price, compared to raw, whole milk that does not contain a surplus of bioactive molecules naturally existing in the liquid volume), processors (by marketing higher quality products at a better price), and end consumers (through the existence of food products with beneficial effects on the body).

Milk and dairy products are important sources of melatonin for the human body; however, the presence of N-acetyl-5-methoxytryptamine has been reported in varying amounts in other raw materials and finished food products of both plant and animal origin [29]. In cereals, large amounts of melatonin have been reported in black rice (182.04 ± 1.62 ng/g dry weight) and in red rice (212.01 ± 1.37 ng/g dry weight); in fruits, large amounts of melatonin have been determined in strawberries *Fragaria ananassa* L. cv. Festival (11.26 ± 0.13 ng/g fresh weight) [29]; and in vegetables, melatonin has been found in large amounts in tomatoes *Lycopersicon esculentum* cv. Gordala (17.1 ± 1.21 ng/g fresh weight) and *Lycopersicon esculentum* cv. Marbonea (18.13 ± 2.24 ng/g fresh weight) [127]. Melatonin is present in varying amounts in animal-origin products, such as lamb meat (1.6 ± 0.14 ng/g), beef (2.1 ± 0.13 ng/g), pork (2.5 ± 0.18 ng/g), chicken meat and skin (2.3 ± 0.23 ng/g), fish meat from salmon (3.7 ± 0.21 ng/g), and in whole eggs, where melatonin is found in a concentration of 1.54 ng/g [29].

The difficulties related to insomnia and the decrease in the amounts of melatonin naturally synthesized in the human body due to aging make the presence of melatonin in plant and animal-origin raw materials a useful technique through which the human body can benefit from exogenous melatonin derived from natural sources. The multiple roles and bioactive functions that melatonin performs in the human body, the evidence reported through clinical studies demonstrating that the presence of higher levels of melatonin in the human body does not appear to have negative effects on human health, and the fact that, up until 2024, no cases have been reported where the consumption of melatonin-rich foods has endangered the health of consumers, make N-acetyl-5-methoxytryptamine an ideal candidate for maintaining human health and improving the performance of physiological processes in the human body.

## 6. The Role of Melatonin in Relation to Other Bioactive Molecules on the Animal and Human Body

Numerous studies have demonstrated that melatonin as a hormone secreted by the pineal and extrapineal pathways, as well as melatonin originating from the body's microflora or brought through the diet of each individual, interacts with a series of other molecules with a biologically active role, to ensure the achievement of various functions and roles in the animal and human body. Studies conducted in this field have reported that melatonin can interact in the animal and human body with serotonin [128], with melatonin receptors MT1, MT2, and MT3 [42,129,130], with lactoferrin [131–137], with vitamins B3, B6, and B12 [138,139], with the stress hormone cortisol [5,140], and with insulin [141–145], as shown in Figure 3.

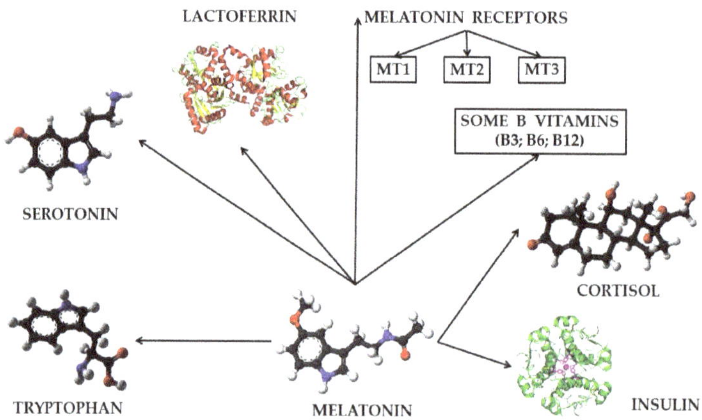


**Figure 3.** Bioactive molecules with which melatonin can interact in the animal and human body. Original processing content. 3D models of Melatonin, Tryptophan, Serotonin, Lactoferrin, Cortisol, and Insulin molecules are taken from online and copyrighted with permission to use them. Graphics: Tryptophan, Serotonin, Lactoferrin, and Cortisol—Copyright © "Creative Commons CC0 1.0 Universal Public Domain Dedication". Graphics: Melatonin—All rights reserved by the Free Software Foundation under the "GNU Free Documentation License". Graphics: Insulin—Copyright © "Creative Commons (CC) Attribution 2.5 Generic".

Serotonin, also known as 5-hydroxytryptamine or 5-HT, acts as a neurotransmitter and a peripheral hormone. Serotonin synthesis is carried out in two steps from the essential amino acid tryptophan. In the first step, tryptophan hydroxylase (TPH) hydrolyzes tryptophan to produce 5-hydroxytryptophan, and in the second step of serotonin synthesis, decarboxylation of L-aromatic amino acids and conversion to 5-hydroxytryptamine are carried out [128]. Any variation in tryptophan and/or serotonin in the body will directly influence the melatonin content in the body and in the milk of mammals.

The MT1 (MTNR1A) and MT2 (MTNR1B) receptors are two membrane receptors for melatonin. These two receptors belong to the G protein-coupled receptor superfamily [119]. Activation of MT1 or MT2 receptors by melatonin leads to inhibition of PKA (kinase A) activity, since melatonin activation of these two receptors decreases the amount of cyclic adenosine monophosphate (cAMP) [42]. In terms of interaction with melatonin, the MT3 receptor shows a lower affinity for this hormone [130].

Lactoferrin (Lf) is a multifunctional glycoprotein, which belongs to the transferrin family and plays a role in iron binding in the animal and human body [131].

Milk proteins are of two types: casein (the main protein in milk, accounting for about 80% of total milk proteins) and serum proteins (accounting for about 20%), which are proteins that pass into whey and buttermilk after milk processing to obtain certain categories of dairy products [132].

Lactoferrin, along with alpha-lactalbumin, beta-lactoglobulin, immunoglobulins, bovine serum albumin, glycomacropeptides, lactoperoxidase, and lysozyme, are part of the serum protein category [133].

Numerous studies have highlighted the role of lactoferrin as an antioxidant, antibacterial, and antiviral factor [134], antimicrobial and anticancer [135], and antiparasitic and antifungal [133], as well as a role in maintaining intestinal health [136].

Lactoferrin performs an antibacterial role in the body (being also the first discovered function of this protein) through two different mechanisms. A first mechanism for achieving the antibacterial function of lactoferrin involves sequestering free iron, thus depriving bacteria of an essential substrate necessary for the growth and development of these microorganisms. The second mechanism for fulfilling the antibacterial role involves the binding of lactoferrin to the lipopolysaccharide that enters the structure of the bacterial cell walls, thus degrading the bacteria by forming peroxides catalyzed by iron (III) ions bound to lactoferrin. Thus, the permeability of the bacterial membrane is affected, resulting in bacterial cell lysis [137].

Bovine lactoferrin (bLF) has the ability to control the production of reactive oxygen species (ROS) and the rate of their elimination by sequestering iron [131].

Pyridoxine (vitamin B6) acts as a coenzyme in the synthesis of melatonin, and any deficiency in this vitamin will inevitably lead to sleep disorders [138,139], and therefore to various conditions with a negative impact, caused by the lack of melatonin in the body.

Some studies have reported that vitamin B3 (niacin) can have a tryptophan-sparing effect and that vitamin B12 (cobalamin) contributes directly to the secretion process of melatonin [139].

Cortisol is a glucocorticoid hormone that is produced by the adrenal glands, and the release of this hormone in the body follows a circadian rhythm regulated by the internal clock located in the suprachiasmatic nucleus [140]. Melatonin can regulate the secretion of certain hormones, in particular inhibiting the release of corticotropin (CRH) from the hypothalamus. By inhibiting CRH secretion by melatonin, it results in decreased levels of adrenocorticotropic hormone (ACTH) and cortisol during the night [5].

Insulin is a polypeptide hormone, composed of 51 amino acids, and secreted mainly by β cells located in the islets of Langerhans of the pancreas. The main role of this hormone in the body is to modulate blood glucose levels, also having a role in glucose homeostasis, metabolism and cell growth [141]. Some clinical studies have highlighted the role that melatonin MT1 and MT2 receptors have on the process of insulin secretion [142].

Research carried out in the field of knowledge of the interactions that can be achieved between melatonin and insulin has demonstrated that there is a direct and inversely proportional relationship between the amount of melatonin synthesized in the body, and the processes of inhibition or stimulation (as appropriate) of insulin synthesis and secretion. Thus, it has been demonstrated that a large amount of melatonin secreted in the body can inhibit insulin secretion, through melatonin receptors located on pancreatic β cells. Melatonin receptors are coupled to three different signaling pathways (cAMP, cGMP, and IP3) and have their own unique and different influences on insulin secretion [143–145]. Binding of melatonin to the MT1 receptor can lead to inhibition of insulin secretion by decreasing cAMP. This phenomenon occurs due to MT1 receptors binding to Gi (inhibitory

G) proteins, which in turn leads to decreasing cAMP (cyclic adenosine monophosphate) and inhibiting PKA activity [145].

Melatonin may also influence the cGMP (cyclic guanosine monophosphate) signaling pathway by stimulating vasodilator signals via nitric oxide (NO). The cGMP signaling pathway may promote insulin secretion, as it may increase pancreatic blood flow by activating protein kinase (which is dependent on the cGMP signaling pathway). Thus, NO may activate guanylate cyclase, thereby increasing cGMP levels and stimulating insulin secretion. The IP3 (inositol triphosphate) pathway plays a role in regulating intracellular calcium concentration, and can be influenced by the presence of melatonin that binds to MT1 and MT2 receptors, receptors that can activate Gq-type G proteins. Thus, when the two melatonin receptors (MT1 and MT2) are activated, they allow the Gq protein to which they are bound to activate PLC (phospholipase C). Subsequently, PLC catalyzes the decomposition of PIP2 (phosphatidylinositol 4,5-bisphosphate) into two other secondary messengers (IP3 and diacylglycerol). Finally, the IP3 pathway stimulates the release of $Ca^{2+}$ ions from the endoplasmic reticulum into the cytoplasm, thus favoring insulin secretion, due to the increase in intracellular calcium levels [145].

Therefore, melatonin can play a role in modulating insulin synthesis and secretion, through the MT1 and MT2 receptors to which melatonin binds. The presence of melatonin in the vertebrate body, and its interaction (of melatonin) with the MT1 and MT2 receptors can modulate insulin synthesis and secretion, through stimulation or inhibition, due to the three signaling pathways (cAMP, cGMP, and IP3) through which melatonin works.

## 7. Conclusions

The multiple roles and functions that melatonin performs in the animal and human body, both individually and in relation to other bioactive molecules, make this hormone a valuable compound that actively participates in improving living conditions by maintaining optimal health of the body.

The proper management of technological and nutritional factors, which can positively influence the melatonin content in cow's milk, has major implications for the implementation of high-quality agricultural and zootechnical practices, due to a higher level of farm animal welfare. This level is achieved by regulating the circadian rhythm, improving sleep quality, and regulating basic physiological processes such as reproduction.

This paper contains valuable information regarding the synthesis and secretion processes of melatonin in vertebrate organisms, its multiple roles and functions in both animal and human organisms, and the main factors that can directly influence melatonin synthesis in the bovine organism, as well as the subsequent release of this hormone into the milk of cows. Specialized studies conducted in this field up until 2024 have provided valuable information, resulting in a deeper understanding of melatonin as a bioactive molecule. However, it is necessary to conduct more experiments focused on studying the impact of cow maintenance conditions on the melatonin content in milk, and some future research should also address the influence of milk processing technological processes on the melatonin content found in finished dairy products.

**Author Contributions:** Conceptualization, V.-C.A. and D.S.; methodology, C.S. and V.-C.A.; software, M.M. and R.-M.R.-R.; writing—original draft preparation, V.-C.A., C.S. and R.-M.R.-R.; writing—review and editing, V.-C.A., M.M. and R.-M.R.-R.; visualization, V.-C.A. and C.S.; supervision, D.S. All authors have read and agreed to the published version of the manuscript.

**Funding:** This research received no external funding.

**Conflicts of Interest:** The authors declare no conflicts of interest.

## References


1. Reiter, R.J.; Sharma, R.; Tan, D.-X.; Chuffa, L.G.d.A.; da Silva, D.G.H.; Slominski, A.T.; Steinbrink, K.; Kleszczynski, K. Dual sources of melatonin and evidence for different primary functions. *Front. Endocrinol.* **2024**, *15*, 1414463. [CrossRef]
2. Chlubek, D.; Sikora, M. Fluoride and Pineal Gland. *Appl. Sci.* **2020**, *10*, 2885. [CrossRef]
3. Tan, D.-X.; Zheng, X.; Kong, J.; Manchester, L.C.; Hardeland, R.; Kim, S.J.; Xu, X.; Reiter, R.J. Fundamental issues related to the origin of melatonin and melatonin isomers during evolution: Relation to their biological functions. *Int. J. Mol. Sci.* **2014**, *15*, 15858–15890. [CrossRef] [PubMed]
4. Tan, D.-X.; Reiter, R.J.; Zimmerman, S.; Hardeland, R. Melatonin: Both a Messenger of Darkness and a Participant in the Cellular Actions of Non-Visible Solar Radiation of Near Infrared Light. *Biology* **2023**, *12*, 89. [CrossRef]
5. Andrani, M.; Dall'Olio, E.; De Rensis, F.; Tummaruk, P.; Saleri, R. Bioactive Peptides in Dairy Milk: Highlighting the Role of Melatonin. *Biomolecules* **2024**, *14*, 934. [CrossRef]
6. Lerner, A.B.; Case, J.D.; Takahashi, Y.; Lee, T.H.; Mori, W. Isolation of melatonin, the pineal gland factor that lightens melanocytes. *J. Am. Chem. Soc.* **1958**, *80*, 2587. [CrossRef]
7. Mannino, G.; Pernici, C.; Serio, G.; Gentile, C.; Bertea, C.M. Melatonin and Phytomelatonin: Chemistry, Biosynthesis, Metabolism, Distribution and Bioactivity in Plants and Animals—An Overview. *Int. J. Mol. Sci.* **2021**, *22*, 9996. [CrossRef]
8. Masters, A.; Pandi-Perumal, S.R.; Seixas, A.; Girardin, J.L.; McFarlane, S.I. Melatonin, the Hormone of Darkness: From Sleep Promotion to Ebola Treatment. *Brain Disord. Ther.* **2014**, *4*, 1000151. [CrossRef]
9. Srinivasan, V.; Spence, W.D.; Pandi-Perumal, S.R.; Zakharia, R.; Bhatnagar, K.P.; Brzezinski, A. Melatonin and human reproduction: Shedding light on the darkness hormone. *Gynecol. Endocrinol.* **2009**, *25*, 779–785. [CrossRef]
10. Patel, S.; Rahmani, B.; Gandhi, J.; Seyam, O.; Joshi, G.; Reid, I.; Smith, N.L.; Waltzer, W.C.; Khan, S.A. Revisiting the pineal gland: A review of calcification, masses, precocious puberty, and melatonin functions. *Int. J. Neurosci.* **2020**, *130*, 464–475. [CrossRef] [PubMed]
11. Tan, D.-X.; Manchester, L.C.; Esteban-Zubero, E.; Zhou, Z.; Reiter, R.J. Melatonin as a Potent and Inducible Endogenous Antioxidant: Synthesis and Metabolism. *Molecules* **2015**, *20*, 18886–18906. [CrossRef]
12. Manchester, L.C.; Coto-Montes, A.; Boga, J.A.; Andersen, L.P.H.; Zhou, Z.; Galano, A.; Vriend, J.; Tan, D.-X.; Reiter, R.J. Melatonin: An ancient molecule that makes oxygen metabolically tolerable. *J. Pineal Res.* **2015**, *59*, 403–419. [CrossRef] [PubMed]
13. Horodincu, L.; Solcan, C. Influence of Different Light Spectra on Melatonin Synthesis by the Pineal Gland and Influence on the Immune System in Chickens. *Animals* **2023**, *13*, 2095. [CrossRef] [PubMed]
14. Polakovičová, S.; Líška, J.; Varga, I.; Gálfiová, P. Morphology of the Human Pineal Gland Studied by Freeze-Fracturing in Scanning Electron Microscopy. *Life* **2024**, *14*, 1617. [CrossRef] [PubMed]
15. Cassone, V.M. Avian circadian organization: A chorus of clocks. *Front. Neuroendocrinol.* **2014**, *35*, 76–88. [CrossRef] [PubMed]
16. Esteban, M.Á.; Cuesta, A.; Chaves-Pozo, E.; Meseguer, J. Influence of Melatonin on the Immune System of Fish: A Review. *Int. J. Mol. Sci.* **2013**, *14*, 7979–7999. [CrossRef]
17. Bisquert, R.; Planells-Cárcel, A.; Alonso-del-Real, J.; Muñiz-Calvo, S.; Guillamón, J.M. The Role of the *PAA1* Gene on Melatonin Biosynthesis in *Saccharomyces cerevisiae*: A Search of New Arylalkylamine N-Acetyltransferases. *Microorganisms* **2023**, *11*, 1115. [CrossRef] [PubMed]
18. Zisapel, N. New perspectives on the role of melatonin in human sleep, circadian rhythms and their regulation. *Br. J. Pharmacol.* **2018**, *175*, 3190–3199. [CrossRef] [PubMed]
19. Li, Z.; Zhang, K.; Zhou, Y.; Zhao, J.; Wang, J.; Lu, W. Role of Melatonin in Bovine Reproductive Biotechnology (Review). *Molecules* **2023**, *28*, 4940. [CrossRef]
20. Kopustinskiene, D.M.; Bernatoniene, J. Molecular Mechanisms of Melatonin-Mediated Cell Protection and Signaling in Health and Disease. *Pharmaceutics* **2021**, *13*, 129. [CrossRef] [PubMed]
21. Mauriz, J.L.; Collado, P.S.; Veneroso, C.; Reiter, R.J.; Gonzalez-Gallego, J. A review of the molecular aspects of melatonin's anti-inflammatory actions: Recent insights and new perspectives. *J. Pineal Res.* **2013**, *54*, 1–14. [CrossRef] [PubMed]
22. Hardeland, R. Aging, Melatonin, and the Pro- and Anti-Inflammatory Networks. *Int. J. Mol. Sci.* **2019**, *20*, 1223. [CrossRef]
23. Mendes, L.; Queiroz, M.; Sena, C.M. Melatonin and Vascular Function. *Antioxidants* **2024**, *13*, 747. [CrossRef] [PubMed]
24. Asghari, M.H.; Moloudizargari, M.; Ghobadi, E.; Fallah, M.; Abdollahi, M. Melatonin as a Multifunctional Anti-Cancer Molecule: Implications in Gastric Cancer. *Life Sci.* **2017**, *185*, 38–45. [CrossRef]
25. Di Bella, G.; Mascia, F.; Gualano, L.; Di Bella, L. Melatonin Anticancer Effects: Review. *Int. J. Mol. Sci.* **2013**, *14*, 2410–2430. [CrossRef]
26. Fernández, A.; Ordóñez, R.; Reiter, R.J.; González-Gallego, J.; Mauriz, J.L. Melatonin and endoplasmic reticulum stress: Relation to autophagy and apoptosis. *J. Pineal Res.* **2015**, *59*, 292–307. [CrossRef]
27. Zhi, S.M.; Fang, G.X.; Xie, X.M.; Liu, L.H.; Yan, J.; Liu, D.B.; Yu, H.Y. Melatonin reduces OGD/R-induced neuron injury by regulating redox/inflammation/apoptosis signaling. *Eur. Rev. Med. Pharmacol. Sci.* **2020**, *24*, 1524–1536. [CrossRef] [PubMed]

28. Lee, J.H.; Yoon, Y.M.; Song, K.-H.; Noh, H.; Lee, S.H. Melatonin suppresses senescence-derived mitochondrial dysfunction in mesenchymal stem cells via the HSPA1L-mitophagy pathway. *Aging Cell* **2020**, *19*, e13111. [CrossRef]
29. Meng, X.; Li, Y.; Li, S.; Zhou, Y.; Gan, R.-Y.; Xu, D.-P.; Li, H.-B. Dietary Sources and Bioactivities of Melatonin. *Nutrients* **2017**, *9*, 367. [CrossRef]
30. Guaadaoui, A.; Bellaoui, M.; Elmajdoub, N.; Bellaoui, M.; Hamal, A. What is a bioactive compound? A combined definition for a preliminary consensus. *Int. J. Nutr. Food Sci.* **2014**, *3*, 174–179. [CrossRef]
31. Vilas-Boas, A.A.; Pintado, M.; Oliveira, A.L.S. Natural Bioactive Compounds from Food Waste: Toxicity and Safety Concerns. *Foods* **2021**, *10*, 1564. [CrossRef]
32. Sorrenti, V.; Burò, I.; Consoli, V.; Vanella, L. Recent Advances in Health Benefits of Bioactive Compounds from Food Wastes and By-Products: Biochemical Aspects. *Int. J. Mol. Sci.* **2023**, *24*, 2019. [CrossRef] [PubMed]
33. Acuña-Castroviejo, D.; Escames, G.; Venegas, C.; Díaz-Casado, M.E.; Lima-Cabello, E.; López, L.C.; Rosales-Corral, S.; Tan, D.X.; Reiter, R.J. Extrapineal melatonin: Sources, regulation, and potential functions. *Cell Mol. Life Sci.* **2014**, *71*, 2997–3025. [CrossRef]
34. Markus, R.P.; Sousa, K.S.; da Silveira Cruz-Machado, S.; Fernandes, P.A.; Ferreira, Z.S. Possible Role of Pineal and Extra-Pineal Melatonin in Surveillance, Immunity, and First-Line Defense. *Int. J. Mol. Sci.* **2021**, *22*, 12143. [CrossRef] [PubMed]
35. Bonnefont-Rousselot, D.; Collin, F. Melatonin: Action as antioxidant and potential applications in human disease and aging. *Toxicology* **2010**, *278*, 55–67. [CrossRef] [PubMed]
36. Lumsden, S.C.; Clarkson, A.N.; Cakmak, Y.O. Neuromodulation of the Pineal Gland via Electrical Stimulation of Its Sympathetic Innervation Pathway. *Front. Neurosci.* **2020**, *14*, 264. [CrossRef]
37. Lee, K.; Back, K. Functional Characterization of the Ciliate *Stylonychia lemnae* Serotonin N-Acetyltransferase, a Pivotal Enzyme in Melatonin Biosynthesis and Its Overexpression Leads to Peroxidizing Herbicide Tolerance in Rice. *Antioxidants* **2024**, *13*, 1177. [CrossRef] [PubMed]
38. Xie, X.; Ding, D.; Bai, D.; Zhu, Y.; Sun, W.; Sun, Y.; Zhang, D. Melatonin biosynthesis pathways in nature and its production in engineered microorganisms. *Synth. Syst. Biotechnol.* **2022**, *7*, 544–553. [CrossRef] [PubMed]
39. Hwang, O.J.; Back, K. Functional Characterization of Arylalkylamine N-Acetyltransferase, a Pivotal Gene in Antioxidant Melatonin Biosynthesis from *Chlamydomonas reinhardtii*. *Antioxidants* **2022**, *11*, 1531. [CrossRef] [PubMed]
40. Zhao, D.; Yu, Y.; Shen, Y.; Liu, Q.; Zhao, Z.; Sharma, R.; Reiter, R.J. Melatonin Synthesis and Function: Evolutionary History in Animals and Plants. *Front. Endocrinol.* **2019**, *10*, 249. [CrossRef] [PubMed]
41. Suofu, Y.; Li, W.; Jean-Alphonse, F.G.; Jia, J.; Khattar, N.K.; Li, J.; Baranov, S.V.; Leronni, D.; Mihalik, A.C.; He, Y.; et al. Dual role of mitochondria in producing melatonin and driving GPCR signaling to block cytochrome c release. *Proc. Natl. Acad. Sci. USA* **2017**, *114*, E7997–E8006. [CrossRef] [PubMed]
42. Wang, L.; Wang, C.; Choi, W.S. Use of Melatonin in Cancer Treatment: Where Are We? *Int. J. Mol. Sci.* **2022**, *23*, 3779. [CrossRef] [PubMed]
43. Tordjman, S.; Chokron, S.; Delorme, R.; Charrier, A.; Bellissant, E.; Jaafari, N.; Fougerou, C. Melatonin: Pharmacology, Functions and Therapeutic Benefits. *Curr. Neuropharmacol.* **2017**, *15*, 434–443. [CrossRef] [PubMed]
44. Miranda-Riestra, A.; Estrada-Reyes, R.; Torres-Sanchez, E.D.; Carreño-García, S.; Ortiz, G.G.; Benítez-King, G. Melatonin: A Neurotrophic Factor? *Molecules* **2022**, *27*, 7742. [CrossRef]
45. Slominski, A.T.; Zmijewski, M.A.; Semak, I.; Kim, T.K.; Janjetovic, Z.; Slominski, R.M.; Zmijewski, J.W. Melatonin, mitochondria, and the skin. *Cell. Mol. Life Sci.* **2017**, *74*, 3913–3925. [CrossRef] [PubMed]
46. Slominski, A.; Tobin, D.J.; Zmijewski, M.A.; Wortsman, J.; Paus, R. Melatonin in the skin: Synthesis, metabolism and functions. *Trends Endocrinol. Metab. TEM* **2008**, *19*, 17–24. [CrossRef]
47. Felder-Schmittbuhl, M.P.; Hicks, D.; Ribelayga, C.P.; Tosini, G. Melatonin in the mammalian retina: Synthesis, mechanisms of action and neuroprotection. *J. Pineal Res.* **2024**, *76*, e12951. [CrossRef] [PubMed]
48. Iuvone, P.M.; Tosini, G.; Pozdeyev, N.; Haque, R.; Klein, D.C.; Chaurasia, S.S. Circadian clocks, clock networks, arylalkylamine N-acetyltransferase, and melatonin in the retina. *Prog. Retin. Eye Res.* **2005**, *24*, 433–456. [CrossRef] [PubMed]
49. Sanchez-Hidalgo, M.; de la Lastra, C.A.; Carrascosa-Salmoral, M.P.; Naranjo, M.C.; Gomez-Corvera, A.; Caballero, B.; Guerrero, J.M. Age-related changes in melatonin synthesis in rat extrapineal tissues. *Exp. Gerontol.* **2009**, *44*, 328–334. [CrossRef] [PubMed]
50. Bocheva, G.; Bakalov, D.; Iliev, P.; Tafradjiiska-Hadjiolova, R. The Vital Role of Melatonin and Its Metabolites in the Neuroprotection and Retardation of Brain Aging. *Int. J. Mol. Sci.* **2024**, *25*, 5122. [CrossRef] [PubMed]
51. Alkozi, H.A.; Pintor, J. TRPV4 activation triggers the release of melatonin from human non-pigmented ciliary epithelial cells. *Exp. Eye Res.* **2015**, *136*, 34–37. [CrossRef] [PubMed]
52. Naranjo, M.C.; Guerrero, J.M.; Rubio, A.; Lardone, P.J.; Carrillo-Vico, A.; Carrascosa-Salmoral, M.P.; Jiménez-Jorge, S.; Arellano, M.V.; Leal-Noval, S.R.; Leal, M.; et al. Melatonin biosynthesis in the thymus of humans and rats. *Cell Mol. Life Sci.* **2007**, *64*, 781–790. [CrossRef] [PubMed]
53. Bonmatí-Carrión, M.-Á.; Tomas-Loba, A. Melatonin and Cancer: A Polyhedral Network Where the Source Matters. *Antioxidants* **2021**, *10*, 210. [CrossRef] [PubMed]

54. Reiter, R.J.; Rosales-Corral, S.; Boga, J.A.; Tan, D.-X.; Davis, J.M.; Konturek, P.C.; Konturek, S.J.; Brzozowski, T. The photoperiod, circadian regulation and chronodisruption: The requisite interplay between the suprachiasmatic nuclei and the pineal and gut melatonin. *J. Physiol. Pharmacol.* **2011**, *62*, 269–274.
55. Bonmatí-Carrión, M.-Á.; Rol, M.-A. Melatonin as a Mediator of the Gut Microbiota–Host Interaction: Implications for Health and Disease. *Antioxidants* **2024**, *13*, 34. [CrossRef] [PubMed]
56. Holtkamp, C.E.; Warmus, D.; Bonowicz, K.; Gagat, M.; Linowiecka, K.; Wolnicka-Glubisz, A.; Reiter, R.J.; Böhm, M.; Slominski, A.T.; Steinbrink, K.; et al. Ultraviolet Radiation-Induced Mitochondrial Disturbances Are Attenuated by Metabolites of Melatonin in Human Epidermal Keratinocytes. *Metabolites* **2023**, *13*, 861. [CrossRef]
57. Bocheva, G.; Slominski, R.M.; Janjetovic, Z.; Kim, T.-K.; Böhm, M.; Steinbrink, K.; Reiter, R.J.; Kleszczyński, K.; Slominski, A.T. Protective Role of Melatonin and Its Metabolites in Skin Aging. *Int. J. Mol. Sci.* **2022**, *23*, 1238. [CrossRef] [PubMed]
58. Galano, A.; Reiter, R.J. Melatonin and Its Metabolites vs Oxidative Stress: From Individual Actions to Collective Protection. *J. Pineal Res.* **2018**, *65*, e12514. [CrossRef]
59. Bae, S.M.; Jeong, J.; Jeon, H.J.; Bang, Y.R.; Yoon, I.Y. Effects of melatonin-rich milk on mild insomnia symptoms. *Sleep. Med. Res.* **2016**, *7*, 60–67. [CrossRef]
60. Brzezinski, A.; Rai, S.; Purohit, A.; Pandi-Perumal, S.R. Melatonin, Clock Genes, and Mammalian Reproduction: What Is the Link? *Int. J. Mol. Sci.* **2021**, *22*, 13240. [CrossRef] [PubMed]
61. Li, Q.; Tang, Y.; Chen, Y.; Li, B.; Wang, H.; Liu, S.; Adeniran, S.O.; Zheng, P. Melatonin Regulates the Expression of VEGF and HOXA10 in Bovine Endometrial Epithelial Cells through the SIRT1/PI3K/AKT Pathway. *Animals* **2024**, *14*, 2771. [CrossRef] [PubMed]
62. Cipolla-Neto, J.; Amaral, F.G.; Afeche, S.C.; Tan, D.X.; Reiter, R.J. Melatonin, energy metabolism, and obesity: A review. *J. Pineal Res.* **2014**, *56*, 371–381. [CrossRef] [PubMed]
63. Suzen, S.; Atayik, M.C.; Sirinzade, H.; Entezari, B.; Gurer-Orhan, H.; Cakatay, U. Melatonin and redox homeostasis. *Melatonin Res.* **2022**, *5*, 304–324. [CrossRef]
64. Rzepka-Migut, B.; Paprocka, J. Melatonin-Measurement Methods and the Factors Modifying the Results. A Systematic Review of the Literature. *Int. J. Environ. Res. Public Health* **2020**, *17*, 1916. [CrossRef] [PubMed]
65. Reiter, R.J.; Tan, D.X.; Kim, S.J.; Cruz, M.H.C. Delivery of pineal melatonin to the brain and SCN: Role of canaliculi, cerebrospinal fluid, tanycytes and Virchow–Robin perivascular spaces. *Brain Struct. Funct.* **2014**, *219*, 1873–1887. [CrossRef] [PubMed]
66. Mao, L.; Dauchy, R.T.; Blask, D.E.; Slakey, L.M.; Xiang, S.; Yuan, L.; Dauchy, E.M.; Shan, B.; Brainard, G.C.; Hanifin, J.P.; et al. Circadian gating of epithelial-to-mesenchymal transition in breast cancer cells via melatonin-regulation of GSK3β. *Mol. Endocrinol.* **2012**, *26*, 1808–1820. [CrossRef] [PubMed]
67. Gao, Y.; Zhao, S.; Zhang, Y.; Zhang, Q. Melatonin Receptors: A Key Mediator in Animal Reproduction. *Vet. Sci.* **2022**, *9*, 309. [CrossRef] [PubMed]
68. Li, D.Y.; Smith, D.G.; Hardeland, R.; Yang, M.Y.; Xu, H.L.; Zhang, L.; Yin, H.D.; Zhu, Q. Melatonin Receptor Genes in Vertebrates. *Int. J. Mol. Sci.* **2013**, *14*, 11208–11223. [CrossRef]
69. Peña-Delgado, V.; Carvajal-Serna, M.; Fondevila, M.; Martín-Cabrejas, M.A.; Aguilera, Y.; Álvarez-Rivera, G.; Abecia, J.A.; Casao, A.; Pérez-Pe, R. Improvement of the Seminal Characteristics in Rams Using Agri-Food By-Products Rich in Phytomelatonin. *Animals* **2023**, *13*, 905. [CrossRef] [PubMed]
70. Pereira, N.; Naufel, M.F.; Ribeiro, E.B.; Tufik, S.; Hachul, H. Influence of Dietary Sources of Melatonin on Sleep Quality: A Review. *J. Food Sci.* **2020**, *85*, 5–13. [CrossRef]
71. Karolczak, K.; Watala, C. Melatonin as a Reducer of Neuro- and Vasculotoxic Oxidative Stress Induced by Homocysteine. *Antioxidants* **2021**, *10*, 1178. [CrossRef] [PubMed]
72. Reiter, R.J.; Tan, D.X.; Korkmaz, A.; Rosales-Corral, S.A. Melatonin and stable circadian rhythms optimize maternal, placental and fetal physiology. *Hum. Reprod. Update* **2014**, *20*, 293–307. [CrossRef]
73. Ma, N.; Zhang, J.; Reiter, R.J.; Ma, X. Melatonin mediates mucosal immune cells, microbial metabolism, and rhythm crosstalk: A therapeutic target to reduce intestinal inflammation. *Med. Res. Rev.* **2020**, *40*, 606–632. [CrossRef] [PubMed]
74. Gombert, M.; Codoñer-Franch, P. Melatonin in Early Nutrition: Long-Term Effects on Cardiovascular System. *Int. J. Mol. Sci.* **2021**, *22*, 6809. [CrossRef] [PubMed]
75. Tamtaji, O.R.; Mirhosseini, N.; Reiter, R.J.; Azami, A.; Asemi, Z. Melatonin, a calpain inhibitor in the central nervous system: Current status and future perspectives. *J. Cell Physiol.* **2018**, *234*, 1001–1007. [CrossRef] [PubMed]
76. Tan, D.-X.; Xu, B.; Zhou, X.; Reiter, R.J. Pineal Calcification, Melatonin Production, Aging, Associated Health Consequences and Rejuvenation of the Pineal Gland. *Molecules* **2018**, *23*, 301. [CrossRef]
77. Carrillo-Vico, A.; Lardone, P.J.; Alvarez-Sánchez, N.; Rodríguez-Rodríguez, A.; Guerrero, J.M. Melatonin: Buffering the immune system. *Int. J. Mol. Sci.* **2013**, *14*, 8638–8683. [CrossRef] [PubMed]
78. Vorster, A.P.A.; van Someren, E.J.W.; Pack, A.I.; Huber, R.; Schmidt, M.H.; Bassetti, C.L.A. Sleep Health. *Clin. Transl. Neurosci.* **2024**, *8*, 8. [CrossRef]

79. Chattu, V.K.; Manzar, M.D.; Kumary, S.; Burman, D.; Spence, D.W.; Pandi-Perumal, S.R. The Global Problem of Insufficient Sleep and Its Serious Public Health Implications. *Healthcare* **2019**, *7*, E1. [CrossRef] [PubMed]
80. Sejbuk, M.; Mirończuk-Chodakowska, I.; Witkowska, A.M. Sleep Quality: A Narrative Review on Nutrition, Stimulants, and Physical Activity as Important Factors. *Nutrients* **2022**, *14*, 1912. [CrossRef] [PubMed]
81. Paditz, E. Postnatal Development of the Circadian Rhythmicity of Human Pineal Melatonin Synthesis and Secretion (Systematic Review). *Children* **2024**, *11*, 1197. [CrossRef]
82. Paditz, E. Melatonin in infants-Physiology, pathophysiology and intervention options. *Somnologie* **2024**, *28*, 103–109. [CrossRef]
83. Honorio-França, A.C.; Hara, C.C.P.; Ormonde, J.V.S.; Nunes, G.T.; França, E.L. Human colostrum melatonin exhibits a day-night variation and modulates the activity of colostral phagocytes. *J. Appl. Biomed.* **2013**, *11*, 153–162. [CrossRef]
84. Italianer, M.F.; Naninck, E.F.G.; Roelants, J.A.; van der Horst, G.T.J.; Reiss, I.K.M.; Goudoever, J.B.v.; Joosten, K.F.M.; Chaves, I.; Vermeulen, M.J. Circadian Variation in Human Milk Composition, a Systematic Review. *Nutrients* **2020**, *12*, 2328. [CrossRef]
85. Wang, X. The antiapoptotic activity of melatonin in neurodegenerative diseases. *CNS Neurosci. Ther.* **2009**, *15*, 345–357. [CrossRef] [PubMed]
86. Biggio, G.; Biggio, F.; Talani, G.; Mostallino, M.C.; Aguglia, A.; Aguglia, E.; Palagini, L. Melatonin: From Neurobiology to Treatment. *Brain Sci.* **2021**, *11*, 1121. [CrossRef]
87. Logan, R.W.; McClung, C.A. Rhythms of life: Circadian disruption and brain disorders across the lifespan. *Nat. Rev. Neurosci.* **2019**, *20*, 49–65. [CrossRef] [PubMed]
88. Mander, B.A.; Winer, J.R.; Walker, M.P. Sleep and Human Aging. *Neuron* **2017**, *94*, 19–36. [CrossRef]
89. Milagres, M.P.; Minim, V.P.R.; A Minim, L.; A Simiqueli, A.; Moraes, L.E.S.; Martino, H.S.D. Night milking adds value to cow's milk. *J. Sci. Food Agric.* **2013**, *94*, 1688–1692. [CrossRef] [PubMed]
90. Usturoi, M.G. *The Technology of Milk and Dairy Products*; Alfa Publishing House: Iasi, Romania, 2007.
91. Rațu, R.N.; Cârlescu, P.M.; Usturoi, M.G.; Lipșa, F.D.; Veleșcu, I.D.; Arsenoaia, V.N.; Florea, A.M.; Ciobanu, M.M.; Radu-Rusu, R.-M.; Postolache, A.N.; et al. Effects of Dairy Cows Management Systems on the Physicochemical and Nutritional Quality of Milk and Yogurt, in a North-Eastern Romanian Farm. *Agriculture* **2023**, *13*, 1295. [CrossRef]
92. Matei, M.; Zaharia, R.; Petrescu, S.-I.; Radu-Rusu, C.G.; Simeanu, D.; Mierliță, D.; Pop, I.M. Persistent Organic Pollutants (POPs): A Review Focused on Occurrence and Incidence in Animal Feed and Cow Milk. *Agriculture* **2023**, *13*, 873. [CrossRef]
93. Robinson, R.C. Structures and metabolic properties of bovine milk oligosaccharides and their potential in the development of novel therapeutics. *Front. Nutr.* **2019**, *6*, 50. [CrossRef]
94. Yuzbashian, E.; Berg, E.; de Campos Zani, S.C.; Chan, C.B. Cow's Milk Bioactive Molecules in the Regulation of Glucose Homeostasis in Human and Animal Studies. *Foods* **2024**, *13*, 2837. [CrossRef]
95. Usturoi, A.; Usturoi, M.-G.; Avarvarei, B.-V.; Pânzaru, C.; Simeanu, C.; Usturoi, M.-I.; Spătaru, M.; Radu-Rusu, R.-M.; Doliş, M.-G.; Simeanu, D. Research Regarding Correlation between the Assured Health State for Laying Hens and Their Productivity. *Agriculture* **2023**, *13*, 86. [CrossRef]
96. Helmreich, S.; Wechsler, B.; Hauser, R.; Gygax, L. Effects of milking frequency in automatic milking systems on salivary cortisol, immunoglobulin A, somatic cell count and melatonin. *Schweiz Arch Tierheilkd* **2016**, *158*, 179–186. [CrossRef] [PubMed]
97. Muthuramalingam, P.; Kennedy, A.D.; Berry, R.J. Plasma melatonin and insulin-like growth factor-1 responses to dim light at night in dairy heifers. *J. Pineal Res.* **2006**, *40*, 225–229. [CrossRef] [PubMed]
98. Bal, M.A.; Penner, G.B.; Oba, M.; Kennedy, A.D. Effects of dim light at night on milk yield, milk composition and endocrine profile of lactating dairy cows. *Can. J. Anim. Sci.* **2008**, *88*, 609–612. [CrossRef]
99. Boztepe, S.; Keskin, I.; Semacan, A.; Akyürek, F.; Aytekin, I.; Sahin, Ö. Melatonin Differences Between Day and Night Milk in Primiparous Holstein Friesian and Jersey Dairy Cattle. *Selcuk. J. Agric. Food Sci.* **2022**, *36*, 27–30. [CrossRef]
100. Keskin, M.; Gül, S.; Karaaslan, İ.; Yakan, A. Controlling the photoperiod to raise the melatonin content of sheep milk. Photoperiod control and milk melatonin content. *J. Hell. Vet. Med. Soc.* **2023**, *74*, 6641–6648. [CrossRef]
101. Şahin, Ö.; Akyürek, F.; Boztepe, S.; Aytekin, İ.; Keskin, İ. Determination of Melatonin Differences between Day and Night Milk in Dairy Cattle. *J. Agric. Sci. Bilim. Derg.* **2021**, *27*, 449–453. [CrossRef]
102. Romanini, E.B.; Marchi Volpato, A.; Dos Santos, J.S.; De Santana, E.H.W.; De Souza, C.H.B.; Ludovico, A. Melatonin concentration in cow's milk and sources of its variation. *J. Appl. Anim. Res.* **2019**, *47*, 140–145. [CrossRef]
103. Asher, A.; Shabtay, A.; Brosh, A.; Eitam, H.; Agmon, R.; Cohen-Zinder, M.; Zubidat, A.E.; Haim, A. "Chrono-functional milk": The difference between melatonin concentrations in night-milk versus day-milk under different night illumination conditions. *Chronobiol. Int.* **2015**, *32*, 1409–1416. [CrossRef] [PubMed]
104. Teng, Z.W.; Yang, G.Q.; Wang, L.F.; Fu, T.; Lian, H.X.; Sun, Y.; Han, L.Q.; Zhang, L.Y.; Gao, T.Y. Effects of the Circadian Rhythm on Milk Composition in Dairy Cows: Does Day Milk Differ from Night Milk? *J. Dairy Sci.* **2021**, *104*, 8301–8313. [CrossRef]
105. Jo, J.-H.; Jalil, G.N.; Kim, W.-S.; Moon, J.-O.; Lee, S.-D.; Kwon, C.-H.; Lee, H.-G. Effects of Rumen-Protected L-Tryptophan Supplementation on Productivity, Physiological Indicators, Blood Profiles, and Heat Shock Protein Gene Expression in Lactating Holstein Cows under Heat Stress Conditions. *Int. J. Mol. Sci.* **2024**, *25*, 1217. [CrossRef]

106. Stelwagen, K.; Phyn, C.V.; Davis, S.R.; Guinard-Flament, J.; Pomiès, D.; Roche, J.R.; Kay, J.K. Invited review: Reduced milking frequency: Milk production and management implications. *J. Dairy. Sci.* **2013**, *96*, 3401–3413. [CrossRef] [PubMed]
107. Wall, E.H.; Bond, J.P.; McFadden, T.B. Milk yield responses to changes in milking frequency during early lactation are associated with coordinated and persistent changes in mammary gene expression. *BMC Genom.* **2013**, *14*, 296. [CrossRef]
108. Asher, A.; Fialko, M.; Fares, F.; Moallem, U.; Yaacoby, S.; Gutman, R. The Effect of Short-Wavelength White LED Illumination throughout the Night on the Milk Fatty Acid Profile of High-Yielding Dairy Cows. *Biology* **2022**, *11*, 1799. [CrossRef] [PubMed]
109. Murphy, B.A.; Herlihy, M.M.; Nolan, M.B.; O'Brien, C.; Furlong, J.G.; Butler, S.T. Identification of the blue light intensity administered to one eye required to suppress bovine plasma melatonin and investigation into effects on milk production in grazing dairy cows. *J. Dairy. Sci.* **2021**, *104*, 12127–12138. [CrossRef]
110. Adamczyk, K.; Herbut, P.; Godyń, D.; Angrecka, S.; Kupczyński, R.; Corrêa Vieira, F.M. Effect of light on dairy cattle in farm conditions—A review. *Ann. Anim. Sci.* **2024**, *24*, 1139–1151. [CrossRef]
111. Elsabagh, M.; Mon, M.; Takao, Y.; Shinoda, A.; Watanabe, T.; Kushibiki, S.; Obitsu, T.; Sugino, T. Exposure to blue LED light before the onset of darkness under a long-day photoperiod alters melatonin secretion, feeding behaviour and growth in female dairy calves. *Jpn. Soc. Anim. Sci.* **2020**, *91*, e13353. [CrossRef] [PubMed]
112. Simeanu, D. *Nutrition and Feeding of Animals*; "Ion Ionescu de la Brad" Publishing House: Iasi, Romania, 2018.
113. Back, K.; Tan, D.X.; Reiter, R.J. Melatonin biosynthesis in plants: Multiple pathways catalyzetryptophan to melatonin in the cytoplasm or chloroplasts (Review article). *J. Pineal Res.* **2016**, *61*, 426–437. [CrossRef] [PubMed]
114. Fan, J.; Xie, Y.; Zhang, Z.; Chen, L. Melatonin: A Multifunctional Factor in Plants. *Int. J. Mol. Sci.* **2018**, *19*, 1528. [CrossRef]
115. Arnao, M.B.; Cano, A.; Hernández-Ruiz, J. Phytomelatonin: An unexpected molecule with amazing performances in plants. *J. Exp. Bot.* **2022**, *73*, 5779–5800. [CrossRef] [PubMed]
116. Niu, T.; Ding, Z.; Zeng, J.; Yan, Z.; Duan, H.; Lv, J.; Zhang, Y.; Zhang, L.; Hu, J. Melatonin Sources in Sheep Rumen and Its Role in Reproductive Physiology. *Animals* **2024**, *14*, 3451. [CrossRef]
117. Holzmann, V.M.M.; Trentin, M.; De Almeida Rego, F.C.; Coelho Cunha Filho, L.F.; Ludovico, A. Melatonin concentration in the milk of cows supplemented with vitamins and milked twice daily. *Semin. Agrar.* **2019**, *40*, 2017–2026. [CrossRef]
118. Hernandez-Ruiz, J.; Cano, A.; Arnao, M.B. Melatonin acts as a growth-stimulating compound in some monocot species. *J. Pineal Res.* **2005**, *39*, 137–142. [CrossRef] [PubMed]
119. Barik, S. The Uniqueness of Tryptophan in Biology: Properties, Metabolism, Interactions and Localization in Proteins. *Int. J. Mol. Sci.* **2020**, *21*, 8776. [CrossRef] [PubMed]
120. Kim, I.-S.; Kim, C.-H.; Yang, W.-S. Physiologically Active Molecules and Functional Properties of Soybeans in Human Health-A Current Perspective. *Int. J. Mol. Sci.* **2021**, *22*, 4054. [CrossRef] [PubMed]
121. Luo, C.; Wang, D.; Lu, N.; Li, H.; Liu, G.; Cao, Z.; Yang, H.; Li, S.; Yu, X.; Shao, W.; et al. Analysis of Chemical Composition, Amino Acid Content, and Rumen Degradation Characteristics of Six Organic Feeds. *Animals* **2022**, *12*, 682. [CrossRef] [PubMed]
122. Kim, Y.-L.; Lee, S.-H.; Son, G.-H.; Shin, J.-S.; Kim, M.-J.; Park, B.-K. Effect of Rumen-Protected L-Tryptophan or L-Ascorbic Acid on Plasma Metabolites and Milk Production Characteristics of Lactating Holstein Cows during Summer Conditions. *Animals* **2024**, *14*, 1820. [CrossRef] [PubMed]
123. Liu, X.; Yao, S.; Liu, Y.; Han, H.; Wang, W.; Yi, Q.; Yan, L.; Ji, P.; Zhang, L.; Liu, G. Effects of Prepartum L-Tryptophan Supplementation on the Postpartum Performance of Holstein Cows. *Animals* **2024**, *14*, 1278. [CrossRef]
124. Kollmann, M.T.; Locher, M.; Hirche, F.; Eder, K.; Meyer, H.H.; Bruckmaier, R.M. Effects of tryptophan supplementation on plasma tryptophan and related hormone levels in heifers and dairy cows. *Domest. Anim. Endocrinol.* **2008**, *34*, 14–24. [CrossRef] [PubMed]
125. Karunanithi, D.; Radhakrishna, A.; Sivaraman, K.P.; Biju, V.M. Quantitative determination of melatonin in milk by LC-MS/MS. *J. Food Sci. Technol.* **2014**, *51*, 805–812. [CrossRef] [PubMed]
126. Kocadağı, T.; Yılmaz, C.; Gökmen, V. Determination of melatonin and its isomer in foods by liquid chromatography tandem mass spectrometry. *Food Chem.* **2014**, *153*, 151–156. [CrossRef] [PubMed]
127. Sturtz, M.; Cerezo, A.B.; Cantos-Villar, E.; Garcia-Parrilla, M.C. Determination of the melatonin content of different varieties of tomatoes (*Lycopersicon esculentum*) and strawberries (*Fragaria ananassa*). *Food Chem.* **2011**, *127*, 1329–1334. [CrossRef] [PubMed]
128. Kanova, M.; Kohout, P. Serotonin—Its Synthesis and Roles in the Healthy and the Critically Ill. *Int. J. Mol. Sci.* **2021**, *22*, 4837. [CrossRef]
129. Dubocovich, M.L. Melatonin receptors: Role on sleep and circadian rhythm regulation. *Sleep. Med.* **2007**, *8* (Suppl. S3), 34–42. [CrossRef] [PubMed]
130. Tan, D.-X.; Manchester, L.C.; Terron, M.P.; Flores, L.J.; Tamura, H.; Reiter, R.J. Melatonin as a naturally occurring co-substrate of quinone reductase-2, the putative MT3melatonin membrane receptor: Hypothesis and significance. *J. Pineal Res.* **2007**, *43*, 317–320. [CrossRef]
131. Superti, F. Lactoferrin from Bovine Milk: A Protective Companion for Life. *Nutrients* **2020**, *12*, 2562. [CrossRef]
132. Flis, Z.; Molik, E. Importance of Bioactive Substances in Sheep's Milk in Human Health. *Int. J. Mol. Sci.* **2021**, *22*, 4364. [CrossRef] [PubMed]

133. Niaz, B.; Saeed, F.; Ahmad, A.; Imran, M.; Maan, A.; Khan, M.; Tufail, T.; Anjum, F.; Hussain, S.; Suleria, H. Lactoferrin (LF): A Natural Antimicrobial Protein. *Int. J. Food Prop.* **2019**, *22*, 1626–1641. [CrossRef]
134. Redwan, E.; Uversky, V.; El-Fakharany, E.; Al-Mehdar, H. Potential lactoferrin activity against pathogenic viruses. *Comptes Rendus Biol.* **2015**, *337*, 581–595. [CrossRef] [PubMed]
135. Cui, S.; Lv, X.; Sun, G.; Wu, W.; Xu, H.; Li, Y.; Liu, Y.; Li, J.; Du, G.; Wang, M.; et al. Recent advances and prospects in purification and heterologous expression of lactoferrin. *Food Bioeng.* **2022**, *1*, 58–67. [CrossRef]
136. Conesa, C.; Bellés, A.; Grasa, L.; Sánchez, L. The Role of Lactoferrin in Intestinal Health. *Pharmaceutics* **2023**, *15*, 1569. [CrossRef] [PubMed]
137. Giansanti, F.; Panella, G.; Leboffe, L.; Antonini, G. Lactoferrin from Milk: Nutraceutical and Pharmacological Properties. *Pharmaceuticals* **2016**, *9*, 61. [CrossRef] [PubMed]
138. Lemoine, P.; Bablon, J.C.; Da Silva, C. A combination of melatonin, vitamin B6 and medicinal plants in the treatment of mild-to-moderate insomnia: A prospective pilot study. *Complement. Ther. Med.* **2019**, *45*, 104–108. [CrossRef] [PubMed]
139. Peukhuri, K.; Sihvola, N.; Korpela, R. Diet promotes sleep duration and quality. *Nutr. Res.* **2012**, *32*, 309–319. [CrossRef]
140. Knezevic, E.; Nenic, K.; Milanovic, V.; Knezevic, N.N. The Role of Cortisol in Chronic Stress, Neurodegenerative Diseases, and Psychological Disorders. *Cells* **2023**, *12*, 2726. [CrossRef]
141. Rahman, M.S.; Hossain, K.S.; Das, S.; Kundu, S.; Adegoke, E.O.; Rahman, M.A.; Hannan, M.A.; Uddin, M.J.; Pang, M.-G. Role of Insulin in Health and Disease: An Update. *Int. J. Mol. Sci.* **2021**, *22*, 6403. [CrossRef]
142. Xia, A.-Y.; Zhu, H.; Zhao, Z.-J.; Liu, H.-Y.; Wang, P.-H.; Ji, L.-D.; Xu, J. Molecular Mechanisms of the Melatonin Receptor Pathway Linking Circadian Rhythm to Type 2 Diabetes Mellitus. *Nutrients* **2023**, *15*, 1406. [CrossRef]
143. Peschke, E.; Bähr, I.; Mühlbauer, E. Melatonin and Pancreatic Islets: Interrelationships between Melatonin, Insulin and Glucagon. *Int. J. Mol. Sci.* **2013**, *14*, 6981–7015. [CrossRef] [PubMed]
144. Peschke, E. Melatonin, endocrine pancreas and diabetes. *J. Pineal Res.* **2008**, *44*, 26–40. [CrossRef]
145. Mulder, H.; Nagorny, C.L.; Lyssenko, V.; Groop, L. Melatonin receptors in pancreatic islets: Good morning to a novel type 2 diabetes gene. *Diabetologia* **2009**, *52*, 1240–1249. [CrossRef] [PubMed]

agriculture

MDPI

Review

# Unlocking the Power of Eggs: Nutritional Insights, Bioactive Compounds, and the Advantages of Omega-3 and Omega-6 Enriched Varieties

**Marius Giorgi Usturoi [1], Roxana Nicoleta Rațu [2,*], Ioana Cristina Crivei [2], Ionuț Dumitru Veleșcu [2], Alexandru Usturoi [3,*], Florina Stoica [4,5] and Răzvan-Mihail Radu Rusu [1]**

1 Department of Management of Animal Resources and Technologies, Faculty of Food and Animal Sciences, Iasi University of Life Sciences, 3 Mihail Sadoveanu Alley, 700489 Iasi, Romania; marius.usturoi@iuls.ro (M.G.U.); razvan.radu@iuls.ro (R.-M.R.R.)
2 Department of Food Technologies, Faculty of Agriculture, Iasi University of Life Sciences, 3 Mihail Sadoveanu Alley, 700489 Iasi, Romania; ioana.crivei@iuls.ro (I.C.C.); ionut.velescu@iuls.ro (I.D.V.)
3 Department of Control, Expertise and Services, Faculty of Food and Animal Sciences, Iasi University of Life Sciences, 3 Mihail Sadoveanu Alley, 700489 Iasi, Romania
4 Department of Pedotechnics, Faculty of Agriculture, Iasi University of Life Sciences, 3 Mihail Sadoveanu Alley, 700489 Iasi, Romania; florina.stoica@iuls.ro
5 Research Institute for Agriculture and Environment, Iasi University of Life Sciences, 14 Sadoveanu Alley, 700490 Iasi, Romania
* Correspondence: roxana.ratu@iuls.ro (R.N.R.); alexandru.usturoi@iuls.ro (A.U.)

**Abstract:** This study explores the nutritional benefits and health implications of omega-3- and omega-6-enriched eggs, positioning them within the context of functional foods aimed at improving public health outcomes. With rising consumer interest in nutritionally fortified foods, omega-enriched eggs have emerged as a viable source of essential fatty acids, offering potential benefits for cardiovascular health, inflammation reduction, and cognitive function. This research examines enrichment techniques, particularly dietary modifications for laying hens, such as the inclusion of flaxseed and algae, to enhance omega-3 content and balance the omega-6-to-omega-3 ratio in eggs. The findings indicate that enriched eggs provide significantly higher levels of essential fatty acids and bioactive compounds than conventional eggs, aligning with dietary needs in populations with limited access to traditional omega-3 sources like fish. This study further addresses consumer perception challenges, regulatory constraints, and environmental considerations related to sustainable production practices. The conclusions underscore the value of omega-enriched eggs as a functional food that aligns with health-conscious dietary trends and recommend ongoing research to refine enrichment methods and expand market accessibility.

**Keywords:** omega-3-enriched eggs; functional foods; public health nutrition; sustainable farming; consumer perception

## 1. Introduction

Eggs are an affordable and highly nutritious food source, widely recognised as a healthy component of human diets [1–3]. According to Réhault-Godbert [4], eggs are rich in high-quality proteins, vitamins, and minerals, making them a valuable dietary recommendation from nutritionists for enhancing health. Beyond their basic nutritional value, eggs also contain bioactive compounds with potential health benefits, which have garnered increasing scientific interest in recent decades.

Historically, egg consumption was controversially associated with high cholesterol levels, leading to negative public perceptions. For instance, in 1968, the American Heart Association (AHA) recommended limiting egg consumption to fewer than three per week due to the prevailing belief that dietary cholesterol contributed significantly to elevated blood cholesterol levels and the development of cardiovascular diseases (CVD) [5,6]. This guidance dramatically shifted public attitudes, with many people avoiding eggs despite their high nutritional value.

However, extensive research conducted over the past 50 years has demonstrated that dietary cholesterol, including cholesterol from eggs, has minimal impact on blood cholesterol levels for most individuals [7]. It is now well established that saturated fats, rather than dietary cholesterol, play a more significant role in elevating blood cholesterol. Consequently, organisations such as the AHA and the British Heart Foundation have updated their recommendations, clarifying that eggs do not inherently contribute to cardiovascular disease. These updated guidelines suggest that individuals with healthy hearts can safely include eggs in their diets, while those with specific cholesterol-related conditions should consult healthcare professionals for tailored advice [8–10].

Recent scientific advancements have further elevated the status of eggs, highlighting their role as a source of bioactive compounds, such as peptides, free radical scavengers, and other biologically active molecules. These compounds, primarily concentrated in the yolk, offer anti-inflammatory, antimicrobial, and antioxidant properties, positioning eggs as a functional food with benefits beyond basic nutrition [11]. Today, eggs are recognised as a valuable dietary component for both disease prevention and health enhancement [12].

One promising area of egg-based nutrition involves the enrichment of eggs with polyunsaturated fatty acids, particularly Omega-3 and Omega-6. These essential nutrients play critical roles in cellular maintenance, regulating inflammatory processes, and supporting cardiovascular health [13]. The Western diet is characterised by an imbalance, with an excess of Omega-6 and a deficiency in Omega-3 fatty acids, contributing to chronic inflammatory disorders, cardiovascular disease, and other health conditions [14]. Enriched eggs, particularly those fortified with Omega-3 fatty acids, help address this imbalance by providing a healthier Omega-6-to-Omega-3 ratio.

The enrichment process involves modifying the diet of laying hens by incorporating Omega-3-rich feed sources, such as flaxseed or fish oil. This strategy enhances the fatty acid profile of eggs, increasing the levels of long-chain polyunsaturated fatty acids, including eicosapentaenoic acid (EPA) and docosahexaenoic acid (DHA) [15–17]. These enriched eggs serve as a convenient dietary source of Omega-3 fatty acids, particularly for individuals who consume insufficient fish or other sources of healthy fats.

By combining bioactive compounds and advancements in Omega-3 and Omega-6 enrichment, eggs are now positioned as a functional food capable of addressing some of the most pressing health challenges of our time. This development underscores the potential of enriched eggs to improve public health by offering a practical approach to incorporating functional foods into daily diets [18].

The primary objective of this research is to provide a comprehensive review of the nutritional composition and bioactive elements of eggs, with a focus on recent advancements in Omega-3 and Omega-6 fatty acid supplementation. This study aims to elucidate the nutritional significance of eggs in human diets, particularly their potential for promoting health through regular consumption. It seeks to summarise and analyse the current literature on the bioactive compounds present in eggs and their effects on health, while exploring strategies to enhance the nutritional profile of eggs through dietary enrichment. Additionally, this study examines the health implications of egg consumption, focusing on its role in cardiovascular health, weight management, and muscle development, and addresses misconceptions surrounding dietary cholesterol. Emerging trends and advances in egg production, including their potential in functional and personalised nutrition, are also discussed. Finally, this study considers ethical and sustainability issues associated with large-scale egg production and offers recommendations for future research to address existing gaps. By adopting a structured approach, this research provides valuable insights into the evolving role of eggs in nutrition and health.

## 2. Nutritional Profile of Eggs

Global egg consumption continues to rise, driven by economic growth and their universal appeal across all age groups, genders, and regions, including both rural and urban settings. Eggs are an integral part of a balanced diet and a staple in numerous international cuisines. Recognised by nutritionists and healthcare professionals worldwide for their exceptional nutritional value, eggs are a highly recommended source of protein. Health and nutrition institutions, such as the European Food Information Council (EUFIC), support the inclusion of eggs as part of a healthy and balanced diet. In Romania, dietary guidelines align with European requirements, which emphasise the importance of eggs as a valuable component of daily nutrition. Studies further highlight that eggs form the foundation of a balanced diet and contribute to maintaining a healthy body weight [19].

Today's consumers demand high-quality eggs with robust shells, free from microbiological contamination, and produced using ethically acceptable farming systems. Breeding programmes for egg-laying hens prioritise egg quality, focusing on shell durability, egg weight stability, and the quality of egg white and yolk. Key indicators of egg-white quality include the albumen height, which reflects freshness, and its ability to inhibit microbial growth, reducing the risk of foodborne illnesses such as salmonellosis [20,21].

According to Valverde et al. [22], eggs are predominantly consumed in fried or prepared forms, but they are also versatile ingredients that enhance a variety of dishes. Enrichment is an innovative approach to improving the biological value of foods by incorporating amino acids, fibres, vitamins, minerals, and antioxidants. The high utilisation of eggs is largely due to their complex protein content, which, alongside lipids and carbohydrates, forms one of the three primary macronutrient categories. Eggs provide 100% of dietary energy and 90% of dry weight [23]. Egg proteins are evenly distributed between the egg white and yolk, while lipids, vitamins, and minerals are primarily concentrated in the yolk. The macronutrient composition of a fresh, uncooked egg consists of approximately 76.1% water, 12.6% protein, 9.5% fat, 0.7% carbohydrates, and 1.1% ash [4,24].

Table 1 provides a comprehensive examination of the nutritional value, categorised by macronutrients and micronutrients, for both the white and yolk components.

**Table 1.** Comprehensive nutritional breakdown of chicken eggs by components (per 100 g of egg white and yolk).

| Nutrient | Egg White | Egg Yolk | Whole Egg | References |
|---|---|---|---|---|
| Energy (kcal) | 52 | 322 | 155 | [4,24,25] |
| Protein (g) | 10.9 | 15.9 | 12.6 | [7,10] |
| Total lipids (fat) (g) | 0.2 | 27.0 | 10.6 | [4,10,26] |
| Saturated fat (g) | 0.0 | 9.6 | 3.3 | [4,27,28] |
| Monounsaturated fat (g) | 0.0 | 12.7 | 4.0 | [4,27,29] |
| Polyunsaturated fat (g) | 0.0 | 4.2 | 1.2 | [25,30,31] |
| Cholesterol (mg) | 0 | 1085 | 373 | [25,27,30] |

**Table 1.** *Cont.*

| Nutrient | Egg White | Egg Yolk | Whole Egg | References |
|---|---|---|---|---|
| Carbohydrates (g) | 0.3 | 0.4 | 0.7 | [4,27,30] |
| Sugars (g) | 0.4 | 0.5 | 1.0 | [23,27,30] |
| Fiber (g) | 0.0 | 0.0 | 0.0 | [25,30,31] |
| Vitamins | | | | [4,27,30,31] |
| Vitamin A (μg) | 0 | 770 | 160 | [4,26,29,31] |
| Vitamin D (IU) | 0 | 145 | 87 | [26,29,31] |
| Vitamin E (mg) | 0 | 3.0 | 1.05 | [4,26,29,31] |
| Vitamin B2 (Riboflavin) (mg) | 0.44 | 0.528 | 0.457 | [4,25,32,33] |
| Vitamin B12 (μg) | 0 | 2.0 | 1.11 | [4,26,29] |
| Folate (μg) | 4 | 146 | 47 | [27,30,33] |
| Minerals | | | | [25,30,32] |
| Calcium (mg) | 7 | 129 | 50 | [26,29,30,33] |
| Iron (mg) | 0.1 | 7.0 | 1.2 | [29,31,33] |
| Potassium (mg) | 163 | 109 | 126 | [26,29,30] |
| Magnesium (mg) | 11 | 3.0 | 7.0 | [26,29,31] |
| Phosphorus (mg) | 15 | 586 | 172 | [25,27,30] |
| Sodium (mg) | 166 | 48 | 124 | [4,10,26,30] |
| Zinc (mg) | 0.0 | 5.3 | 1.0 | [26,29,30] |
| Selenium (mg) | 20 | 60 | 30.7 | [20,26,30,31] |

Note: Percent Daily Value (DV) based on a 2000-calorie diet.

Egg proteins are recognised as one of the best sources of high-quality protein for humans, with only breast milk offering comparable benefits [34,35]. These proteins include ovalbumin, which enhances antioxidant activity through covalent bonding with polysaccharides, ovotransferrin, which chelates $Fe^{3+}$, and phosvitin, rich in phosphoserines. Proteins such as apolipoproteins and riboflavin-binding proteins are specific to the yolk, while compartment-specific proteins support embryonic development [36]. Compared to other protein sources, egg proteins, particularly those in the yolk, demonstrate higher satiety effects, contributing to a reduced energy intake and promoting weight loss during hypocaloric diets [37].

Proteomic analyses have identified nearly one thousand unique proteins in chicken eggs, including peptides with antioxidant, anticancer, and antimicrobial properties. The average protein content in a whole, raw egg is 12.5 g per 100 g, with egg white and yolk contributing 10.9 g and 15.9 g, respectively. Minor variations occur due to age and genetics [38–40].

The lipid content of eggs is predominantly found in the yolk, ranging from 8.7 to 11.2 g per 100 g. These lipids are essential components of yolk lipoproteins, comprising cholesterol, phospholipids, and triglycerides. While the yolk-to-egg-white ratio determines the fat content, dietary modifications significantly influence the fatty acid profile. Studies have shown that feeding hens with marigold powder increases yolk lutein content and enhances its fatty acid composition, contributing to its nutritional value [41–43].

Eggs contain approximately 0.7% carbohydrates, primarily as glucose distributed between the yolk and albumen. Glycoproteins, resulting from post-translational glycosylation during egg formation, represent a significant portion of egg carbohydrates. Additionally, eggs are rich in micronutrients, including zinc, selenium, and fat-soluble vitamins. Their proteins are highly digestible, with cooked egg proteins exhibiting a digestibility rate of 90.9% in humans. Branched-chain amino acids such as leucine, isoleucine, and valine further support muscle protein synthesis [44,45].

Vitamins in eggs include A, D, E, K, and the B-complex, with the yolk containing the majority [46]. The hen's diet directly affects the levels of fat-soluble vitamins such as A, D, E, and K in the yolk. For instance, dietary supplementation with retinol or vitamin D3 can significantly enhance their concentration in eggs [47,48]. Similarly, carotenoids, such as lutein and zeaxanthin, derived from feed sources like marigold petals or corn, contribute to the yolk's orange–yellow hue while promoting visual health and reducing the risk of age-related macular degeneration [49].

Eggs are also a primary source of choline, a nutrient essential for cellular maintenance, brain development, and neurotransmission. The yolk contains approximately 680 mg of choline per 100 g, making eggs a leading dietary source. Supplementation through hen nutrition further enhances egg nutrient profiles, addressing specific consumer needs [50].

Minerals such as calcium, phosphorus, and potassium are present in moderate amounts in eggs, with the yolk serving as the primary reservoir for iron and zinc. Trace elements like selenium and iodine can be increased through dietary supplementation of hens. Selenium-enriched eggs, for example, provide up to 40 μg of selenium per egg, supporting antioxidant functions and reducing deficiency risks [51,52].

Eggs are highly valued for their nutritional benefits and play a critical role in diverse diets, particularly for vulnerable groups such as children, the elderly, and athletes. They are increasingly used in food manufacturing due to their affordability, versatility, and ease of storage. Pasteurised egg products are preferred in the food industry for their safety and convenience. The growing interest in functional foods, driven by an ageing population and dietary trends, highlights the continued relevance of eggs in addressing global nutritional needs [1,4,34].

While concerns regarding cholesterol and cardiovascular risk persist, recent studies confirm that egg consumption does not significantly impact cardiovascular health in most individuals. For high-risk groups, moderation and personalised dietary recommendations remain essential. Overall, eggs offer an affordable, sustainable, and nutrient-dense food source with broad health benefits, underscoring their role in human nutrition [49,50].

## 3. Bioactive Compounds in Eggs

Eggs contain a diverse range of nutrients and bioactive compounds that significantly contribute to human health and nutrition. These include proteins, peptides, phospholipids, omega-3 fatty acids, vitamins, and minerals, all of which have been recognised for their potential health benefits and functional properties [12]. The growing interest in these compounds stems from their dual roles in supporting nutrition and their potential applications in biomedical fields, establishing eggs as a key focus in functional food research [11].

Egg proteins, such as ovalbumin and ovotransferrin, have demonstrated several health-promoting properties, including antioxidant, antimicrobial, and anticancer effects [53]. Additionally, omega-3-enriched eggs have been associated with improvements in cardiovascular health and cognitive function, further enhancing their value as a functional food [54,55].

These findings reinforce the perception of eggs as functional foods capable of improving overall well-being, especially for health-conscious consumers. Bioactive compounds in egg white and yolk are classified based on their antioxidant, antimicrobial, immunomodulatory, anti-inflammatory, and antihypertensive properties. This highlights the extensive nutritional and health benefits of eggs [56–58].

Despite their benefits, eggs have been the subject of ongoing debate due to concerns about cholesterol content and its potential impact on cardiovascular health. Some studies indicate that moderate egg consumption does not significantly increase the risk of cardiovascular diseases and may even confer protective effects against certain conditions [59]. However, uncertainties remain regarding the influence of saturated fats in eggs on heart health, emphasising the need for further investigation [60,61].

Recent developments in egg enrichment, such as incorporating bioactive compounds from herbs and employing genetic modifications, have underscored the innovative potential of eggs in functional food development. These advancements aim to enhance the nutritional benefits of eggs while addressing dietary challenges, solidifying their position as an essential source of bioactive compounds in modern diets [62,63].

### 3.1. *Types of Bioactive Compounds in Eggs*

Eggs are a rich source of bioactive compounds, which provide notable health benefits and possess functional properties. These compounds are essential not only for human nutrition but also for their potential applications in biomedical research and therapeutic interventions [57,64].

#### 3.1.1. Proteins and Peptides

Eggs contain a variety of proteins, including ovalbumin and ovotransferrin, both of which have been extensively studied for their health-promoting effects. Ovotransferrin exhibits antioxidant, antibacterial, and anticancer properties, making it vital for embryonic development and overall health [58]. Hydrolysed ovotransferrin, in particular, has shown significant anticancer activity, particularly against colon and breast cancer cells, underscoring its therapeutic potential [65–67].

Additionally, lysozyme, the first protein identified from hen eggs, has demonstrated the ability to enhance immune responses and inhibit the proliferation of cancer cells, further reinforcing the importance of egg-derived peptides in health and disease management [68–71].

#### 3.1.2. Phospholipids and Lecithin

Phospholipids, a major component of egg yolks, play a critical role in lipid metabolism and the absorption of fat-soluble vitamins. They also function as emulsifiers, making them essential in both the food and pharmaceutical industries [40]. Lecithin, another key component extracted from eggs, is widely used as a natural emulsifier and is in high demand globally due to its economic and industrial significance [40,72].

#### 3.1.3. Omega-3 Fatty Acids

Eggs can be enriched with omega-3 fatty acids, which are recognised for their cholesterol-lowering and anti-inflammatory effects. Omega-3-enriched eggs have been associated with improved cardiovascular health and enhanced cognitive function, making them a functional food option for health-conscious individuals [73,74]. Studies indicate that dietary modifications in hens, such as the inclusion of flaxseed, can elevate alpha-linolenic acid (ALA) to 200 mg per egg and docosahexaenoic acid (DHA) to 90 mg per egg, further enhancing their nutritional value [75]. Furthermore, achieving an optimal omega-6-to-omega-3 ratio, which is critical for balancing pro-inflammatory and anti-inflammatory effects, is an essential consideration in the production of these enriched eggs, ensuring their alignment with modern dietary recommendations [76].

#### 3.1.4. Vitamins and Minerals

Eggs provide essential vitamins and minerals, including selenium and vitamin D, which support immune function, bone health, and other physiological processes. These micronutrients make eggs a vital component of a balanced diet [3,4,77].

#### 3.1.5. Immunomodulatory Compounds

Bioactive compounds in eggs have shown immunomodulatory effects, aiding in the management of diseases by influencing immune pathways and enhancing the body's defence mechanisms against infections [78].

Recent advancements in egg enrichment have focused on the incorporation of specific bioactive compounds derived from herbs. Such modifications have demonstrated promising outcomes, including reductions in triglyceride levels and improvements in immunity among human volunteers [79]. Additionally, genetic engineering techniques have been employed to produce eggs enriched with proteins or peptides of pharmaceutical interest, offering innovative opportunities for functional food development [80].

### 3.2. *Health Benefits of Bioactive Compounds in Eggs*

Eggs are recognised for their comprehensive nutritional profile, containing high-quality proteins, vitamins, and minerals. Increasingly, they are also regarded as a significant source of bioactive compounds with the potential to deliver substantial health benefits. The bio-accessibility of these compounds, even in cooked eggs, highlights their potential as functional foods capable of improving various health outcomes [4,12].

#### 3.2.1. Immune System Modulation

Egg-derived compounds have demonstrated immunomodulatory effects, which enhance the body's defence mechanisms against pathogens. Research highlights that proteins such as lactoperoxidase can inhibit viral infectivity, including HIV (Human Immunodeficiency Virus) and the herpes simplex virus, suggesting a critical role for eggs in supporting immune health [4,11,81].

#### 3.2.2. Anti-Cancer Properties

Eggs possess pharmacological properties, including anti-cancer effects. Bioactive peptides in eggs are linked to anti-inflammatory and antibacterial activities, which may contribute to disease prevention and treatment. These properties, combined with the nutrient density and high-quality protein content of eggs, underscore their potential as functional foods in both general and therapeutic contexts [1,57,82].

#### 3.2.3. Cardiometabolic Health

Egg consumption has been associated with various aspects of cardiometabolic health. A meta-analysis indicated that a moderate egg intake does not significantly increase cardiovascular disease risk and may even provide protective benefits. Furthermore, dietary components such as choline, lutein, and zeaxanthin present in eggs are thought to support cognitive function and protect against neurodegenerative diseases [83–85].

Egg consumption has also been shown to improve dietary quality. Studies report that individuals consuming eggs experience increased postprandial satiety, reduced ghrelin responses, and lower intake of total and added sugars compared to non-consumers [34,36,86]. Findings from Andersen et al. [87] further demonstrate that the proportion of kilocalories derived from carbohydrates decreased during the whole-egg diet phase compared to egg-free or egg white-only diets, while the overall energy intake remained stable. Additionally, whole eggs contributed to the intake of nutrients associated with cardiometabolic health, such as total fats, arachidonic acid, and sodium.

#### 3.2.4. Muscle Protein Synthesis

Although limited research exists on the effects of eggs on muscle protein synthesis, existing studies suggest promising outcomes. Research by Moore et al. [88], cited by Puglisi et al. [89], demonstrated that consuming 20 g of whole-egg protein post-resistance training optimised muscle protein synthesis in young men. This aligns with findings from Witard et al. [90], who reported similar results with whey protein, and with the recommendations of the International Society of Sports Nutrition advocating 20–40 g of quality protein per serving [91].

#### 3.2.5. Weight Management and Body Composition

Epidemiological and intervention studies present mixed findings regarding egg consumption and body weight. A cross-sectional study by Garrido-Miguel et al. [92] found that consuming more than five eggs per week was associated with a lower body mass index and body fat percentage in young adults (aged 18–30), compared to those consuming less than one egg per week. Similarly, a study involving 2241 Chinese adults (aged 18–80 years) showed that consuming over 50 g of eggs daily reduced the risk of central obesity and body fat in women, with stronger protective effects observed in men [93,94].

Research by Emrani et al. [95] further suggests that incorporating whole eggs into an energy-restricted diet can enhance weight loss in healthy individuals. However, the lack of direct long-term studies on the effects of whole-egg consumption on weight and body composition highlights the need for further clinical trials, particularly to examine the impact of a sustained egg intake on anthropometric indices.

### *3.3. Clarifying Misconceptions About Egg Consumption and Its Impact on Cardiovascular and Bone Health*

One of the primary concerns in studies evaluating egg consumption has been its cholesterol content, with a single egg containing approximately 200 mg of cholesterol. This level has often been considered potentially significant for cardiovascular disease (CVD) risk [58]. However, research has shown that egg consumption only slightly elevates LDL cholesterol levels when included in a balanced diet that already contains dietary cholesterol. This contrasts with lactovegetarian diets, which typically feature a low cholesterol intake. The findings suggest that individuals with a higher baseline cholesterol intake exhibit a diminished response to additional dietary cholesterol, such as that from eggs [96].

Meta-analyses of controlled, randomised studies have reported minimal increases in plasma LDL and HDL cholesterol levels following egg consumption. Importantly, these studies indicate no significant changes in the total cholesterol ratio or the LDL:HDL ratio, both of which are key markers for cardiovascular disease risk [97]. Additionally, experimental research conducted by McDonald et al. [98] demonstrated that egg consumption does not adversely affect endothelial function in human subjects. As noted by Godos et al. [99], the observed correlation between egg consumption and a reduced risk of stroke may reflect the interplay of dietary cholesterol with other bioactive constituents present in eggs.

Beyond cardiovascular health, eggs provide several nutrients that may support bone health, particularly in older adults. Rich in high-quality protein, eggs supply essential amino acids necessary for the formation and maintenance of the bone matrix [100]. As Rizzoli et al. [101] explain, eggs also serve as a natural source of vitamin D, a critical nutrient that enhances calcium absorption and facilitates bone mineralisation. This unique nutritional composition positions eggs as an important dietary component for preserving bone density and reducing fracture risk among the elderly [85].

Adopting a holistic approach that integrates eggs into the diet, alongside adequate physical activity and lifestyle adjustments, may significantly contribute to healthier ageing [100,101]. A study conducted by Olagunju et al. [102] on a cohort of 176 individuals aged 65 and older investigated the impact of egg consumption on bone density. The study found that eggs' bioactive components positively influenced bone mineral density, with a statistically significant association between a higher egg intake and increased whole-body T-scores. Participants who consumed more eggs exhibited improved bone density and a lower incidence of fractures [103]. The analysis also highlighted that gender and the body mass index (BMI) played critical roles in bone health outcomes, with females showing higher T-scores and individuals with a higher BMI displaying enhanced bone density [104].

Interestingly, the study also revealed a negative correlation between daily egg consumption and the incidence of fractures, suggesting that regular egg consumption may reduce fracture risk. A positive association between HDL cholesterol levels and multiple fractures was also identified, indicating that cholesterol may influence bone health. These findings provide novel insights into the relationship between egg consumption and bone strength in older adults, suggesting that incorporating eggs into the diet may reduce the risk of osteoporosis and fractures in this population [5,8,85].

### *3.4. Factors Affecting Bioactive Compound Levels Dietary Influences*

The levels of bioactive compounds in eggs are influenced by multiple factors, including dietary practices, oxidative stress, and environmental conditions, which collectively affect their nutritional and functional properties.

Dietary influences. The concentration of bioactive substances in eggs is significantly influenced by the dietary practices of laying hens. The inclusion of specific feed additives, such as phytogenic substances, has been shown to enhance the production of bioactive peptides. For example, bioactive components derived from herbs, such as menthol from peppermint and chlorogenic acid from various plant extracts, exhibit antimicrobial, antifungal, and antioxidant properties. These additions have been demonstrated to improve protein synthesis and enhance albumen quality in eggs, particularly through increasing the ovomucin content, which is critical for the functional properties of the thick albumen [105].

Furthermore, the use of fermented feeds has been reported to increase nutrient availability, thereby improving the overall protein content and enhancing the levels of bioactive compounds in eggs. Fermentation processes contribute to the breakdown of antinutritional factors, maximising nutrient absorption and utilisation in hens [106,107].

Oxidative stress and its impact. Oxidative stress is another significant factor affecting the levels of bioactive compounds in eggs. Elevated oxidative products in the body can impair the activities of antioxidant enzymes, which are vital for cellular protection. High levels of reactive oxygen species (ROS) can cause damage to cellular structures, leading to diminished egg quality, particularly with regard to albumen properties. Such oxidative damage not only compromises the physical characteristics of eggs but also reduces their nutritional and antioxidant capacities, ultimately impacting consumer health [108].

Environmental factors. The environmental conditions in which hens are raised, including factors such as temperature and housing, also play a critical role in determining the levels of bioactive compounds in eggs. Hens raised in suboptimal conditions are more likely to experience heightened stress, which contributes to increased oxidative stress and a subsequent decline in egg quality. Ensuring balanced environmental conditions can help to alleviate these stressors and promote higher levels of bioactive compounds in eggs [109].

## 4. Omega-3 and Omega-6 Enrichment in Eggs

The enrichment of eggs with omega-3 and omega-6 fatty acids involves enhancing their fatty acid profile by incorporating specific omega-3 and omega-6 sources into the diets of laying hens [110,111]. This nutritional intervention has garnered considerable attention due to the increasing consumer interest in functional foods that provide health benefits beyond basic nutrition. Omega-3 fatty acids, particularly eicosapentaenoic acid (EPA) and docosahexaenoic acid (DHA), are well known for their positive effects on cardiovascular health, cognitive function, and inflammation reduction. In contrast, omega-6 fatty acids play a critical role in cell signalling and inflammatory responses [108].

Research indicates that omega-3-enriched eggs can contain significantly elevated levels of these beneficial fatty acids compared to conventional eggs, with some studies reporting up to a twelvefold increase in omega-3 content [112]. These enriched eggs have gained popularity among health-conscious consumers, particularly those seeking alternative dietary sources of omega-3s, such as individuals who do not consume fish, a traditional source of these nutrients [14,113].

However, the optimal omega-3-to-omega-6 ratio for health remains a topic of debate among researchers, with ongoing discussions surrounding the ideal balance to achieve favourable health outcomes [114,115]. The production of omega-3-enriched eggs typically involves feeding hens diets supplemented with sources such as flaxseed or fish oil, which significantly increases the levels of these beneficial fatty acids in the eggs [116,117].

Despite their enhanced nutritional profile, omega-3-enriched eggs are often more expensive than standard eggs, raising accessibility concerns for price-sensitive consumers. Nevertheless, the market for these speciality eggs continues to expand, driven by the growing consumer demand for healthier food options. This trend reflects a broader shift towards dietary awareness and an increasing preference for functional nutrition [118,119].

In addition to their health benefits, omega-3-enriched eggs are subject to regulatory oversight to ensure their safety and proper labelling. Regulatory bodies, such as the U.S. Food and Drug Administration (FDA) and the U.S. Department of Agriculture (USDA), enforce compliance with specific standards to safeguard consumer interests and maintain market credibility [120].

As awareness of the health advantages of omega-3 fatty acids continues to rise, the demand for omega-3- and omega-6-enriched eggs is expected to grow further. This presents both opportunities and challenges for producers operating within the dynamic and evolving food market [121].

Omega-3 and omega-6 fatty acids are indispensable components of a balanced diet, playing vital roles in numerous physiological processes and overall health. Both types of fatty acids are crucial for maintaining the structural integrity and functionality of cell membranes, with omega-3 fatty acids particularly recognised for their extensive health benefits [107,118,119].

### 4.1. *Methods of Enrichment*

The enrichment of eggs with omega-3 fatty acids involves strategic nutritional interventions for laying hens, primarily through the inclusion of alpha-linolenic acid (ALA) in their diet. Research has demonstrated that varying the dietary concentration of ALA, ranging from 0.3% to 6% of energy, significantly alters the fatty acid profile of eggs. Diets containing 6% ALA, for instance, have been shown to increase total n-3 fatty acids in eggs nearly ninefold while simultaneously reducing the n-6/n-3 ratio from 7.17 to 1.29, thereby enhancing the nutritional quality of the eggs produced [110,122].

#### 4.1.1. Oil Sources for Enrichment

Various oils are used as ALA sources in the diets of laying hens, including macadamia oil, vegetable oil blends, and flaxseed oil. A diet composed of 60% canola oil and 40% flaxseed oil has been employed to achieve moderate ALA levels, whereas pure flaxseed oil is utilised for higher ALA content [122]. The composition and fatty acid profiles of these oils are crucial in optimising the nutritional value of enriched eggs [123].

To evaluate the effectiveness of these dietary interventions, Oliveira et al. [124] employed a controlled experimental design. Hens were allocated to cages, and dietary treatments were applied randomly, allowing for a rigorous statistical analysis via One-Way ANOVA. Significant differences in fatty acid profiles across treatments were identified using Tukey's multiple comparison test, with a significance threshold of $p < 0.05$. This methodological rigour ensures that findings on omega-3 enrichment are scientifically reliable.

4.1.2. Considerations for Eggshell Quality

While enriching eggs with omega-3 fatty acids, it is essential to address the impact of dietary interventions on eggshell integrity. An imbalanced diet may cause eggshell abnormalities, which are not solely attributable to calcium deficiency. Providing hens with a well-balanced feed is recommended to maintain both optimal fatty acid enrichment and eggshell quality [125,126].

4.1.3. Types of Enriched Eggs

To provide a clear comparison and overview, the different types and characteristics of enriched eggs are presented in Table 2, highlighting their nutritional composition and benefits.

**Table 2.** Types of enriched eggs.

| Type of Egg | Dietary Modification | Key Nutritional Benefits | Additional Notes | Reference |
|---|---|---|---|---|
| Omega-3-enriched eggs | Diets fortified with omega-3 sources, such as flaxseed | Contain significantly higher levels of omega-3 fatty acids, including ALA, DHA, and EPA. Reported to improve cardiovascular health, brain function, and reduce inflammation | Some studies show up to a 12-fold increase in omega-3 content compared to conventional eggs. May have lower cholesterol levels | [74,124–126] |
| Standard eggs | Conventional poultry diets | Lower omega-3 fatty acid levels compared to enriched varieties. | Australian study reports 1.3% omega-3 fatty acids in yolks | [127] |
| Organic eggs | Organic feed and farming practices | Slightly higher nutrient levels than standard eggs but lower omega-3 content than omega-3-enriched eggs | Australian study reports similar omega-3 levels as standard eggs (approximately 1.3%) | [127] |

4.1.4. Other Specialty Eggs

Apart from omega-3-enriched eggs, other types of specialty eggs include organic and cage-free varieties. Organic eggs are produced by hens raised in cage-free environments and fed organic grains free from pesticides and animal by-products [128]. Cage-free eggs, by contrast, are laid by hens that roam freely within a hen house, enhancing animal welfare. While both types offer distinct benefits, omega-3-enriched eggs stand out for their superior nutritional profile, particularly their higher omega-3 content [129].

### *4.2. Regulatory Oversight*

The safety and labelling of omega-3-enriched eggs in the United States fall under the jurisdiction of the U.S. Food and Drug Administration (FDA) and the U.S. Department of Agriculture (USDA). The FDA regulates the production, transport, and storage of shell eggs under the Federal Food, Drug, and Cosmetic Act to ensure consumer protection against unsafe or mislabelled products [21].

#### 4.2.1. Egg Safety Final Rule

The Egg Safety Final Rule, established to mitigate foodborne illnesses such as Salmonella Enteritidis, requires egg producers to implement stringent safety measures throughout production. It is estimated that these regulations prevent approximately 79,000 cases of foodborne illness annually [4]. This rule is particularly pertinent for omega-3-enriched eggs, which must also comply with rigorous safety standards to safeguard consumer health.

#### 4.2.2. Egg Regulatory Program Standards

The Egg Regulatory Program Standards (ERPS) [130] provide a comprehensive framework for the state-level oversight of eggs and egg products in the United States. These standards are designed to harmonise regulatory activities among partner agencies, thereby enhancing overall egg safety, including that of omega-3-enriched varieties [127]. Compliance with these standards is critical for producers seeking to market omega-3-enriched eggs within the U.S. regulatory system.

In contrast, the European Union operates under its own regulatory frameworks for eggs and egg products. Key regulations include Regulation (EC) No 853/2004 [131], which establishes specific hygiene rules for food of animal origin, and Regulation (EC) No 178/2002 [132], which lays down the general principles and requirements of food law. Additionally, the Council Directive 1999/74/EC [133] outlines minimum welfare standards for laying hens, further influencing the production of enriched eggs.

Producers aiming to market omega-3-enriched eggs across different regions must ensure adherence to both ERPS in the U.S. and the corresponding EU regulations. This dual compliance not only supports safety and quality but also ensures alignment with the diverse regulatory requirements of international markets.

#### 4.2.3. Nutritional Claims and Labelling

Producers of omega-3-enriched eggs must comply with stringent labelling guidelines to accurately convey their nutritional benefits and maintain consumer trust. In the United States, the USDA oversees labelling practices for enriched eggs, ensuring that claims regarding omega-3 content are substantiated and not misleading [134]. For instance, eggs derived from hens fed omega-3-enriched diets, such as those containing flaxseed, can be labelled as having higher omega-3 levels than conventional eggs.

Similarly, in the European Union, food labelling is regulated under Regulation (EU) No 1169/2011 [135] on the provision of food information to consumers. This regulation requires that any nutritional claim, such as "high in omega-3 fatty acids", must meet strict compositional criteria as outlined in Regulation (EC) No 1924/2006 [136] on nutrition and health claims made on foods. Furthermore, the labelling must be clear, accurate, and not mislead consumers regarding the nutritional qualities of the product.

Adhering to both USDA and EU labelling standards ensures that producers of omega-3-enriched eggs can effectively market their products in compliance with regulatory requirements across these regions. This dual compliance also enhances transparency and consumer confidence in the functional benefits of enriched eggs.

### *4.3. Market Trends and Consumer Demand*

The market for omega-3- and omega-6-enriched eggs has expanded considerably in recent years, driven by increased consumer awareness of the health benefits associated with omega-3 fatty acids. These enriched eggs are viewed as a convenient and accessible source of essential nutrients, particularly appealing to health-conscious individuals seeking functional food options [137].

#### 4.3.1. Demand Drivers and Challenges

The growing focus on health and wellness has fuelled demand for omega-3-enriched eggs, which offer additional benefits such as enhanced cardiovascular health and cognitive function. This unique value proposition enables producers to differentiate their products in a competitive market [126]. However, the production of omega-3-enriched eggs involves higher costs due to the need for specialised feed, potentially resulting in higher retail prices. This price differential may deter price-sensitive consumers, limiting market growth [6]. Furthermore, while consumer awareness of omega-3 benefits is rising, a significant portion of the population remains unaware, which could hinder further market expansion [138].

#### 4.3.2. Future Opportunities

Despite these challenges, the omega-3-enriched egg market presents abundant growth opportunities. Producers can capitalise on the increasing demand for functional foods by diversifying product offerings, expanding into untapped markets, and collaborating with health professionals to raise consumer awareness. This trend towards health-focused nutrition is expected to create favourable conditions for the omega-3 egg market, ultimately benefiting both consumers and the food industry [139].

## 5. Health Benefits of Regular Egg Consumption

Eggs are recognised as a highly nutritious, accessible, and cost-effective dietary option. Recent studies have demonstrated that their consumption has either a neutral or beneficial effect on various health markers, without posing significant risks when consumed regularly as part of a balanced diet. Eggs are particularly valuable for populations with elevated nutritional requirements, such as athletes, elderly individuals, infants, children, and pregnant women. They provide high-quality protein and essential micronutrients, including vitamin D, iodine, folate, and choline—nutrients that are frequently deficient in standard diets [140,141].

Globally, Europe ranks as the second-largest egg producer, surpassed only by China, which significantly leads in production volume. In 2021, the average annual per capita egg consumption in Europe was estimated at 220–225 eggs, lower than consumption levels in the United States and Canada, which recorded 285 and 253 eggs per capita, respectively [142,143]. Mexico remains among the highest egg-consuming nations, with recent figures estimating annual per capita consumption at 409 eggs [144–146].

Haward et al. [147] reviewed studies suggesting that consuming only egg whites while excluding yolks could serve as a preventive measure against cardiovascular diseases; however, the evidence regarding cardiac outcomes remains inconclusive. A three-month investigation by de'Ogburn et al. [84] involving participants with metabolic syndrome revealed that consuming three eggs daily reduced tumour necrosis factor and C-reactive protein levels compared to a placebo group, highlighting the anti-inflammatory properties of eggs.

Wang et al. [148] reported that consuming more than four eggs per week did not significantly affect blood pressure or lipid profiles compared to an intake of four eggs or fewer weekly. Similarly, Rouhani et al. [97] found that while egg consumption led to increases in total cholesterol (TC), low-density lipoprotein cholesterol (LDL-C), and high-density lipoprotein cholesterol (HDL-C), it had no notable impact on the TC/HDL ratio or triglycerides—key markers of cardiovascular risk. Furthermore, Li et al. [96] observed that a higher egg consumption correlated with elevated LDL-C and an increased LDL-C/HDL-C ratio, though the effect on HDL-C levels was not significant. These findings suggest that while eggs may influence lipid levels, their overall effect on cardiovascular risk factors, such as the TC/HDL ratio and triglycerides, is negligible or insignificant.

A meta-analysis spanning 54 years of research found no significant association between egg consumption and cardiovascular disease risk. Interestingly, consuming more than one egg per day was linked to a reduced risk of coronary heart disease. Another review of eight observational studies similarly concluded that there was no substantial relationship between egg consumption and cardiovascular diseases [149]. Drouin-Chartier et al. [150] confirmed that a moderate egg intake ($\leq$1 egg per day) is not associated with an increased risk of cardiovascular disorders, aligning with findings by Krittanawong et al. [151], who reported no link between consuming one or more eggs daily and cardiovascular disease.

Additional studies indicate that dietary cholesterol, including that derived from eggs, does not significantly heighten the risk of coronary heart disease or stroke [152]. Notably, substituting 3% of energy from plant-based proteins with egg protein was associated with a reduced risk of cardiovascular disease mortality, particularly from heart disease and stroke in men [153]. Bechthold et al. [154], in a meta-analysis of 12 key food groups, including eggs, demonstrated that moderate egg consumption might reduce coronary heart disease risk. Moreover, Krittanawong et al. [151] suggested that egg consumption could prevent coronary heart disease through carotenoid-enhanced absorption, which improves HDL cholesterol functionality and increases the availability of bioactive compounds, such as lutein and zeaxanthin, thereby providing protection against atherosclerosis.

### 5.1. *Impact of Eggs on Cardiovascular Health, Weight Management, and Muscle Growth*

Regular egg consumption is associated with a broader dietary pattern that often includes a diverse range of protein sources, fish, vegetables, fresh fruits, whole grains, and dairy products, contributing to overall healthier eating habits [155]. Additionally, individuals who consume eggs report greater postprandial satiety, reduced ghrelin responses, and lower consumption of both total and added sugars compared to those who do not include eggs in their diet [156]. These findings are consistent with research by Andersen [87], which highlighted a reduction in the proportion of kilocalories derived from carbohydrates during a whole-egg dietary phase, as opposed to egg-free or egg-white diet phases, while maintaining a consistent total energy intake. Furthermore, the inclusion of whole eggs significantly increased the intake of nutrients associated with cardiometabolic health, such as total fats, arachidonic acid, and sodium.

The role of eggs in muscle protein synthesis has been explored in limited studies; however, available evidence suggests promising outcomes. Research by Moore et al., [88], cited by Puglisi et al. [89], evaluated the impact of consuming 0, 5, 10, 20, and 40 g of protein derived from whole eggs on muscle protein synthesis in healthy young men following leg resistance training [150]. The findings indicated that 20 g of protein were sufficient to maximise muscle protein synthesis. This observation aligns with similar findings by Witard et al. [90], who reported improved muscle protein synthesis with 20 g of whey protein. These results correspond with the International Society of Sports Nutrition's recommendations, which advocate for 20–40 g of high-quality protein per serving to support optimal muscle protein synthesis [84].

The relationship between egg consumption and body weight or composition has yielded mixed results in epidemiological and intervention studies. A cross-sectional study conducted by Garrido-Miguel et al. [92] revealed that consuming more than five eggs per week was associated with a lower body mass index (BMI) and reduced body fat percentage compared to individuals consuming fewer than one egg per week. These findings were based on a young adult population (ages 18–30) and support existing evidence suggesting that eggs enhance satiety and dietary quality [12]. A separate study involving 2241 Chinese adults aged 18 to 80 years demonstrated that consuming more than 50 g of eggs daily was linked to a reduced risk of central obesity and lower body fat levels in women [61]. Conversely, in men, stronger protective associations were observed between egg consumption and a reduced likelihood of being classified as metabolically unhealthy obese [90].

Research by Emrani et al. [95] further suggests that whole-egg consumption, when integrated into an energy-restricted diet, can facilitate greater weight loss in healthy individuals. Given the absence of direct, long-term studies examining the effects of whole-egg consumption on weight and body composition, these findings underscore the need for future clinical trials to explore the impact of eggs on anthropometric measures and overall body weight management.

### 5.2. Clarifying Misconceptions Around Egg Consumption and Cholesterol

A primary concern in research on egg consumption relates to their cholesterol content (approximately 200 mg per egg), which has often been considered a potential contributor to cardiovascular disease (CVD) risk [157,158]. However, studies have demonstrated that egg consumption only marginally increases low-density lipoprotein (LDL) cholesterol levels when incorporated into a well-balanced diet that includes other dietary cholesterol sources. This contrasts with lactovegetarian diets, which typically involve lower cholesterol intakes. These findings suggest that individuals with higher baseline cholesterol intakes exhibit a reduced physiological response to additional dietary cholesterol, such as that derived from eggs [159].

Meta-analyses of controlled, randomised studies indicate minimal elevations in plasma LDL and high-density lipoprotein (HDL) cholesterol levels associated with egg consumption. Furthermore, these studies report negligible changes in critical cardiovascular markers, such as the total HDL and the LDL:HDL cholesterol ratio [94]. According to McDonald et al. [98], experimental research involving human participants has shown that egg consumption does not adversely affect endothelial function. Similarly, Zappala et al. [156] identified a correlation between egg consumption and a reduced risk of stroke, which may be attributed to interactions between cholesterol and other bioactive constituents present in eggs.

Beyond cardiovascular health, eggs represent an important source of nutrients that may positively impact bone health, particularly in elderly populations. Eggs provide high-quality protein and essential amino acids critical for the formation and maintenance of the bone matrix. Moreover, as noted by Rizzoli et al. [101], eggs serve as a natural source of vitamin D, a nutrient essential for calcium absorption and bone mineralisation. This unique nutritional profile positions eggs as a valuable dietary component for addressing the nutritional requirements necessary to support bone health in older adults [85]. A holistic approach, combining egg consumption with physical activity and other lifestyle interventions, could play a significant role in promoting healthier ageing [152].

Olagunju et al. [158] conducted a study involving 176 individuals aged 65 and older to explore the potential effects of egg consumption on bone mineral density. Their findings highlighted the bioactive compounds in eggs as potential contributors to improved bone health. A statistically significant positive correlation was identified between whole-body T-scores and egg intake, with a higher egg consumption linked to greater T-scores, indicative of improved bone density. The study also revealed that gender and the body mass index (BMI) significantly influence bone health [160,161]. Females demonstrated significantly higher T-scores, while individuals with elevated BMIs exhibited superior bone density. Furthermore, a negative correlation was observed between frequent fractures and daily egg consumption, suggesting that those who consumed more eggs were less likely to experience multiple fractures. Interestingly, HDL cholesterol concentrations were positively associated with fracture frequency, suggesting a potential link between cholesterol and bone health [162]. These findings offer new insights into the relationship between egg consumption and bone strength in the elderly, suggesting that regular consumption of whole eggs may enhance bone mineral density and contribute to a reduced risk of fractures and osteoporosis in ageing populations [163].

## 6. Future Directions and Emerging Trends in Omega-3- and Omega-6-Enriched Eggs

The ongoing advancements in the production and marketing of omega-3- and omega-6-enriched eggs highlight their increasing prominence as functional foods. These enriched eggs have gained recognition for their significant health benefits, including improved cardiovascular health, enhanced brain function, and reduced inflammation. As consumers place greater emphasis on nutritional quality and preventive healthcare, omega-enriched eggs have emerged as an essential dietary option, particularly for individuals who do not frequently consume fish, the conventional source of these essential fatty acids [164]. Research has demonstrated that some omega-enriched eggs can contain up to 100 mg of omega-3 fatty acids per egg, making them a substantial contributor to mitigating dietary deficiencies [165].

Innovative approaches in poultry nutrition, such as incorporating flaxseed and algae into chicken feed, enable the precise manipulation of fatty acid profiles in eggs. Additionally, the transition towards sustainable farming practices, including free-range systems and environmentally conscious feed options, addresses not only consumer health preferences but also broader environmental concerns. These efforts respond to increasing warnings regarding the depletion of marine resources often associated with conventional omega-3 sourcing [166].

Despite the growing demand for omega-enriched eggs, several challenges remain. Regulatory restrictions may hinder flexible marketing strategies, while some consumers remain sceptical about the health claims associated with these products [167]. Furthermore, competition from alternative sources of omega-3, such as fish oil and plant-based supplements, may affect the adoption of enriched eggs in consumer diets. As the market for omega-enriched eggs continues to expand, sustained research and development will be critical to overcoming these obstacles. Such efforts will focus on enhancing the nutritional benefits of enriched eggs while ensuring sustainable and environmentally friendly production practices [168].

### *6.1. Impact on Nutritional Profile*

Enhancing chicken feed with omega-3 sources, such as flaxseed, has been shown to significantly improve the nutritional composition of eggs, particularly by increasing the omega-3 fatty acid content. Studies indicate that omega-enriched eggs may provide up to 100 mg of omega-3 fatty acids per egg, thereby serving as a superior dietary source of these vital nutrients. The inclusion of omega-enriched eggs in regular diets is vital for addressing nutritional gaps and supporting overall health, particularly among populations with an inadequate omega-3 fatty acid intake from traditional dietary sources [87].

### *6.2. Production Methods*

#### 6.2.1. Innovative Feeding Practices

Advancements in feeding strategies have emerged as a pivotal approach for enhancing the nutritional quality of eggs [159]. Nutritionists are increasingly formulating specialised feed mixtures rich in polyunsaturated fatty acids (PUFAs), including omega-3 and omega-6 fatty acids, to augment poultry products with these essential nutrients. This involves the strategic inclusion of various lipid sources in the diets of laying hens, enabling precise adjustments to the fatty acid profile of the eggs produced [160].

#### 6.2.2. Tailored Feed Formulations

Recent developments in the poultry industry have seen producers collaborating closely with nutritionists to design bespoke feed formulations tailored to meet specific nutritional objectives. For instance, St Ewe Eggs, a UK-based brand, partnered with Humphrey Feeds & Pullets to develop a unique feed blend enriched with selenium and algae, resulting in nutritionally superior eggs. This focus on customisation reflects a wider trend within the poultry feed sector, wherein manufacturers are transitioning from generic approaches to specialised strategies that cater to the distinct requirements of different bird species [161].

#### 6.2.3. Sustainable Practices

Sustainability has become a central tenet of contemporary poultry farming, shaping both production methodologies and feeding practices. Farmers are increasingly adopting eco-friendly measures that not only promote animal welfare but also align with environmental sustainability goals. For example, free-range systems allow hens to engage in natural behaviours while simultaneously enhancing soil health and biodiversity through their foraging activities. Additionally, the integration of trees and hedgerows into free-range environments supports local ecosystems and minimises reliance on chemical pest control interventions [162].

#### 6.2.4. Impact on Egg Quality

Sustainable farming practices have a notable influence on the quality of eggs produced. Hens reared in enriched environments typically lay eggs with enhanced nutritional profiles, benefiting from diverse diets and opportunities for natural foraging behaviours. This dual approach, prioritising both bird welfare and product quality, addresses the increasing consumer demand for nutritionally enriched food products [163].

#### 6.2.5. Regulatory Considerations

As the demand for omega-enriched eggs continues to grow, regulatory frameworks are evolving to ensure product quality and safety. Emerging regulations are anticipated to emphasise the nutritional composition of poultry products and establish standards for producers to follow in the manufacture of enriched eggs. These quality assurance measures are expected to shape the future of the poultry sector, guiding producers towards innovative and diversified product offerings [126,132,164].

By integrating specialised feeding techniques, sustainable farming methods, and compliance with regulatory standards, the production of omega-enriched eggs is well positioned for significant growth. This aligns with consumer preferences for food products that prioritise health and wellness [165].

### *6.3. Research and Development*

#### 6.3.1. Overview of Omega-Enriched Eggs

An overview of omega-enriched eggs highlights their increasing significance within nutritional science, particularly because of the health benefits associated with omega-3 and omega-6 fatty acids. Research has demonstrated that targeted dietary interventions for laying hens can significantly improve the fatty acid composition of eggs. For instance, "Benefic eggs" are produced by supplementing standard poultry feed with autoclaved linseed, combined with specific vitamins, minerals, and lutein to enhance the nutritional profile of the eggs.

Empirical data supports these advancements. Standard eggs typically contain approximately 0.1% omega-3 fatty acids relative to their total fat content, whereas omega-enriched eggs have been shown to contain between 0.5% and 1.5% omega-3 fatty acids, depending on the dietary modifications applied. For example, a study by Simopoulos et al. [114] demonstrated that eggs from hens fed a linseed-enriched diet contained 300–600 mg of omega-3 fatty acids per egg, compared to 60 mg in conventional eggs. Similarly, the ratio of omega-6 to omega-3 fatty acids, a critical marker of nutritional balance, was reduced from a conventional value of 15:1 to approximately 3:1 in omega-enriched eggs.

Such quantitative improvements underline the efficacy of fortification strategies, making omega-enriched eggs a valuable nutritional option for consumers seeking to address dietary deficiencies in omega-3 and omega-6 fatty acids while benefiting from a holistic nutritional profile. These advancements align with the growing demand for functional foods that address both health and convenience [17,166].

#### 6.3.2. Health Benefits, Consumer Awareness, and Allergy Considerations

The nutritional profile of enriched eggs, particularly Benefic eggs, exhibits marked improvements over conventional eggs, rendering them a valuable dietary option. Benefic eggs contain a significantly higher concentration of essential nutrients, including a sixfold increase in alpha-linolenic acid (ALA) compared to standard eggs, contributing approximately 15% of the recommended daily allowance (RDA) in France. Moreover, the omega-3 fatty acid docosahexaenoic acid (DHA) is enriched to three times the level found in standard eggs, meeting 100% of the RDA. The enrichment process not only boosts omega-3 fatty acid levels but also incorporates key nutrients such as vitamin D, folic acid, vitamin E, lutein, zeaxanthin, iodine, and selenium. These enhancements are achieved while maintaining an optimal omega-6-to-omega-3 fatty acid ratio, aligning with modern nutritional guidelines and promoting a balanced intake of essential fatty acids [18,167].

However, it is also essential to consider potential allergenic risks associated with eggs, as they are a common food allergen. While enrichment enhances the nutritional profile, it is vital to ensure that these modifications do not increase the allergenicity of the product, particularly for sensitive individuals. Further research could explore the allergenic potential of enriched eggs to ensure consumer safety and acceptance [36].

#### 6.3.3. Methodological Considerations

Future research on omega-enriched eggs should prioritise the utilisation of comprehensive and detailed data sets, encompassing both wholesale and retail-level cost analyses. Such an approach would facilitate a more nuanced understanding of the supply and demand dynamics within the omega-enriched egg market. This methodology offers the potential to estimate unconditional elasticities, providing a broader and more accurate representation of market responses. These insights are particularly relevant for policymaking, as they move beyond the current reliance on conditional elasticity estimates, which may inadequately capture the complexities of consumer behaviour and market interactions [14]. By expanding the scope of elasticity estimates, future studies could offer policymakers more robust and actionable data, enabling informed decisions regarding the production and regulation of nutritionally enriched egg products [166,168].

#### 6.3.4. Market Trends and Opportunities

Emerging trends in the global omega-3 market are expected to significantly influence the development and diversification of omega-enriched egg products. Recent market analyses highlight a growing consumer preference for foods with enhanced nutritional value and health-promoting properties, fostering a favourable environment for innovation within this sector [95,169]. This increasing demand emphasises the necessity for continued research and development to optimise the nutritional composition of omega-enriched eggs, aligning with evolving consumer expectations.

To fully exploit these trends, collaboration between researchers and industry stakeholders is essential, leveraging advanced methodologies and technologies. Such cooperative efforts could facilitate the creation of enriched egg products with superior nutritional profiles, while also increasing their accessibility and appeal in an increasingly competitive global marketplace [98].

#### 6.3.5. Environmental Impact

The production and consumption of omega-3- and omega-6-enriched eggs carry significant environmental implications, particularly in terms of sustainability and resource utilisation. Marine ecologists have warned of the potential collapse of global fish stocks by 2050 if current exploitation rates continue, highlighting the urgent need for alternative strategies to meet health and environmental objectives [6]. The increasing popularity of omega-3-enriched eggs is partly driven by heightened consumer awareness of the health benefits of omega-3 fatty acids, traditionally sourced from fish [3]. However, this reliance on marine-derived omega-3s, such as fish oils, poses substantial sustainability challenges, as enhancing hen diets with these ingredients could exacerbate the strain on already overexploited marine ecosystems [170].

In light of these ecological concerns, some producers are turning to more sustainable dietary alternatives for hens, such as plant-based sources like algae, which provide omega-3 fatty acids without directly impacting marine life. This shift is essential not only to reduce dependence on marine resources but also to mitigate the environmental damage caused by pollution and habitat destruction associated with conventional fishing practices [171,172]. Furthermore, the growing consumer preference for organic and cage-free egg options reflects a broader movement towards sustainable agricultural practices that prioritise animal welfare and minimise chemical inputs [173,174].

Recent studies indicate that the demand for omega-3-enriched eggs is influenced not only by health benefits but also by environmental considerations [175,176]. Products perceived as both nutritionally superior and environmentally sustainable are increasingly favoured, offering a competitive advantage in the marketplace [177]. Therefore, addressing the environmental impact of omega-3-enriched egg production is critical for the future of this sector. By integrating health benefits with eco-friendly practices, the industry can safeguard marine ecosystems while advancing sustainable agricultural methods to meet consumer demands [178].

#### 6.3.6. Challenges and Barriers

The regulatory framework governing enriched eggs, particularly in relation to the addition of vitamins and minerals, presents significant challenges for producers. According to Regulation (EC) No. 852/2004 [175] on the hygiene of foodstuffs, the addition of vitamins and minerals to unprocessed foods is explicitly prohibited. This regulation aims to preserve the transparency of the natural nutritional value of fresh food products, preventing potential consumer confusion regarding their nutrient content. However, such restrictions may obstruct marketing strategies for omega-3- and omega-6-enriched eggs, as consumers may perceive the inclusion of added nutrients as misleading, potentially undermining the authenticity of these products.

The regulatory landscape is further complicated by the Nutrition (Amendment etc.) (EU Exit) Regulations 2019, which have reallocated certain regulatory functions from the European Commission to national authorities within the United Kingdom. Responsibilities are now distributed among distinct entities, including the Secretary of State for England, the Scottish Ministers, and the Welsh Ministers, resulting in a fragmented regulatory framework [176]. This decentralisation may lead to inconsistencies in regulatory interpretation and enforcement across the UK, creating challenges for producers seeking to distribute enriched egg products across multiple regions. Consequently, compliance with varying requirements may hinder producers' ability to maintain uniform operations within the UK market [177].

Addressing these regulatory complexities is critical for advancing the development and commercialisation of nutritionally enhanced egg products. Harmonising regulations and improving clarity in compliance requirements will be essential to support producers in meeting both consumer expectations and market demands.

#### 6.3.7. Market Acceptance

The acceptance of omega-3- and omega-6-enriched eggs by consumers presents several challenges that warrant consideration. Although public awareness of the health benefits associated with these essential fatty acids is on the rise, certain consumer segments remain sceptical regarding the claims made by producers. This scepticism is often rooted in a limited understanding of the nutritional science underlying these enhancements, which can lead to consumer resistance, especially in markets where traditional egg consumption habits are firmly established [102]. Such resistance underscores the importance of effective communication strategies that can demystify the benefits of enriched eggs and align consumer perceptions with scientifically substantiated health advantages [178].

Economic factors are also critical in determining the market adoption of enriched eggs. Research indicates that tariff reductions and other economic measures can have a tangible impact on consumer welfare, as fluctuations in pricing directly influence demand [179]. Enriching eggs with additional nutrients entails higher production costs, which are typically reflected in elevated retail prices. This price increase may deter consumers who are particularly sensitive to cost, which is a significant consideration in regions where purchasing decisions are heavily influenced by price rather than by nutritional content. Consequently, the higher cost of enriched eggs could limit their market penetration, particularly in price-sensitive demographics, thereby posing a challenge to the overall economic viability of these products [74].

Additionally, the market for omega-3- and omega-6-enriched eggs faces competition from alternative sources of these fatty acids, such as fish oil supplements and plant-based omega-3 products. These alternatives often appeal to consumers due to their convenience and the comparable health benefits they offer, without requiring major adjustments in dietary habits. The ready availability of such alternatives could potentially erode the market share of enriched eggs unless producers can effectively communicate the distinct benefits that their products provide. For enriched eggs to secure a competitive position, producers must emphasise unique selling points, such as the integration of these nutrients into a familiar food format and the potential for broader nutritional benefits beyond those provided by isolated supplements [172].

In summary, the successful market adoption of omega-3- and omega-6-enriched eggs depends on addressing consumer scepticism through education, managing the economic factors that influence purchasing decisions, and differentiating enriched eggs from alternative omega-3 sources. These strategies are essential to overcome the barriers posed by consumer acceptance, economic constraints, and competitive pressures within the omega-enriched food market.

#### 6.3.8. Production Challenges

The production of omega-3- and omega-6-enriched eggs requires a meticulous and methodical approach to feed formulations and farming procedures, as these elements are essential for attaining the needed nutritional enhancements. To enhance the concentrations of omega-3 and omega-6 fatty acids in eggs, chickens require a meticulously formulated diet that fulfils their nutritional requirements while maximising the incorporation of these critical fatty acids into the egg yolk [137]. This dietary supplementation often includes sources rich in omega-3, such as flaxseed or algae, which must be incorporated in a way that maintains animal health and productivity while enhancing egg quality. However, designing and maintaining such specialised feed formulations can be complex, requiring substantial investment in research and development to refine and validate feeding protocols that yield consistent results [173].

Furthermore, the scalability of omega-enriched egg production presents logistical issues, especially in maintaining uniformity in product quality across increased production quantities. As the demand for these nutritionally fortified eggs increases, manufacturers must retain the integrity of nutritional profiles while expanding operations. Variability in nutritional composition resulting from disparities in feed formulation, environmental factors, and hen health between production locations may affect the uniformity of enhanced egg products. Confronting these problems is crucial to fulfil consumer expectations and comply with legal criteria for product labelling. Consequently, continuous progress in agricultural and nutritional science is essential for developing dependable procedures that promote both the nutritious improvement of eggs and the scalability of production [180].

## 7. Conclusions

The nutritional profile of omega-3- and omega-6-enriched eggs establishes them as a significant advancement in functional foods, offering health benefits such as improved cardiovascular function, reduced inflammation, and enhanced cognitive performance. Through dietary modifications for poultry, incorporating polyunsaturated fatty acid-rich sources such as flaxseed and algae, producers can effectively address common dietary deficiencies, particularly in populations with limited fish consumption.

The rising demand for functional foods among health-conscious consumers has driven innovation in omega-enriched egg production. This trend is particularly evident in regions such as North America, Europe, and Asia–Pacific, where awareness of the health benefits of omega-3 fatty acids is increasing. Consequently, omega-enriched eggs are positioned to occupy a substantial segment of the functional food market, aligning with the growing consumer preference for preventive health measures and wellness-oriented diets.

Despite their potential, challenges remain in terms of consumer perception and regulatory frameworks. Misunderstandings about the health benefits and science behind enriched eggs may foster scepticism, particularly in regions with deeply rooted traditional diets. Additionally, regulatory restrictions, such as those in the European Union that limit nutrient additions to unprocessed foods, complicate the marketing of these products. Overcoming these barriers will be critical to ensuring the long-term success of omega-enriched eggs in the marketplace.

The environmental impact of omega-enriched egg production must also be addressed. Traditional omega-3 sources, such as fish oils, place significant strain on marine ecosystems. Transitioning to sustainable feed sources, such as algae, presents a viable solution to mitigate ecological pressures while maintaining the nutritional integrity of enriched eggs. Moreover, the increasing demand for organic and cage-free options reflects consumer interest in sustainable and ethical farming practices, aligning with broader environmental goals.

Future research should focus on personalised feed formulations, improving bioactive compound enrichment, and analysing economic implications to facilitate widespread adoption. Studies examining market dynamics, including cost elasticity at wholesale and retail levels, will inform effective pricing and policy strategies. Enhanced feed sources could enable the development of products targeting specific health concerns, such as cardiovascular disease and cognitive decline.

From a public health perspective, omega-enriched eggs offer an affordable, nutrient-dense food option that can contribute to meeting dietary recommendations for essential fatty acids. Their regular consumption could significantly address nutritional gaps and improve health outcomes, particularly for vulnerable groups such as the elderly, athletes, and pregnant women.

In summary, omega-enriched eggs represent a crucial intersection of nutrition, sustainability, and innovation in functional food development. By addressing consumer health needs and environmental sustainability, these products are well positioned within the evolving landscape of personalised and preventive nutrition. Continued growth in this market will depend on overcoming regulatory challenges, fostering consumer trust, and advancing research to demonstrate the comprehensive benefits of these enriched egg products.

**Author Contributions:** Conceptualisation, M.G.U. and R.N.R.; methodology, R.N.R.; software, F.S.; validation, M.G.U., R.-M.R.R. and A.U.; formal analysis, R.N.R. and I.C.C.; investigation, I.D.V. and I.C.C.; resources, I.D.V.; data curation, M.G.U.; writing—original draft preparation, R.N.R.; writing—review and editing, M.G.U.; visualisation, R.-M.R.R.; supervision, A.U.; project administration, R.N.R.; and funding acquisition, R.N.R. All authors have read and agreed to the published version of the manuscript.

**Funding:** This research received no external funding.

**Institutional Review Board Statement:** Not applicable.

**Data Availability Statement:** Data sharing is not applicable to this article.

**Conflicts of Interest:** The authors declare no conflicts of interest.

## References

1. Gavril, R.N.R.; Usturoi, M.G.; Usturoi, A. Table Eggs Quality, According to the Storage Period. *Curr. Opin. Biotechnol.* **2013**, *24*, S83. [CrossRef]
2. Anton, M.; Nau, F.; Guérin-Dubiard, C. Bioactive Fractions of Eggs for Human and Animal Health. In *Improving the Safety and Quality of Eggs and Egg Products*; Elsevier: Amsterdam, The Netherlands, 2011; pp. 321–345.
3. Adeoye, A.A.; Oyeleye, O.O.; Olorunsola, R.A.; Udoh, J.E.; Oladepo, A.D. Table Egg Quality and Nutritional Composition Assessements of Different Breeds and Ages of Laying Hens. *Slovak J. Anim. Sci.* **2023**, *56*, 38–45. [CrossRef]
4. Réhault-Godbert, S.; Guyot, N.; Nys, Y. The Golden Egg: Nutritional Value, Bioactivities, and Emerging Benefits for Human Health. *Nutrients* **2019**, *11*, 684. [CrossRef]
5. Sugano, M.; Matsuoka, R. Nutritional Viewpoints on Eggs and Cholesterol. *Foods* **2021**, *10*, 494. [CrossRef] [PubMed]
6. McNamara, D.J. The Fifty Year Rehabilitation of the Egg. *Nutrients* **2015**, *7*, 8716–8722. [CrossRef]
7. Usturoi, A.; Simeanu, C.; Usturoi, M.G.; Dolis, M.G.; Ratu, R.N.; Simeanu, D. Influence of Packaging Type on the Dynamics of Powdered Eggs Chemical Composition. *Mater. Plast.* **2017**, *54*, 380–385. [CrossRef]
8. Antoni, R. Dietary Saturated Fat and Cholesterol: Cracking the Myths around Eggs and Cardiovascular Disease. *J. Nutr. Sci.* **2023**, *12*, e97. [CrossRef] [PubMed]
9. Fabro, C.; Romanzin, A.; Spanghero, M. Fatty Acid Profile of Table Eggs from Laying Hens Fed Hempseed Products: A Meta-Analysis. *Livest. Sci.* **2021**, *254*, 104748. [CrossRef]
10. Usturoi, M.G.; Rațu, R.N.; Mihail Radu-Rusu, R.; Ivancia, M.; Usturoi, A. Fatty Acid Profile in Eggs and Eggs Products. *Sci. Papers. Ser. D Anim. Sci.* **2021**, *64*, 399–403.
11. Kovacs-Nolan, J.; Phillips, M.; Mine, Y. Advances in the Value of Eggs and Egg Components for Human Health. *J. Agric. Food Chem.* **2005**, *53*, 8421–8431. [CrossRef] [PubMed]
12. Andersen, C.J. Bioactive Egg Components and Inflammation. *Nutrients* **2015**, *7*, 7889–7913. [CrossRef] [PubMed]

13. Mens, A.J.W.; van Krimpen, M.M.; Kar, S.K.; Guiscafre, F.J.; Sijtsma, L. Enriching Table Eggs with N-3 Polyunsaturated Fatty Acids through Dietary Supplementation with the Phototrophically Grown Green Algae Nannochloropsis Limnetica: Effects of Microalgae on Nutrient Retention, Performance, Egg Characteristics and Health Parameters. *Poult. Sci.* **2022**, *101*, 101869. [CrossRef] [PubMed]
14. Fraeye, I.; Bruneel, C.; Lemahieu, C.; Buyse, J.; Muylaert, K.; Foubert, I. Dietary Enrichment of Eggs with Omega-3 Fatty Acids: A Review. *Food Res. Int.* **2012**, *48*, 961–969. [CrossRef]
15. Nistor, L.I.; Albu, A.; Nistor, A.C.; Usturoi, M.G. Aspects of Eggs Quality Provided from Free Range and Conventional Systems. *J. Microbiol. Biotechnol. Food Sci.* **2015**, *5*, 186–189. [CrossRef]
16. Criste, F.L.; Mierlita, D.; Simeanu, D.; Boisteanu, P.C.; Pop, I.M.; Georgescu, B.; Nacu, G. Study of Fatty Acids Profile and Oxidative Stability of Egg Yolk from Hens Fed a Diet Containing White Lupine Seeds Meal. *Rev. Chim.* **2018**, *69*, 2454–2460. [CrossRef]
17. Ciobanu, M.M.; Boisteanu, P.C.; Simeanu, D.; Postolache, A.N.; Lazar, R.; Vintu, C.R. Study on the Profile of Fatty Acids of Broiler Chicken Raised and Slaughtered in Industrial System. *Rev. Chim.* **2019**, *70*, 4089–4094. [CrossRef]
18. Herranz, B.; Romero, C.; Sánchez-Román, I.; López-Torres, M.; Viveros, A.; Arija, I.; Álvarez, M.D.; de Pascual-Teresa, S.; Chamorro, S. Enriching Eggs with Bioactive Compounds through the Inclusion of Grape Pomace in Laying Hens Diet: Effect on Internal and External Egg Quality Parameters. *Foods* **2024**, *13*, 1553. [CrossRef]
19. Pal, M.; Molnár, J. The Role of Eggs as an Important Source of Nutrition in Human Health. *Int. J. Food Sci. Agric.* **2021**, *5*, 180–182. [CrossRef]
20. Bílková, B.; Świderská, Z.; Zita, L.; Laloë, D.; Charles, M.; Beneš, V.; Stopka, P.; Vinkler, M. Domestic Fowl Breed Variation in Egg White Protein Expression: Application of Proteomics and Transcriptomics. *J. Agric. Food Chem.* **2018**, *66*, 11854–11863. [CrossRef]
21. Krawczyk, J.; Lewko, L.; Sokołowicz, Z.; Koseniuk, A.; Kraus, A. Effect of Hen Genotype and Laying Time on Egg Quality and Albumen Lysozyme Content and Activity. *Animals* **2023**, *13*, 1611. [CrossRef] [PubMed]
22. Valverde, D.; Laca, A.; Estrada, L.N.; Paredes, B.; Rendueles, M.; Díaz, M. Egg Yolk Fractions as Basic Ingredient in the Development of New Snack Products. *Int. J. Gastron. Food Sci.* **2016**, *3*, 23–29. [CrossRef]
23. Raţu, gavril; Usturoi, R.N. Effects of Temperature and Storage Time on Hen Eggs Quality. *J. Biotechnol.* 2012.
24. Houchins, J.; Fulgoni, V.L., III; Papanikolaou, Y. Eggs Contribute Essential Nutrients to the Diet of American Children across Ethnicities. *Curr. Dev. Nutr.* **2024**, *8*, 103339. [CrossRef]
25. Ruxton, C.H.S.; Derbyshire, E.; Gibson, S. The Nutritional Properties and Health Benefits of Eggs. *Nutr. Food Sci.* **2010**, *40*, 263–279. [CrossRef]
26. Stadelman, W.J.; Cotterill, O. *Egg Science and Technology*; Haworth Press: New York, NY, USA, 1995.
27. Radu-Rusu, R.M.; Usturoi, M.G.; Leahu, A.; Amariei, S.; Radu-Rusu, C.G.; Vacaru-Opriş, I. Chemical Features, Cholesterol and Energy Content of Table Hen Eggs from Conventional and Alternative Farming Systems. *S. Afr. J. Anim. Sci.* **2014**, *44*, 33. [CrossRef]
28. McCance, R.A.; Widdowson, E.M. *The Composition of Foods*; The Royal Society of Chemistry: London, UK, 2006.
29. Zeisel, S.H.; Mar, M.H.; Howe, J.C.; Holden, J.M. Concentrations of Choline-Containing compounds and Betaine in Common Foods. *J. Nutr.* **2003**, *133*, 1302–1307. [CrossRef]
30. Nistor, A.C.; Nistor, L.I.; Usturoi, M.G. A Review of Fatty Acid and Amino Acids Profile from Pasteurized Egg Liquids Produced in Romania. *Bull. Univ. Agric. Sci. Vet. Med. Cluj-Napoca Food Sci. Technol.* **2018**, *75*, 143–148. [CrossRef] [PubMed]
31. Ariza, A.G.; González, F.J.; Arbulu, A.A.; Bermejo, J.V.; Vallejo, M.E. Hen Breed and Variety Factors as a Source of Variability for the Chemical Composition of Eggs. *J. Food Compost. Anal.* **2021**, *95*, 103673. [CrossRef]
32. Chen, G.; Cai, Y.; Su, Y.; Gao, B.; Wu, H.; Cheng, J. Effects of Spirulina Algae as a Feed Supplement on Nutritional Value and Flavour Components of Silkie Hens Eggs. *J. Anim. Physiol. Anim. Nutr.* **2019**, *103*, 1408–1417. [CrossRef]
33. Raţu, R.N.; Ciobanu, M.M.; Radu-Rusu, R.M.; Usturoi, M.G.; Ivancia, M.; Doliș, M.G. Study on the chemical composition and nitrogen fraction of milk from different animal species. *Sci. Papers. Ser. D Anim. Sci.* **2021**, *64*, 374–379.
34. Escamilla Rosales, M.F.; Olvera Rosales, L.; Jara Gutiérrez, C.E.; Jaimez Ordaz, J.; Santana Sepúlveda, P.A.; González Olivares, L.G. Proteins of Milk, Egg and Fish as a Source of Antioxidant Peptides: Production, Mechanism of Action and Health Benefits. *Food Rev. Int.* **2024**, *40*, 1600–1620. [CrossRef]
35. Kuang, H.; Yang, F.; Zhang, Y.; Wang, T.; Chen, G. The Impact of Egg Nutrient Composition and Its Consumption on Cholesterol Homeostasis. *Cholesterol* **2018**, *2018*, 6303810. [CrossRef] [PubMed]
36. Chang, C.; Lahti, T.; Tanaka, T.; Nickerson, M.T. Egg Proteins: Fractionation, Bioactive Peptides and Allergenicity. *J. Sci. Food Agric.* **2018**, *98*, 5547–5558. [CrossRef] [PubMed]
37. Miranda, J.M.; Anton, X.; Redondo-Valbuena, C.; Roca-Saavedra, P.; Rodriguez, J.A.; Lamas, A.; Franco, C.M.; Cepeda, A. Egg and Egg-Derived Foods: Effects on Human Health and Use as Functional Foods. *Nutrients* **2015**, *7*, 706–729. [CrossRef] [PubMed]
38. Anderson, K.E. Comparison of Fatty Acid, Cholesterol, Vitamin A and E Composition, and Trans Fats in Eggs from Brown and White Egg Strains That Were Molted or Nonmolted. *Poult. Sci.* **2013**, *92*, 3259–3265. [CrossRef] [PubMed]

39. Lopez-Rodriguez, E.; Roldan, N.; Garcia-Alvarez, B.; Pérez-Gill, J. Protein and Lipid Fingerprinting of Native-like Membrane Complexes by Combining TLC and Protein Electrophoresis. *J. Lipid Res.* **2019**, *60*, 430–435. [CrossRef] [PubMed]
40. Seuss-Baum, I.; Nau, F.; Guérin-Dubiard, C.; Nys, Y.; Bain, M. The Nutritional Quality of Eggs. In *Improving the Safety and Quality of Egg and Egg Products*; Egg Safety and Nutritional Quality; Woodhead Publishing: Cambridge, UK, 2011; Volume 2, pp. 201–236.
41. Abeyrathne, E.D.N.S.; Nam, K.-C.; Huang, X.; Ahn, D.U. Egg Yolk Lipids: Separation, Characterization, and Utilization. *Food Sci. Biotechnol.* **2022**, *31*, 1243–1256. [CrossRef]
42. Grčević, M.; Kralik, Z.; Kralik, G.; Galović, O. Effects of Dietary Marigold Extract on Lutein Content, Yolk Color and Fatty Acid Profile of Omega-3 Eggs: Lutein Content in Omega-3 Eggs. *J. Sci. Food Agric.* **2019**, *99*, 2292–2299. [CrossRef]
43. Sanlier, N.; Üstün, D. Egg Consumption and Health Effects: A Narrative Review. *J. Food Sci.* **2021**, *86*, 4250–4261. [CrossRef] [PubMed]
44. Somaratne, G.; Nau, F.; Ferrua, M.J.; Singh, J.; Ye, A.; Dupont, D.; Singh, R.P.; Floury, J. Characterization of Egg White Gel Microstructure and Its Relationship with Pepsin Diffusivity. *Food Hydrocoll.* **2020**, *98*, 105258. [CrossRef]
45. Patil, S.; Rao, B.; Matondkar, M.; Bhushette, P.; Sonawane, S.K. Review on understanding of egg yolk as functional ingredients. *J. Microbiol. Biotechnol. Food Sci.* **2022**, *11*, e4627. [CrossRef]
46. Manzoor, S.; Fayaz, U.; Dar, A.H.; Dash, K.K.; Shams, R.; Bashir, I.; Pandey, V.K.; Abdi, G. Sustainable Development Goals through Reducing Food Loss and Food Waste: A Comprehensive Review. *Future Foods* **2024**, *9*, 100362. [CrossRef]
47. Vizzoto, F.; Testa, F.; Iraldo, F. Strategies to Reduce Food Waste in the Foodservices Sector: A Systematic Review. *Int. J. Hosp. Manag.* **2021**, *95*, 102933. [CrossRef]
48. Schiavone, A.; Barroeta, A.C. *Improving the Safety and Quality of Eggs and Egg Products*; Van, I.F., Nys, Y., Bain, M., Eds.; Woodhead Publishing: Cambridge, UK, 2011; Volume 2.
49. Tang, S.G.H.; Sieo, C.C.; Kalavathy, R.; Saad, W.Z.; Yong, S.T.; Wong, H.K.; Ho, Y.W. Chemical Compositions of Egg Yolks and Egg Quality of Laying Hens Fed Prebiotic, Probiotic, and Synbiotic Diets: Chemical Compositions and Egg Quality. *J. Food Sci.* **2015**, *80*, C1686–C1695. [CrossRef]
50. Naber, E.C. Modifying Vitamin Composition of Eggs: A Review. *J. Appl. Poult. Res.* **1993**, *2*, 385–393. [CrossRef]
51. Wallace, T.C.; Fulgoni, V.L. Usual Choline Intakes Are Associated with Egg and Protein Food Consumption in the United States. *Nutrients* **2017**, *9*, 839. [CrossRef]
52. Cho, C.E.; Aardema, N.D.J.; Bunnell, M.L.; Larson, D.P.; Aguilar, S.S.; Bergeson, J.R.; Malysheva, O.V.; Caudill, M.A.; Lefevre, M. Effect of Choline Forms and Gut Microbiota Composition on Trimethylamine-N-Oxide Response in Healthy Men. *Nutrients* **2020**, *12*, 2220. [CrossRef] [PubMed]
53. Obeid, R.; Karlsson, T. Choline—A Scoping Review for Nordic Nutrition Recommendations 2023. *Food Nutr. Res.* **2023**, *67*, 10303. [CrossRef]
54. Ranard, K.M.; Jeon, S.; Mohn, E.S.; Griffiths, J.C.; Johnson, E.J.; Erdman, J.W., Jr. Dietary Guidance for Lutein: Consideration for Intake Recommendations Is Scientifically Supported. *Eur. J. Nutr.* **2017**, *56* (Suppl. S3), 37–42. [CrossRef]
55. Zampiga, M.; Calini, F.; Sirri, F. Importance of Feed Efficiency for Sustainable Intensification of Chicken Meat Production: Implications and Role for Amino Acids, Feed Enzymes and Organic Trace Minerals. *Worlds. Poult. Sci. J.* **2021**, *77*, 639–659. [CrossRef]
56. Nys, Y.; Schlegel, P.; Durosoy, S.; Jondreville, C.; Narcy, A. Adapting Trace Mineral Nutrition of Birds for Optimising the Environment and Poultry Product Quality. *Worlds. Poult. Sci. J.* **2018**, *74*, 225–238. [CrossRef]
57. Sarantidi, E.; Ainatzoglou, A.; Papadimitriou, C.; Stamoula, E.; Maghiorou, K.; Miflidi, A.; Trichopoulou, A.; Mountzouris, K.C.; Anagnostopoulos, A.K. Egg White and Yolk Protein Atlas: New Protein Insights of a Global Landmark Food. *Foods* **2023**, *12*, 3470. [CrossRef]
58. Anton, M.; Nau, F.; Nys, Y. Bioactive Egg Components and Their Potential Uses. *Worlds. Poult. Sci. J.* **2006**, *62*, 429–438. [CrossRef]
59. Nolasco, E.; Yang, J.; Ciftci, O.; Vu, D.C.; Alvarez, S.; Purdum, S.; Majumder, K. Evaluating the Effect of Cooking and Gastrointestinal Digestion in Modulating the Bio-Accessibility of Different Bioactive Compounds of Eggs. *Food Chem.* **2021**, *344*, 128623. [CrossRef] [PubMed]
60. Zhang, X.; Chelliappan, B.; Antonysamy, M. Recent Advances in Applications of Bioactive Egg Compounds in Nonfood Sectors. *Front. Bioeng. Biotechnol.* **2021**, *9*, 738993. [CrossRef] [PubMed]
61. Lee, J.H.; Paik, H.-D. Anticancer and Immunomodulatory Activity of Egg Proteins and Peptides: A Review. *Poult. Sci.* **2019**, *98*, 6505–6516. [CrossRef]
62. Carter, S.; Connole, E.S.; Hill, A.M.; Buckley, J.D.; Coates, A.M. Eggs and Cardiovascular Disease Risk: An Update of Recent Evidence. *Curr. Atheroscler. Rep.* **2023**, *25*, 373–380. [CrossRef]
63. Schade, D.S.; Gonzales, K.; Kaminsky, N.; Adolphe, A.; Shey, L.; Eaton, R.P. Resolving the Egg and Cholesterol Intake Controversy: New Clinical Insights into Cholesterol Regulation by the Liver and Intestine. *Endocr. Pract.* **2022**, *28*, 102–109. [CrossRef] [PubMed]

64. Miguel, M.; Vassallo, D.V.; Wiggers, G.A. Bioactive Peptides and Hydrolysates from Egg Proteins as a New Tool for Protection against Cardiovascular Problems. *Curr. Pharm. Des.* **2020**, *26*, 3676–3683. [CrossRef]
65. Liu, Y.-F.; Oey, I.; Bremer, P.; Carne, A.; Silcock, P. Bioactive Peptides Derived from Egg Proteins: A Review. *Crit. Rev. Food Sci. Nutr.* **2018**, *58*, 2508–2530. [CrossRef] [PubMed]
66. Bhat, Z.F.; Kumar, S.; Bhat, H.F. Bioactive Peptides from Egg: A Review. *Nutr. Food Sci.* **2015**, *45*, 190–212. [CrossRef]
67. Trziszka, T.; Różański, H.; Polanowski, A. Eggs as a Very Promising Source of Biomedical and Nutraceutical Preparations: A Review. *J. Life Sci.* **2013**, *8*, 862. [CrossRef]
68. Legros, J.; Jan, S.; Bonnassie, S.; Gautier, M.; Croguennec, T.; Pezennec, S.; Cochet, M.-F.; Nau, F.; Andrews, S.C.; Baron, F. The Role of Ovotransferrin in Egg-White Antimicrobial Activity: A Review. *Foods* **2021**, *10*, 823. [CrossRef]
69. Wang, J.; Wu, J. Proteomic Analysis of Fertilized Egg White during Early Incubation. *EuPA Open Proteom.* **2014**, *2*, 38–59. [CrossRef]
70. Mann, K.; Mann, M. In-Depth Analysis of the Chicken Egg White Proteome Using an LTQ Orbitrap Velos. *Proteome Sci.* **2011**, *9*, 7. [CrossRef]
71. Anton, M. Egg Yolk: Structures, Functionalities, and Processes. *J. Sci. Food Agric.* **2013**, *93*, 2871–2880. [CrossRef]
72. Palacios, L.E. *Egg Yolk Lecithin Fractionation and Characterization*; Department of Food Science and Human Nutrition, Center for Crops Iowa State University: Ames, IA, USA, 2024.
73. Ohman, M.; Akerfeldt, T.; Nilsson, I.; Rosen, C.; Hansson, L.-O.; Carlsson, M.; Larsson, A. Biochemical Effects of Consumption of Eggs Containing Omega-3 Polyunsaturated Fatty Acids. *Ups. J. Med. Sci.* **2008**, *113*, 315–323. [CrossRef]
74. Shakoor, H.; Khan, M.I.; Sahar, A.; Khan, M.K.I.; Faiz, F.; Basheer Ahmad, H. Development of Omega-3 Rich Eggs through Dietary Flaxseed and Bio-Evaluation in Metabolic Syndrome. *Food Sci. Nutr.* **2020**, *8*, 2619–2626. [CrossRef]
75. Lemahieu, C.; Bruneel, C.; Ryckebosch, E.; Muylaert, K.; Buyse, J.; Foubert, I. Impact of Different Omega-3 Polyunsaturated Fatty Acid (n-3 PUFA) Sources (Flaxseed, Isochrysis Galbana, Fish Oil and DHA Gold) on n-3 LC-PUFA Enrichment (Efficiency) in the Egg Yolk. *J. Funct. Foods* **2015**, *19*, 821–827. [CrossRef]
76. Attia, Y.; Sagan, A.A.A.; Hussein, E.-S.O.S.; Olal, M.J.; Ebeid, T.A.; Alabdullatif, A.A.; Alhotan, R.A.; Qaid, M.M.; Tufarelli, V.; Pugliese, G.; et al. Enhancing the Nutritional Values of Egg Yolks of Laying Hens by Different Dietary Sources of Omega-3 Fatty Acids, Vitamin e and Trace Elements. *Livest. Sci.* **2024**, *289*, 105573. [CrossRef]
77. Batiha, G.E.-S.; Alqarni, M.; Awad, D.A.B.; Algammal, A.M.; Nyamota, R.; Wahed, M.I.I.; Shah, M.A.; Amin, M.N.; Adetuyi, B.O.; Hetta, H.F.; et al. Dairy-Derived and Egg White Proteins in Enhancing Immune System against COVID-19. *Front. Nutr.* **2021**, *8*, 629440. [CrossRef] [PubMed]
78. Elkin, R.G.; Harvatine, K.J. A Review of Recent Studies on the Enrichment of Eggs and Poultry Meat with Omega-3 Polyunsaturated Fatty Acids: Novel Findings and Unanswered Questions. *Poult. Sci.* **2023**, *102*, 102938. [CrossRef] [PubMed]
79. Dullius, A.; Fassina, P.; Giroldi, M.; Goettert, M.I.; Volken de Souza, C.F. A Biotechnological Approach for the Production of Branched Chain Amino Acid Containing Bioactive Peptides to Improve Human Health: A Review. *Food Res. Int.* **2020**, *131*, 109002. [CrossRef] [PubMed]
80. Choi, M.; Lee, J.-H.; Lee, Y.-J.; Paik, H.-D.; Park, E. Egg Yolk Protein Water Extracts Modulate the Immune Response in BALB/c Mice with Immune Dysfunction Caused by Forced Swimming. *Foods* **2022**, *11*, 121. [CrossRef]
81. Meram, C.; Wu, J. Anti-Inflammatory Effects of Egg Yolk Livetins (α, β, and γ-Livetin) Fraction and Its Enzymatic Hydrolysates in Lipopolysaccharide-Induced RAW 264.7 Macrophages. *Food Res. Int.* **2017**, *100*, 449–459. [CrossRef] [PubMed]
82. Mah, E.; Chen, C.-Y.O.; Liska, D.J. The Effect of Egg Consumption on Cardiometabolic Health Outcomes: An Umbrella Review. *Public Health Nutr.* **2020**, *23*, 935–955. [CrossRef]
83. Kouvari, M.; Damigou, E.; Florentin, M.; Kosti, R.I.; Chrysohoou, C.; Pitsavos, C.S.; Panagiotakos, D.B. Egg Consumption, Cardiovascular Disease and Cardiometabolic Risk Factors: The Interaction with Saturated Fatty Acids. Results from the ATTICA Cohort Study (2002–2012). *Nutrients* **2022**, *14*, 5291. [CrossRef] [PubMed]
84. Ratliff, J.; Leite, J.O.; de Ogburn, R.; Puglisi, M.J.; VanHeest, J.; Fernandez, M.L. Consuming Eggs for Breakfast Influences Plasma Glucose and Ghrelin, While Reducing Energy Intake during the next 24 Hours in Adult Men. *Nutr. Res.* **2010**, *30*, 96–103. [CrossRef] [PubMed]
85. Missimer, A.; Dimarco, D.M.; Andersen, C.J.; Murillo, A.G.; Vergara-Jimenez, M.; Fernandez, M.L. Consuming Two Eggs per Day, as Compared to an Oatmeal Breakfast, Increases Plasma Ghrelin While Maintaining the LDL/HDL Ratio. *Nutrients* **2017**, *9*, 89. [CrossRef] [PubMed]
86. Mason, P. The Importance of Eggs in an Environmentally Sustainable Diet. *Nutr. Bull.* **2023**, *48*, 400–410. [CrossRef]
87. Andersen, C.J.; Huang, L.; Zhai, F.; Esposito, C.P.; Greco, J.M.; Zhang, R.; Woodruff, R.; Sloan, A.; Van Dyke, A.R. Consumption of Different Egg-Based Diets Alters Clinical Metabolic and Hematological Parameters in Young, Healthy Men and Women. *Nutrients* **2023**, *15*, 3747. [CrossRef] [PubMed]

88. Moore, D.R.; Robinson, M.J.; Fry, J.L.; Tang, J.E.; Glover, E.I.; Wilkinson, S.B.; Prior, T.; Tarnopolsky, M.A.; Phillips, S.M. Ingested Protein Dose Response of Muscle and Albumin Protein Synthesis after Resistance Exercise in Young Men. *Am. J. Clin. Nutr.* **2009**, *89*, 161–168. [CrossRef] [PubMed]
89. Puglisi, M.J.; Fernandez, M.L. The Health Benefits of Egg Protein. *Nutrients* **2022**, *14*, 2904. [CrossRef] [PubMed]
90. Witard, O.C.; Jackman, S.R.; Breen, L.; Smith, K.; Selby, A.; Tipton, K.D. Myofibrillar Muscle Protein Synthesis Rates Subsequent to a Meal in Response to Increasing Doses of Whey Protein at Rest and after Resistance Exercise. *Am. J. Clin. Nutr.* **2014**, *99*, 86–95. [CrossRef] [PubMed]
91. Jäger, R.; Kerksick, C.M.; Campbell, B.I.; Cribb, P.J.; Wells, S.D.; Skwiat, T.M.; Purpura, M.; Ziegenfuss, T.N.; Ferrando, A.A.; Arent, S.M.; et al. International Society of Sports Nutrition Position Stand: Protein and Exercise. *J. Int. Soc. Sports Nutr.* **2017**, *14*, 20. [CrossRef]
92. Garrido-Miguel, M.; Mesas, A.E.; Fernández-Rodríguez, R.; Fernández-Franco, S.; Pozuelo-Carrascosa, D.P.; López-Gil, J.F.; Martínez-Vizcaíno, V. The Role of Protein Intake in the Relationship between Egg Consumption and Body Composition in Young Adults. A Mediation Analysis. *Clin. Nutr.* **2022**, *41*, 2356–2363. [CrossRef]
93. Liu, R.; Zhao, Y.; Li, Q.; Dang, S.; Yan, H. Body Fat Mass, Fat Distribution and Egg Consumption: A Population-Based Study in Chinese Adults: Egg Consumption and Body Fat in Rural Chinese. *J. Am. Coll. Nutr.* **2020**, *39*, 528–536. [CrossRef]
94. Tabatabaeyan, A.; Lotfi, K.; Mirzaei, S.; Asadi, A.; Akhlaghi, M.; Saneei, P. The Association between Egg Consumption and Metabolic Health Status in Overweight and Obese Adolescents. *Sci. Rep.* **2023**, *13*, 2778. [CrossRef]
95. Emrani, A.S.; Beigrezaei, S.; Zademohammadi, F.; Salehi-Abargouei, A. The Effect of Whole Egg Consumption on Weight and Body Composition in Adults: A Systematic Review and Meta-Analysis of Clinical Trials. *Syst. Rev.* **2023**, *12*, 125. [CrossRef]
96. Li, M.-Y.; Chen, J.-H.; Chen, C.; Kang, Y.-N. Association between Egg Consumption and Cholesterol Concentration: A Systematic Review and Meta-Analysis of Randomized Controlled Trials. *Nutrients* **2020**, *12*, 1995. [CrossRef] [PubMed]
97. Rouhani, M.H.; Rashidi-Pourfard, N.; Salehi-Abargouei, A.; Karimi, M.; Haghighatdoost, F. Effects of Egg Consumption on Blood Lipids: A Systematic Review and Meta-Analysis of Randomized Clinical Trials. *J. Am. Coll. Nutr.* **2018**, *37*, 99–110. [CrossRef] [PubMed]
98. McDonald, J.D.; Chitchumroonchokchai, C.; Li, J.; Mah, E.; Labyk, A.N.; Reverri, E.J.; Ballard, K.D.; Volek, J.S.; Bruno, R.S. Replacing Carbohydrate during a Glucose Challenge with the Egg White Portion or Whole Eggs Protects against Postprandial Impairments in Vascular Endothelial Function in Prediabetic Men by Limiting Increases in Glycaemia and Lipid Peroxidation. *Br. J. Nutr.* **2018**, *119*, 259–270. [CrossRef]
99. Godos, J.; Micek, A.; Brzostek, T.; Toledo, E.; Iacoviello, L.; Astrup, A.; Franco, O.H.; Galvano, F.; Martinez-Gonzalez, M.A.; Grosso, G. Egg Consumption and Cardiovascular Risk: A Dose-Response Meta-Analysis of Prospective Cohort Studies. *Eur. J. Nutr.* **2021**, *60*, 1833–1862. [CrossRef]
100. Fratoni, V.; Brandi, M.L. B Vitamins, Homocysteine and Bone Health. *Nutrients* **2015**, *7*, 2176–2192. [CrossRef] [PubMed]
101. Rizzoli, R.; Biver, E.; Brennan-Speranza, T.C. Nutritional Intake and Bone Health. *Lancet Diabetes Endocrinol.* **2021**, *9*, 606–621. [CrossRef]
102. Olagunju, M.T.; Abodunrin, O.R.; Omotoso, I.O.; Adewole, I.E.; Ola, O.M.; Abel, C.; Akinsolu, F.T. Egg Consumption and Bone Mass Density among the Elderly: A Scoping Review. PLOS Glob. *Public Health* **2024**, *4*, e0002519. [CrossRef] [PubMed]
103. Obianwuna, U.E.; Oleforuh-Okoleh, V.U.; Wang, J.; Zhang, H.-J.; Qi, G.-H.; Qiu, K.; Wu, S.-G. Natural Products of Plants and Animal Origin Improve Albumen Quality of Chicken Eggs. *Front. Nutr.* **2022**, *9*, 875270. [CrossRef] [PubMed]
104. Sugiharto, S.; Ranjitkar, S. Recent Advances in Fermented Feeds towards Improved Broiler Chicken Performance, Gastrointestinal Tract Microecology and Immune Responses: A Review. *Anim. Nutr.* **2019**, *5*, 1–10. [CrossRef]
105. Hussain, T.; Murtaza, G.; Metwally, E.; Kalhoro, D.H.; Kalhoro, M.S.; Rahu, B.A.; Sahito, R.G.A.; Yin, Y.; Yang, H.; Chughtai, M.I.; et al. The Role of Oxidative Stress and Antioxidant Balance in Pregnancy. *Mediators Inflamm.* **2021**, *2021*, 9962860. [CrossRef]
106. Oke, O.E.; Akosile, O.A.; Oni, A.I.; Opowoye, I.O.; Ishola, C.A.; Adebiyi, J.O.; Odeyemi, A.J.; Adjei-Mensah, B.; Uyanga, V.A.; Abioja, M.O. Oxidative Stress in Poultry Production. *Poult. Sci.* **2024**, *103*, 104003. [CrossRef] [PubMed]
107. Nienaber, J.A.; Hahn, G.L. Livestock Production System Management Responses to Thermal Challenges. *Int. J. Biometeorol.* **2007**, *52*, 149–157. [CrossRef] [PubMed]
108. Kpomasse, C.C.; Kouame, Y.A.E.; N'nanle, O.; Houndonougbo, F.M.; Tona, K.; Oke, O.E. The Productivity and Resilience of the Indigenous Chickens in the Tropical Environments: Improvement and Future Perspectives. *J. Appl. Anim. Res.* **2023**, *51*, 456–469. [CrossRef]
109. Pizzino, G.; Irrera, N.; Cucinotta, M.; Pallio, G.; Mannino, F.; Arcoraci, V.; Squadrito, F.; Altavilla, D.; Bitto, A. Oxidative Stress: Harms and Benefits for Human Health. *Oxid. Med. Cell. Longev.* **2017**, *2017*, 8416763. [CrossRef] [PubMed]
110. Djuricic, I.; Calder, P.C. Beneficial Outcomes of Omega-6 and Omega-3 Polyunsaturated Fatty Acids on Human Health: An Update for 2021. *Nutrients* **2021**, *13*, 2421. [CrossRef] [PubMed]

111. Cachaldora, P.; Garcia-Rebollar, P.; Alvarez, C.; De Blas, J.C.; Mendez, J. Effect of Type and Level of Basal Fat and Level of Fish Oil Supplementation on Yolk Fat Composition and N−3 Fatty Acids Deposition Efficiency in Laying Hens. *Anim. Feed. Sci. Technol.* **2008**, *141*, 104–114. [CrossRef]
112. Ahmed, M.E.; Abdelati, K.A. Effect of Dietary Graded Levels of Leucaena Leucocepahala Seeds on Layers Performance, Egg Quality and Bird Parameters. *Int. J. Poult. Sci.* **2009**, *8*, 475–479. [CrossRef]
113. Milinsk, M.C.; Murakami, A.E.; Gomes, S.T.M.; Matsushita, M.; de Souza, N.E. Fatty Acid Profile of Egg Yolk Lipids from Hens Fed Diets Rich in N-3 Fatty Acids. *Food Chem.* **2003**, *83*, 287–292. [CrossRef]
114. Simopoulos, A.P. An Increase in the Omega-6/Omega-3 Fatty Acid Ratio Increases the Risk for Obesity. *Nutrients* **2016**, *8*, 128. [CrossRef]
115. Simopoulos, A.P. The Omega-6/Omega-3 Fatty Acid Ratio: Health Implications. *OCL* **2010**, *17*, 267–275. [CrossRef]
116. Coorey, R.; Novinda, A.; Williams, H.; Jayasena, V. Omega-3 Fatty Acid Profile of Eggs from Laying Hens Fed Diets Supplemented with Chia, Fish Oil, and Flaxseed: Omega-3 in Eggs from Chia Fed Hens. *J. Food Sci.* **2015**, *80*, S180–S187. [CrossRef] [PubMed]
117. Ehr, I.J.; Persia, M.E.; Bobeck, E.A. Comparative Omega-3 Fatty Acid Enrichment of Egg Yolks from First-Cycle Laying Hens Fed Flaxseed Oil or Ground Flaxseed. *Poult. Sci.* **2017**, *96*, 1791–1799. [CrossRef] [PubMed]
118. Baltzer, K. Consumers' Willingness to Pay for Food Quality—The Case of Eggs. *Food Econ. Acta Agric. Scand. Sect. C* **2004**, *1*, 78–90. [CrossRef]
119. Zamani, O.; Bittmann, T.; Ortega, D.L. The Effect of Avian Influenza Outbreaks on Retail Price Premiums in the United States Poultry Market. *Poult. Sci.* **2024**, *103*, 104102. [CrossRef]
120. Eggersdorfer, N.S.A. Is the World Supply of Omega-3 Fatty Acids Adequate for Optimal Human Health? *Curr. Opin. Clin. Nutr. Metab. Care* **2015**, *18*, 147–154.
121. Balić, A.; Vlašić, D.; Žužul, K.; Marinović, B.; Bukvić Mokos, Z. Omega-3 versus Omega-6 Polyunsaturated Fatty Acids in the Prevention and Treatment of Inflammatory Skin Diseases. *Int. J. Mol. Sci.* **2020**, *21*, 741. [CrossRef]
122. Li, D.; Weisinger, H.S.; Weisinger, R.S.; Mathai, M.; Armitage, J.A.; Vingrys, A.J.; Sinclair, A.J. Omega 6 to Omega 3 Fatty Acid Imbalance Early in Life Leads to Persistent Reductions in DHA Levels in Glycerophospholipids in Rat Hypothalamus Even after Long-Term Omega 3 Fatty Acid Repletion. Prostaglandins Leukot. *Essent. Fatty Acids* **2006**, *74*, 391–399. [CrossRef] [PubMed]
123. Kartikasari, L.R.; Geier, M.S.; Hughes, R.J.; Bastian, S.E.; Gibson, R.A. Assessment of Omega-3 and Omega-6 Fatty Acid Profiles and Ratio of Omega-6/Omega-3 of White Eggs Produced by Laying Hens Fed Diets Enriched with Omega-3 Rich Vegetable Oil. *Open Agric.* **2024**, *9*, 20220274. [CrossRef]
124. Oliveira, D.D.; Baião, N.C.; Cançado, S.V.; Grimaldi, R.; Souza, M.R.; Lara, L.J.C.; Lana, A.M.Q. Effects of Lipid Sources in the Diet of Laying Hens on the Fatty Acid Profiles of Egg Yolks. *Poult. Sci.* **2010**, *89*, 2484–2490. [CrossRef] [PubMed]
125. Maina, A.N.; Lewis, E.; Kiarie, E.G. Egg Production, Egg Quality, and Fatty Acids Profiles in Eggs and Tissues in Lohmann LSL Lite Hens Fed Algal Oils Rich in Docosahexaenoic Acid (DHA). *Poult. Sci.* **2023**, *102*, 102921. [CrossRef] [PubMed]
126. Świątkiewicz, S.; Arczewska-Włosek, A.; Krawczyk, J.; Szczurek, W.; Puchała, M.; Józefiak, D. Effect of Selected Feed Additives on Egg Performance and Eggshell Quality in Laying Hens Fed a Diet with Standard or Decreased Calcium Content. *Ann. Anim. Sci.* **2018**, *18*, 167–183. [CrossRef]
127. Yalçin, H.; Unal, M.K. The Enrichment of Hen Eggs with Omega-3 Fatty Acids. *J. Med. Food* **2010**, *13*, 610–614. [CrossRef] [PubMed]
128. Kralik, Z.; Kralik, G.; Košević, M.; Galović, O.; Samardžić, M. Natural Multi-Enriched Eggs with n-3 Polyunsaturated Fatty Acids, Selenium, Vitamin E, and Lutein. *Animals* **2023**, *13*, 321. [CrossRef]
129. Soesanto, I.R.H.; Yani, A.; Tanti, A. Identification of Market Demand for Omega-3 Eggs in Bogor Regency. *J. Ilmu-Ilmu Peternak.* **2024**, *34*, 171–178. [CrossRef]
130. United States Food and Drug Administration (FDA). Egg Regulatory Program Standards (ERPS). 2017. Available online: https://www.fda.gov/food/regulatory-program-standards-food/egg-regulatory-program-standards-erps (accessed on 8 December 2024).
131. *Regulation (EC) No 853/2004 of the European Parliament and of the Council of 29 April 2004 Laying Down Specific Hygiene Rules for Food of Animal Origin*; Official Journal of the European Union Law: Luxembourg, 2004; pp. 55–205.
132. *Regulation (EC) No 178/2002 of the European Parliament and of the Council of 28 January 2002 Laying Down the General Principles and Requirements of Food Law, Establishing the European Food Safety Authority, and Laying Down Procedures in Matters of Food Safety*; Official Journal of the European Union Law: Luxembourg, 2002; pp. 1–24.
133. *Council Directive 1999/74/EC of 19 July 1999 Laying Down Minimum Standards for the Protection of Laying Hens*; Official Journal of the European Communities Law: Luxembourg, 1999; pp. 53–57.
134. United States Department of Agriculture (USDA). USDA Guidelines. Available online: https://www.dietaryguidelines.gov/sites/default/files/2020-12/Dietary_Guidelines_for_Americans_2020-2025.pdf (accessed on 8 December 2024).
135. *Regulation (EU) No 1169/2011 of the European Parliament and of the Council of 25 October 2011 on the Provision of Food Information to Consumers*; Official Journal of the European Union Law: Luxembourg, 2011; pp. 18–63.

136. *Regulation (EC) No 1924/2006 of the European Parliament and of the Council of 20 December 2006 on Nutrition and Health Claims Made on Foods*; Official Journal of the European Union Law: Luxembourg, 2006; pp. 9–25.
137. Khan, S.A.; Khan, A.; Khan, S.A.; Beg, M.A.; Ali, A.; Damanhouri, G. Comparative Study of Fatty-Acid Composition of Table Eggs from the Jeddah Food Market and Effect of Value Addition in Omega-3 Bio-Fortified Eggs. *Saudi J. Biol. Sci.* **2017**, *24*, 929–935. [CrossRef] [PubMed]
138. Bakhtavoryan, R.; Hovhannisyan, V.; Devadoss, S.; Lopez, J. An Empirical Evaluation of Egg Demand in the United States. *J. Agric. Appl. Econ.* **2021**, *53*, 280–300. [CrossRef]
139. Medina-Cruz, M.F.; Zárate-Contreras, D.; Pérez-Ruiz, R.V.; Aguilar-Toalá, J.E.; Rosas-Espejel, M.; Cruz-Monterrosa, R.G. Nutritional Aspects, Production and Viability in the Market of Organic Chicken Eggs: Review. *Food Chem. Adv.* **2024**, *4*, 100595. [CrossRef]
140. Egg Guidance Documents & Regulatory Information. Available online: https://www.fda.gov/food/guidance-documents-regulatory-information-topic-food-and-dietary-supplements/egg-guidance-regulation-and-other-information (accessed on 2 December 2024).
141. United States Congress Senate. *To Amend the Federal Meat Inspection ACT, the Poultry Products Inspection ACT, the Egg Products Inspection ACT, and the Federal Food, Drug, and Cosmetic ACT to Provide for Improved Public Health and Food Safety Through Enhanced Enforcement*; Bibliogov: Secaucus, NJ, USA, 2011.
142. Mulatsih, S.; Soesanto, I.R.; Retnani, Y.; Yani, A.; Mutia, R.; Tanti, A. Exploring the Potential of Omega-3 Enriched Egg Industry in Indonesia: Production, Consumer Demand, and Competitiveness. *J. Ilmu Produksi Teknol. Has. Peternak.* **2024**, *12*, 75–81. [CrossRef]
143. Bourre, J.M.; Galea, F. An Important Source of Omega-3 Fatty Acids, Vitamins D and E, Carotenoids, Iodine and Selenium: A New Natural Multi-Enriched Egg. *J. Nutr. Health Aging* **2006**, *10*, 371–376. [PubMed]
144. Palmieri, N.; Stefanoni, W.; Latterini, F.; Pari, L. Factors Influencing Italian Consumers' Willingness to Pay for Eggs Enriched with Omega-3-Fatty Acids. *Foods* **2022**, *11*, 545. [CrossRef] [PubMed]
145. Magette, W.L. *Development and Testing of an Environmental Management System for Organic Wastes from Intensive Pig and Poultry Units: Literature Review*; Environmental Protection Agency: County Wexford, Ireland, 2001.
146. Myers, M.; Ruxton, C.H.S. Eggs: Healthy or Risky? A Review of Evidence from High Quality Studies on Hen's Eggs. *Nutrients* **2023**, *15*, 2657. [CrossRef] [PubMed]
147. Haward, R.; Chacko, J.; Konjeti, S.; Metri, G.R.; Binoy, B.K.; Haward, R.; Raju, S. Debunking the Myth: Eggs and Heart Disease. *Cureus* **2024**, *16*, e59952. [CrossRef]
148. Wang, M.X.; Wong, C.H.; Kim, J.E. Impact of whole egg intake on blood pressure, lipids and lipoproteins in middle-aged and older population: A systematic review and meta-analysis of randomized controlled trials. *Nutr. Metab. Cardiovasc. Dis.* **2019**, *29*, 653–664. [CrossRef]
149. Rong, Y.; Chen, L.; Zhu, T.; Song, Y.; Yu, M.; Shan, Z.; Sands, A.; Hu, F.B.; Liu, L. Egg Consumption and Risk of Coronary Heart Disease and Stroke: Dose-Response Meta-Analysis of Prospective Cohort Studies. *BMJ* **2013**, *346*, e8539. [CrossRef] [PubMed]
150. Drouin-Chartier, J.P.; Chen, S.; Li, Y.; Schwab, A.L.; Stampfer, M.J.; Sacks, F.M.; Rosner, B.; Willett, W.C.; Hu, F.B.; Bhupathiraju, S.N. Egg Consumption and Risk of Cardiovascular Disease: Three Large Prospective US Cohort Studies, Systematic Review, and Updated Meta Analysis. *BMJ* **2020**, *368*, m513. [CrossRef] [PubMed]
151. Krittanawong, C.; Narasimhan, B.; Wang, Z.; Virk, H.U.H.; Farrell, A.M.; Zhang, H.; Tang, W.W. Association Between Egg Consumption and Risk of Cardiovascular Outcomes: A Systematic Review and Meta-Analysis. *Am. J. Med.* **2020**, *134*, 76–83.e2. [CrossRef] [PubMed]
152. Berger, S.; Raman, G.; Vishwanathan, R.; Jacques, P.F.; Johnson, E.J. Dietary Cholesterol and Cardiovascular Disease: A Systematic Review and Meta-Analysis. Am. *J. Clin. Nutr.* **2015**, *102*, 276–294. [CrossRef]
153. Huang, J.; Liao, L.M.; Weinstein, S.J.; Sinha, R.; Graubard, B.I.; Albanes, D. Association between Plant and Animal Protein Intake and Overall and Cause-Specific Mortality. *JAMA Intern. Med.* **2020**, *180*, 1173–1184. [CrossRef] [PubMed]
154. Bechthold, A.; Boeing, H.; Schwedhelm, C.; Hoffmann, G.; Knüppel, S.; Iqbal, K.; De Henauw, S.; Michels, N.; Devleesschauwer, B.; Schlesinger, S.; et al. Food Groups and Risk of Coronary Heart Disease, Stroke and Heart Failure: A Systematic Review and Dose-Response Meta-Analysis of Prospective Studies. *Crit. Rev. Food Sci. Nutr.* **2019**, *59*, 1071–1090. [CrossRef] [PubMed]
155. Calle, M.C.; Andersen, C.J. Assessment of Dietary Patterns Represents a Potential, yet Variable, Measure of Inflammatory Status: A Review and Update. *Dis. Markers* **2019**, *2019*, 3102870. [CrossRef]
156. Zappala, G.; Platania, A.; Paladino, G.; Nicolosi, L.K.; Ragusa, R.; Marranzano, M. Meal Habits and Metabolic Status in Southern Italian Adults. *Nutr. Healthy Aging* **2019**, *5*, 199–207. [CrossRef]
157. Marventano, S.; Godos, J.; Tieri, M.; Ghelfi, F.; Titta, L.; Lafranconi, A.; Gambera, A.; Alonzo, E.; Sciacca, S.; Buscemi, S.; et al. Egg Consumption and Human Health: An Umbrella Review of Observational Studies. *Int. J. Food Sci. Nutr.* **2020**, *71*, 325–331. [CrossRef]

158. Olagunju, M.; Aleru, E.O.; Abodunrin, O.R.; Adedini, C.B.; Ola, O.M.; Abel, C.; Adewole, I.E.; Okunbor, H.N.; Akinsolu, F.T. Association between Meal Skipping and the Double Burden of Malnutrition among University Students. *N. Afr. J. Food Nutr. Res.* **2024**, *8*, 167. [CrossRef]
159. Blasbalg, T.L.; Hibbeln, J.R.; Ramsden, C.E.; Majchrzak, S.F.; Rawlings, R.R. Changes in Consumption of Omega-3 and Omega-6 Fatty Acids in the United States during the 20th Century. *Am. J. Clin. Nutr.* **2011**, *93*, 950–962. [CrossRef] [PubMed]
160. Cherian, G.; Sim, J.S. Omega-3 Fatty Acids in Table Eggs. World's Poult. *Sci. J.* **2021**, *77*, 65–78.
161. Koppenol, A.; Van Harn, J.; Hendriks, W.H.; Dekker, R.A.; Van Krimpen, M.M. Enriching Eggs with Omega-3 Fatty Acids: A Review. *Anim. Feed Sci. Technol.* **2022**, *292*, 115432. [CrossRef]
162. Hargis, P.S.; Van Elswyk, M.E. Modifying Egg Yolk Lipids with Dietary Omega-3 Fatty Acids. *Poult. Sci.* **2020**, *99*, 4042–4050.
163. Beynen, A.C.; Van Der Horst, H.; Van Hoek, M. Omega-6/Omega-3 Balance in Enriched Eggs: Effects of Linseed Oil Feeding to Hens. *Food Chem.* **2021**, *343*, 128501. [CrossRef]
164. Kokoszyński, D. Guinea Fowl, Goose, Turkey, Ostrich, and Emu Eggs. In *Egg Innovations and Strategies for Improvements*; Elsevier: Amsterdam, The Netherlands, 2017; pp. 33–43.
165. Newberry, R.C. Commercial Free-Range Egg Production Practices. In *Egg Innovations and Strategies for Improvements*; Elsevier: Amsterdam, The Netherlands, 2017; pp. 89–102.
166. Pesti, G.M.; Choct, M. The Future of Feed Formulation for Poultry: Toward More Sustainable Production of Meat and Eggs. *Anim. Nutr.* **2023**, *15*, 71–87. [CrossRef]
167. Vlaicu, P.A.; Untea, A.E.; Oancea, A.G. Sustainable Poultry Feeding Strategies for Achieving Zero Hunger and Enhancing Food Quality. *Agriculture* **2024**, *14*, 1811. [CrossRef]
168. Philippe, F.X.; Mahmoudi, Y.; Cinq-Mars, D.; Lefrançois, M.; Moula, N.; Palacios, J.; Pelletier, F.; Godbout, S. Comparison of Egg Production, Quality and Composition in Three Production Systems for Laying Hens. *Livest. Sci.* **2020**, *232*, 103917. [CrossRef]
169. Vlaicu, P.A.; Untea, A.E.; Varzaru, I.; Saracila, M.; Oancea, A.G. Designing Nutrition for Health-Incorporating Dietary by-Products into Poultry Feeds to Create Functional Foods with Insights into Health Benefits, Risks, Bioactive Compounds, Food Component Functionality and Safety Regulations. *Foods* **2023**, *12*, 4001. [CrossRef]
170. Ijomah, T.I.; Idemudia, C.; Eyo-Udo, N.L.; Anjorin, K.F. Innovative Digital Marketing Strategies for SMEs: Driving Competitive Advantage and Sustainable Growth. *Int. J. Manag. Entrep. Res.* **2024**, *6*, 2173–2188. [CrossRef]
171. Abdel-Wareth, A.A.A.; Williams, A.N.; Salahuddin, M.; Gadekar, S.; Lohakare, J. Algae as an Alternative Source of Protein in Poultry Diets for Sustainable Production and Disease Resistance: Present Status and Future Considerations. *Front. Vet. Sci.* **2024**, *11*, 1382163. [CrossRef]
172. Muharam, H.; Wandrial, S.; Rumondang, P.R.; Handayani, M.A.; Masruchan, M. Innovative Strategies in Digital Marketing: Enhancing Consumer Engagement and Brand Loyalty. *Global* **2024**, *2*, 1629–1643. [CrossRef]
173. Sgroi, F.; Sciortino, C.; Baviera-Puig, A.; Modica, F. Analyzing Consumer Trends in Functional Foods: A Cluster Analysis Approach. *J. Agric. Food Res.* **2024**, *15*, 101041. [CrossRef]
174. Bist, R.B.; Bist, K.; Poudel, S.; Subedi, D.; Yang, X.; Paneru, B.; Mani, S.; Wang, D.; Chai, L. Sustainable Poultry Farming Practices: A Critical Review of Current Strategies and Future Prospects. *Poult. Sci.* **2024**, *103*, 104295. [CrossRef]
175. *Regulation (EC) No 852/2004 of the European Parliament and of the Council of 29 April 2004 on the Hygiene of Foodstuffs, The Nutrition (Amendment etc.) (EU Exit) Regulations 2019: Practical Changes for Industry*; Official Journal of the European Union Law: Luxembourg, 2004.
176. Barnett, J.; McConnon, A.; Kennedy, J.; Raats, M.; Shepherd, R.; Verbeke, W.; Fletcher, J.; Kuttschreuter, M.; Lima, L.; Wills, J.; et al. Development of Strategies for Effective Communication of Food Risks and Benefits across Europe: Design and Conceptual Framework of the FoodRisC Project. *BMC Public Health* **2011**, *11*, 308. [CrossRef]
177. Doyon, M.; Bergeron, S.; Saulais, L.; Labonté, M.È.; Provencher, V. Do Consumers Value Welfare and Environmental Attributes in Egg Production Similarly in Fresh Eggs and Prepared Meals? *Animals* **2023**, *13*, 324. [CrossRef] [PubMed] [PubMed Central]
178. Rahmani, D.; Kallas, Z.; Pappa, M.; Gil, J.M. Are Consumers' Egg Preferences Influenced by Animal-Welfare Conditions and Environmental Impacts? *Sustainability* **2019**, *11*, 6218. [CrossRef]
179. Wägeli, S.; Janssen, M.; Hamm, U. Organic Consumers' Preferences and Willingness-to-Pay for Locally Produced Animal Products: Preferences for Locally Produced Animal Products. *Int. J. Consum. Stud.* **2016**, *40*, 357–367. [CrossRef]
180. Antony, B.; Benny, M.; Jose, S.; Jacob, S.; Nedumpilly, V.; Ajimol, M.S.; Abraham, G. Development of Omega-3 Enriched Egg Using Fish-Oil Based Fowl Feed Supplement. *J. Appl. Poult. Res.* **2024**, *33*, 100429. [CrossRef]

agriculture

MDPI

Review

# Equine Milk and Meat: Nutritious and Sustainable Alternatives for Global Food Security and Environmental Sustainability—A Review

**Claudia Pânzaru [1], Marius Gheorghe Doliș [1], Răzvan-Mihail Radu-Rusu [1], Constantin Pascal [1], Vasile Maciuc [1] and Mădălina-Alexandra Davidescu [2,*]**

1 Department of Animal Resources and Technology, Faculty of Food and Animal Sciences, "Ion Ionescu de la Brad" University of Life Sciences, 8 Mihail Sadoveanu Alley, 700489 Iasi, Romania; claudia.panzaru@iuls.ro (C.P.); marius.dolis@iuls.ro (M.G.D.); razvan.radu@iuls.ro (R.-M.R.-R.); constantin.pascal@iuls.ro (C.P.); vasile.maciuc@iuls.ro (V.M.)

2 Department of Control, Expertise and Services, Faculty of Food and Animal Sciences, "Ion Ionescu de la Brad" University of Life Sciences, 8 Mihail Sadoveanu Alley, 700489 Iasi, Romania

* Correspondence: madalina.davidescu@iuls.ro

**Simple Summary:** The research dedicated to equine products, such as milk and meat for human consumption, requires further development in response to current global challenges. These products have the potential to fulfill the growing demand for essential nutrients critical to human health. Additionally, equine farming offers an environmentally sustainable alternative to traditional livestock systems. This review aims to highlight the potential of equine milk and meat as sustainable, nutritionally valuable food sources by examining their unique composition and health benefits; it also explores the sustainability of equine raising in the context of supporting the transition to greener agricultural practices.

**Abstract:** Global hunger and malnutrition continue to pose significant challenges, entailing innovative and environmentally responsible practices to improve food security. Equine products offer a valuable opportunity to diversify diets and combat nutritional deficiencies. Equine milk, rich in bioavailable nutrients, essential fatty acids, and hypoallergenic properties, serves as an excellent dietary supplement; this aspect could be applied particularly in regions where conventional dairy is inaccessible or unsuitable due to allergies, lactose intolerance, or other health conditions. Similarly, equine meat, known for its high-quality protein content, low fat content, and essential micronutrients like iron and zinc, provides an affordable and sustainable protein source for food-insecure populations. The ability of equines to thrive on marginal lands, coupled with their lower environmental impact compared to traditional livestock (such as ruminants), underscores their potential within sustainable agricultural systems. This review examines the role of equine products in addressing global hunger, highlighting their nutritional benefits, environmental advantages, and the necessity for further research to tackle challenges related to versatility, cultural acceptance, and policy integration.

**Keywords:** donkey and horse milk and meat; equine potential; global hunger; alternative foods; sustainability

## 1. Introduction

Equine raising is recognized as one of the most sustainable agricultural practices, playing a vital role in supporting rural development, particularly in those regions where it serves as a key element of local economies. Horses (*Equus caballus*), donkeys (*Equus asinus*), and their hybrids (mules, the result of crossing the donkey male with the mare) are integral to this practice, contributing through their adaptability and diverse applications [1]. Equines are increasingly valued for their roles in tourism and agricultural labor, especially in areas where mechanization is limited [2–4]. Their high productivity, along with low feeding costs as part of their efficiency in grazing on pastures, makes them an economical option for rural communities [5]. Beyond these functions, equines are highly versatile, supporting activities such as leisure riding and equine-assisted therapy (hippotherapy) [6], which contribute to physical and mental well-being and in some cases have unexpected effects even applying in very short sessions (a meta-analysis applied in children with spastic palsy showed that 8–10 min of horseback riding had a significant impact in reducing the asymmetrical activity of the hip adductor muscles) [7].

In addition to their traditional uses, equines play a growing role in the production of animal-derived products like milk and meat. Equine milk, known for its hypoallergenic qualities and digestibility, is especially beneficial for individuals with conditions such as lactose intolerance, allergies, or metabolic disorders [8–12]. Similarly, equine meat offers a lean, high-quality protein source enriched with essential micronutrients, making it a valuable dietary option [13–15]. These products not only meet nutritional needs but also represent a sustainable alternative within the context of increasing global food demands [16], allergies [17,18], and digestive sensitivity [19,20]. By 2050, meeting the soaring global demand for animal-derived food products, projected to increase by 58–73%, will require sustainable solutions to enhance production efficiency while addressing resource limitations and balancing economic, social, and environmental sustainability [21].

Equine milk and meat are increasingly recognized for their remarkable nutritional value and health benefits, making them valuable alternatives in the food industry. Unlike cow's milk, equine milk is lower in fat and contains higher levels of several nutrients such as vitamin C (which is absent in bovine milk) [22–24]. This makes horse and donkey milk a suitable option for individuals with lactose intolerance or milk allergies. It is also considered closer in composition to human breast milk, making it easier to digest, particularly for infants with specific dietary needs [25]. It is popular in several areas where it has cultural and nutritional significance as Kazakhstan, Kyrgyzstan, Mongolia, Russia, and China—these countries have an old tradition of consuming mare's milk, as a fermented beverage named *kumis*, known for being nutritional and also medicinal—South Korea, Japan—in these regions, donkey and mare milk is considered a delicacy, being also a part of cosmetic products—Turkey, Arabian Peninsula, and other Middle Eastern countries—*kumis* is also popular—Italy, Spain, France—it started to have popularity, as the idea of being healthy has been spread; it is also preferred as a food supplement and in cosmetic products—it started to gain more interest also in the United States and Canada, since it represents an alternative to cow milk products. In Kazakhstan, Kyrgyzstan, and Mongolia, the consumption of *kumis* ranges between 1 and 2 L/day/person. In Russia, the registered quantities are from 1–2 L/week/person (it is less common than cow's and goat's milk or products). In China, it is considered a niche market in most areas, but the volume is around 1–2 L/week/person for the people who consume it for health benefits. In South Korea and Japan, volumes that range from 0.5–1 L/week/person are registered, and the values are sustained by a small percentage of the population that prefers equine milk regularly. In Turkey and the Arabian Peninsula, while camel's milk is traditional, *kumis* consumption registers 0.5–1 L/week/person. In Europe, equine milk (as *kumis* and supplements) is consumed in quantities smaller than 1 L/week (especially in Italy, Spain, and France). In the USA and Canada, it is preferred by individuals with dairy allergies or those seeking healthier products, the reason for which the consumption does not surpass 1 L/month/person [24].

Equine meat, known for being lean and rich in protein, is also appreciated for its lower fat content and high levels of essential minerals such as iron and zinc. This makes it desirable, particularly in regions where other forms of meat may not be as accessible or affordable. When compared to beef, equine meat offers a healthier alternative, as it is lower in saturated fats and cholesterol, making it particularly advantageous for individuals aiming to manage their heart health [26–28]. Compared to rabbit meat, equine meat is better suited for diets emphasizing higher omega-3 fatty acid intake and offers superior sensory qualities, such as flavor and texture. On the other hand, rabbit meat is a low-fat option that is rich in polyunsaturated fatty acids (PUFAs), making it ideal for heart-healthy and low-calorie diets [29,30]. Compared to poultry meat, equine meat also has higher levels of omega-3. As a result of grazing on pastures, it is also lower in fat content, richer in iron, and more fibrous than poultry meat [31,32]. In comparison to sheep meat, equine meat is much leaner, with a significantly lower fat content and a healthier omega-3 fatty acid profile; while sheep meat, especially lamb, is prized for its tender texture and distinctive gamey flavor, horse meat tends to offer a sweeter, richer taste with a firmer texture, often resembling beef [26,33]. Additionally, equine meat is lower in saturated fats and contains more iron, making it a heart-healthy choice and suitable for treating anemia. In contrast, mutton, especially, tends to have a higher fat content, which may make it less favorable for those seeking leaner options [34–36]. In Europe, horse meat is popular in France (in dishes as horse steak or steak de cheval, cured meat or bresaola and sausages [36]), where consumption of 40,000–60,000 tons is registered annually [37], then in Italy (in traditional dishes as patissada de cavalo or horse stew [38]) in quantities that range between 5000 and 10,000 tons/annually [37] and also in Belgium (preferred as steak de cheval, burgundy horse meat stew, saucisses de cheval or sausages, and tartare de cheval [38] in approximately 4000 tons/annually [37]) and Switzerland (boeuf de cheval or horse beef stew, horse meat carpaccio, steak, sausages, and cured horse meat [38] but only in approximately 2000 tons/annually [37]). In Asia, Mongolia, Kyrgyzstan, Turkmenistan, and Japan are countries with tradition in this sense, but Kazakhstan is still in the first place for horse meat consumption in the whole world (more than 40,000 tons/annually [37] as beshbarmak or boiled horse meat with noodles or sausages [36]). In second place is Kyrgystan, with 10,000–20,000 tons/annually [37], where beshbarmak, stews, and sausages are preferred (the highest consumption is registered during various cultural festivals, especially Nowruz, which is the Persian New Year) [36]. In Mongolia, horse meat consumption is approximately 10,000 tons/annually [36], in traditional dishes such as beshbarmak, buuz (steamed dumplings filled with minced meat), horhog (horse meat with vegetables), shul (soup), tsuivan (noodle dish), khuushuur (deep-fried pastries filled with meat), airag with horse meat (combination between horse meat and milk), and tamir (dried horse meat) [37]. Then, Turkmenistan is well-known for this practice, where döşeme, or a type of meat stew, and aslama, or stew with horse meat and vegetables, are traditional dishes [37], in quantities that range between 5000 and 10,000 tons/annually [36]. In Japan, sakuraniku, basashi (raw horse meat), and uni-kushi (skewers) [38] are consumed in quantities from 3000 to 5000 tons/annually [37]. In South America, Argentina is a country where horse meat is consumed in rural areas, but in small quantities, beef being their first option; in Mexico and Canada, there are some communities where such practices exist, but again to a limited extent. Horse meat consumption is restricted or even forbidden in most states of the USA, United Kingdom, and Ireland (where horses are considered companion animals generally), and in India, there are religious restrictions (in some areas, horses are seen as sacred animals). In Saudi Arabia, Indonesia, and other Muslim-majority countries, even if it is not forbidden through religion, horse meat consumption is uncommon; in Israel, on the other hand, eating horse meat is banned for Jewish people under kosher dietary laws, and in Australia, it is culturally taboo (horses are considered companions or working animals) [39].

On the other hand, equine raising provides advantages when it comes to sustainable farming. In rural and marginal areas, where pastures and natural meadows are predominant, equine farming presents a sustainable, resource-efficient method of food production. The low environmental impact of equines, along with their ability to thrive on natural grasslands, supports eco-sustainable farming practices [21]. This approach not only supports landscape conservation and biodiversity but also strengthens food security by ensuring a reliable supply of both milk and meat. By incorporating equine products into local economies, communities can strengthen their resistance, improve nutritional intake, and foster sustainable agricultural practices that contribute to long-term environmental and social stability [5,32,40–42].

More than that, the adaptability of equines to diverse environments, from arid to temperate regions, further enhances their significance. Their ability to provide ecological, economic, and social benefits underscores their importance in promoting sustainable agricultural systems, addressing food security, and fostering rural livelihoods. This multifunctionality makes equines indispensable in ensuring both economic resilience and environmental sustainability in rural areas. At the same time, sanitary management practices such as disease prevention strategies, sanitary control measures, and animal welfare conditions are important. Through implementing these biosecurity and management practices, the risk of infectious diseases is reduced; this involves quarantining new horses for at least 30 days before introducing them to the herd and also ensuring that all the new horses have health certificates, vaccinations, and negative disease test results. Also, cleaning and disinfecting transportation equipment and clothing after handling all the new horses is crucial; special attention should be given to the foals and older equines due to their susceptibility to getting ill, and good practice is to limit visitor access, to monitor the farm entrance and to use footbaths at their entrance contribute to minimizing the risk of introducing pathogens inside the herd. However, active immunity is a key concept in disease prevention, acquired through vaccination, which helps the body to produce antibodies (this action can prevent infectious diseases in equines if given before exposure to the illness such as tetanus, encephalomyelitis, rabies, West Nile virus—core disease vaccinations—and anthrax, botulism, equine herpesvirus, equine viral arteritis, influenza, Potomac horse fever, rotavirus A, strangles—risk-based disease vaccinations; also, adenovirus, African horse sickness, brucellosis, clostridial enterocolitis, contagious equine metritis, equine infectious anemia, vesicular stomatitis are other diseases that could appear in this case) [39]. When discussing welfare conditions for equine raised for milk and meat, they should benefit from enough space (stalls of 3 × 3 m for foals and small breeds up to 4.2 × 4.2 m for large breeds and 0.6–0.8 hectares of pasture per animal), clean living areas and the possibility to interact with others since this is their natural behavior. The stalls should be equipped with fresh watering machines and comfortable straw beds and should also be well lit (8–12 h of preferably natural light; if the light is artificial, the brightness has to be 100–150 lux); then, the diets must be balanced and sufficient, the equines should be handled with respect, and the daily care is essential, as is regular veterinary checks [40].

In conclusion, equine rearing offers nutritious and health-conscious alternatives through both milk and meat while contributing to rural economies and sustainable agriculture. Equine milk has a lower fat content and higher vitamin C levels compared to cow's milk, being an ideal option for individuals with lactose intolerance or milk allergies. Additionally, its closer composition to human breast milk enhances digestibility, particularly for infants with special dietary needs. More than that, equine meat is a lean, low-fat protein source enriched with omega-3 fatty acids and essential minerals, making it a valuable dietary choice compared to other meats such as beef, poultry, and lamb. In addition to its health benefits, equine farming is environmentally sustainable, as equines are well-suited to diverse climates and contribute to maintaining biodiversity. Their adaptability, low environmental impact, and multifaceted roles underscore their importance in fostering long-term economic and environmental resilience in rural communities.

## 2. Equine Population and Productions Worldwide

### 2.1. The Total Number of Equines Raised Worldwide

Data on the accurate number of equines raised worldwide are missing in many countries due to a lack of communication between organizations and governments. Still, it is estimated that the total number of horses, donkeys, and mules is 116 million (36 million are in 38 of the lowest-income countries). Data are essential not only for disease control and epidemiological research but also for finding solutions to actual problems such as climate change, water access, and global food challenges. There is a report that refers to 36 countries that managed to explore problems for governments, and it was able to recommend a couple of ideas in this respect. Also, studies that involved the United

Kingdom, China, Mexico, India, Mexico, Kenya, and Senegal are available [37,43]. From the total number of registered equines, 57 million are horses, 50.5 million are donkeys (99% of the total number of the donkey population is in low-to-middle-income countries—LMICs), and 7.9 million are mules. Unfortunately, 22 of 36 countries have not reported their equid populations in the past decade. Governments in lower-income countries face challenges in collecting consistent and accurate livestock data. This results in limited insight into the socio-economic contributions of equids (such as donkeys and horses), emerging threats to their populations, and cross-border movements. The lack of reliable data contributes to the vulnerability of donkey populations, with some national populations in decline, though accurate statistics remain scarce [37,44,45].

According to other sources [41], while in 1980 a number of 59.63 million horses, 38.9 million donkeys, and 12.93 million mules were raised, in 2010, the situation would appear to reveal numbers like 59.76 million horses, 43.47 million donkeys, and 10.47 million mules (Table 1).

**Table 1.** The total number of equines between 1980 and 2010 (worldwide) *.

| Equines | Mil./Year | | | |
|---|---|---|---|---|
| | **1980** | **1990** | **2000** | **2010** |
| Horses | 59.63 | 61 | 57.76 | 59.76 |
| Donkeys | 38.9 | 43.76 | 41.63 | 43.47 |
| Mules | 12.93 | 14.83 | 12.93 | 10.47 |

* Source: [41].

## 2.2. The Equines Raised for Milk Production

Cattle account for 81% of global milk production, followed by buffaloes at 15%, goats at 2%, and sheep at 1%. Camels contribute 0.4%, and the rest comes from other dairy species like equines and yaks. In developing countries, about one-third of milk production comes from buffaloes, goats, camels, and sheep. Conversely, in developed nations, cattle are responsible for nearly all milk production. In several areas, non-cattle dairy species account for 38% of milk production in Asia, 22% in Africa, 3% in Europe, and only 0.5% in the Americas, with minimal production in Oceania [45].

In several regions, horse (*Equus caballus*) and donkey (*Equus asinus*) milk serve as essential sources of nutrition for subsistence farmers. Horses are typically used for dairy production in cooler climates, while donkeys are favored in dry, semi-arid areas. Milking equines is labor-intensive, requiring frequent sessions—usually 5 to 6 times a day. Additionally, a mare will not release milk unless its foal is present to stimulate it. In the steppes of Central Asia, horse milk is consumed in the form of *kumis*, a traditional fermented lactic-alcoholic beverage. Horse milk also plays a vital role as an animal protein source for pastoralists in Mongolia (in this region, yaks are the major source of milk and milk products). In contrast, donkey milk has become less common for consumption, though it is still used for medicinal purposes in some African communities [46].

## 2.3. The Equines Raised for Meat Production

The global horse meat market, while a niche sector compared to larger industries like beef and poultry, holds significant cultural and economic importance in certain regions, especially Kazakhstan, China, and parts of Europe. The industry is influenced by varying factors, including shifting global consumption trends, ethical debates surrounding horse slaughter, and the maintenance of demand in key markets. Kazakhstan is the global leader in horse meat production, slaughtering around 450,000 horses annually, making it the largest producer worldwide. The country's vast pasturelands are ideal for equine farming, driving this high level of production [44]. Horse meat is a central component of traditional diets in Kazakhstan and neighboring countries such as Kyrgyzstan and Turkmenistan, where it is consumed in various forms, including *kumis*, a fermented mare's milk drink. This cultural attachment to horse meat ensures its continued demand and significant role in the region's culinary heritage. In 2021, Kazakhstan exported around 10,000 tons of horse meat, primarily to Russia and other Central Asian nations. However, export figures can fluctuate based on regional trade agreements and global demand. The Kazakh government has made efforts to support the industry, bolstering both domestic consumption and international exports [36,45,46].

In China, the consumption of horse meat is less widespread compared to Kazakhstan or Europe, but it holds a niche market, especially in the northern regions like Inner Mongolia. Though production data are not as readily available, estimates suggest that China produces several thousand tons of horse meat annually. Horse meat is used in traditional dishes and sometimes consumed for medicinal purposes, particularly in rural and pastoral regions. The demand in China remains low compared to the other regions, but it is still an important dietary component in specific areas [36,44,47].

In Europe, countries like France, Italy, and Romania have long-standing traditions of consuming horse meat, albeit to varying degrees. France has historically been one of the largest consumers of horse meat in Europe. Although the demand has waned in recent decades, the country produced about 33,000 tons of horse meat in 2020. Horse meat is still available through specialized butchers and restaurants, where it is marketed as a healthy and lean alternative to beef. In Italy, the interest in horse meat has seen a resurgence, particularly in traditional regions like Lombardy and Tuscany. Italy produced approximately 4000 tons of horse meat in 2020, much of which is used in sausages and premium dishes like horse steaks. Romania produces a steady supply of horse meat, primarily for local consumption. With a horse population exceeding 1 million, Romania's annual production hovers around 10,000 tons [36,44,48].

According to other sources [49,50], the global livestock of horses has increased by 1,762,110 heads and by 83,356 heads when analyzing the donkey herds between 2019 and 2022. The production of donkey meat decreased by 20,965.89 tons in this period, and the carcass weight increased by 20 g/year during the given period (Table 2). For horses, there are not enough data available.

**Table 2.** Equine meat production worldwide (2019–2022) *.

| | | | Year | | | |
|---|---|---|---|---|---|---|
| Item | | Unit | 2019 | 2020 | 2021 | 2022 |
| Livestock | Horses | N | 58,819,179 | 59,592,065 | 59,780,596 | 60,581,289 |
| | Donkeys | N | 51,603,917 | 52,636,569 | 52,075,158 | 51,687,273 |
| Production | Meat of horses | t | - | - | - | - |
| | Meat of donkeys | t | 125,270.58 | 123,654.85 | 114,993.36 | 104,304.69 |
| Yield/carcass weight | Horses | 100 g/An | - | - | - | - |
| | Donkeys | 100 g/An | 876 | 883 | 885 | 896 |

* Source [50]—N = number of heads per year; t = tones; 100 g/An = the amount produced per year.

The global demand for horse meat, while still concentrated in specific regions, is gradually increasing as consumers seek healthier, leaner meat options [15]. Although the market for horse meat remains small in countries like Japan and Belgium, these nations feature niche markets, particularly in urban areas where there is a growing appreciation for diverse culinary experiences [44]. This trend reflects a broader interest in alternative, nutritious meats as part of more varied diets [51].

## 3. Equine Milk—Benefits for Human Consumption

### 3.1. Historical Uses

The medicinal properties of various substances, including equine milk, have been recognized since ancient times by influential figures whose writings still inform our understanding of health and wellness. Herodotus (c. 484–c. 425 BCE), the renowned Greek historian, offers valuable insights into the use of equine milk in his seminal work Histories. He describes the nomadic Scythian tribes and other Central Asian people, who relied on mare's milk as a vital resource. Herodotus highlights the consumption of fermented mare's milk, particularly in the form of *kumis*, a traditional fermented beverage. He underscores not only the nutritional benefits of mare's milk but also its crucial role in the nomadic lifestyle, providing essential nourishment during long journeys and harsh conditions [52]. Moreover, Hippocrates (c. 460–c. 370 BCE), often called the "Father of Medicine", also acknowledged the therapeutic properties of equine milk. In his comprehensive medical writings, he recognized mare's milk for its healing qualities, recommending it for various ailments, including digestive issues and respiratory conditions. It was especially valued as a tonic for overall health and was thought to be beneficial for patients with chronic illnesses such as tuberculosis and other respiratory diseases. This aligns with contemporary views on the potential health benefits of mare's milk, which is appreciated for its unique nutritional composition [53]. Then, Pliny the Elder (c. 23–79 CE), in his *Natural History*, offers additional insights into the use of equine milk, particularly in Central Asia. He emphasizes the health-promoting qualities of mare's milk and its restorative effects on the body, underscoring its importance in the diets and traditions of various cultures in the region [54]. Also, Galen (c. 129–c. 216 CE), a renowned physician of the Roman Empire, acknowledged the therapeutic benefits of mare's milk, particularly in addressing lung ailments and digestive issues. Similar to Hippocrates, he recommended equine milk as a restorative remedy to strengthen the body and support the immune system [55]. Marco Polo (1254–1324 CE), the famed Venetian explorer, recorded the widespread consumption of equine milk during his travels across Asia. He detailed the dietary habits of Mongol tribes, highlighting their reliance on *kumis*, a fermented mare's milk, as a crucial element of their sustenance. Polo noted that *kumis* played a significant role in maintaining their health and vitality, serving as both a nutritional cornerstone and a vital resource for their nomadic lifestyle, which demanded physical endurance and an active way of life [56]. Then, also Avicenna (c. 980–1037 CE), the renowned Persian polymath and a pivotal figure of the Islamic Golden Age, highlighted the benefits of equine milk in his groundbreaking medical treatise, *The Canon of Medicine*. He praised mare's milk for its nutritional value, recommending it for its strengthening effects and its ability to support the treatment of chronic illnesses. Furthermore, Avicenna recognized its effectiveness in alleviating inflammation and mitigating adverse reactions caused by the consumption of specific foods and beverages, emphasizing its therapeutic versatility [57].

Nowadays, many studies have emphasized the nutritional value and health benefits of equine milk, which has become an increasingly appreciated food product, especially in pastoral and rural regions. With a chemical composition that closely resembles human milk, equine milk is rich in lactose, essential fatty acids, and bioactive compounds that promote health [58]. Mare's milk is highly regarded for its immune-enhancing properties, which stem from its distinctive blend of proteins, vitamins, and minerals. It is believed to offer therapeutic benefits for managing digestive disorders, respiratory conditions, and chronic illnesses. Furthermore, equine milk is recognized for its ability to support immune function and contribute to overall health and vitality [59,60].

### 3.2. *The Chemical Composition of Equine Milk*

Horse and donkey milk are considered closer to human milk in terms of protein composition (horse milk has 1.5–2.5%, donkey milk has 1.5–2.0%, and human milk has 1.0–1.5%) and lactose content (horse and donkey milk have 6.0–7.0% and human milk has 1.0–7.5%) when compared to cow's milk (3.0–3.5% protein and 4.5–5.0% lactose) [23,60]. Studies show that most people can tolerate up to 14 g of lactose per day, and the fermentation process of horse milk further reduces its lactose content, enhancing its digestibility for those with sensitivities. Additionally, horse milk is low in calories, making it an ideal choice for weight-loss diets and reducing the risk of weight gain. However, due to its relatively low energy content, equine milk is not recommended as the sole source of nutrition for infants unless supplemented with vegetable oil (approximately 40 g per liter) to ensure sufficient caloric intake and essential nutrients for growth and development [61–67] (Table 3).

**Table 3.** The chemical composition of milk from various species *.

| **Item (%)** | **Species** | | | | | |
|---|---|---|---|---|---|---|
| | **Horses** [a] | **Donkeys** [b] | **Mules** [c] | **Cows** [d] | **Sheep** [e] | **Goat** [f] |
| Total solids | 10–12 | 9–11 | 10–12 | 12–13 | 18–20 | 12–14 |
| Fat | 1.0–2.5 | 1.0–1.5 | 1.0–2.0 | 3.5–4.5 | 6.0–7.0 | 4.0–5.0 |
| Protein | 1.5–2.5 | 1.5–2.0 | 1.5–2.5 | 3.0–3.5 | 5.5–6.0 | 3.0–4.0 |
| Lactose (carbohydrates) | 6.0–7.0 | 6.0–7.0 | 6.5–7 | 4.5–5.0 | 4.5–5.0 | 4.1–4.5 |
| Ash | 0.3–0.5 | 0.3–0.6 | 0.9–1.2 | 0.7–1.0 | 1.0–1.2 | 0.9–1.1 |
| Water content | 88–90 | 89–91 | 88–90 | 87–88 | 80–82 | 88–86 |

* Sources: [a] [68]; [b] [69]; [c] [70]; [d] [69] [e] [70]; [f] [71].

Sheep milk, with its higher fat content, is an excellent choice for those seeking a richer source of energy and nutrition. This makes it particularly beneficial for individuals with increased caloric needs, such as athletes, the elderly, or those recovering from illness. Its fat-rich composition is also valuable for people dealing with conditions like malnutrition or cachexia, where maintaining weight and muscle mass is crucial [69]. Moreover, the protein and fat content in sheep milk helps strengthen the immune system, aiding in recovery from infections or post-surgery convalescence. Goat milk, while having a more moderate fat content than sheep milk, still contains more fat than cow's milk [68–70]. Goat milk has been recommended for individuals with sensitive digestive systems or food allergies. Due to its smaller fat globules, goat milk is easier to digest, making it a suitable option for individuals who experience difficulty processing cow's milk. The reduced fat content makes them a preferred choice for people with cardiovascular concerns, as they contribute less to saturated fat intake, potentially lowering the risk of heart disease [71–75].

Individuals aiming to increase their intake without consuming high levels of saturated fats may opt for horse or donkey milk, as these offer a light yet nutritious choice, especially for those managing weight loss or caloric restriction in a healthy way. For people with sensitive digestion, horse and donkey milk are ideal choices, offering a gentler option without the added fat [76].

Clinical studies on the effects of consuming horse and donkey milk in children with various allergies have been documented since 1992. Donkey milk, in particular, has been the focus of several studies due to its high palatability, attributed to its lactose content of 6.0–7.0%, and its low allergenicity. Moreover, when produced under high hygiene standards and properly supplemented for nutrition, equine milk has been shown to be a promising alternative in the dietary management of children with conditions such as immunoglobulin E-mediated cow's milk protein allergy and food protein-induced enterocolitis, which typically manifests within the first 6 months of life [77]. However, the evidence regarding the effectiveness of equine milk in meeting the nutritional needs of children is not yet conclusive, as further research with larger studies is required to validate these findings. Despite its potential, donkey milk is used cautiously and is generally recommended for children over the age of one, as part of a solid food diet, or as an added ingredient [78–83].

### 3.3. *Bioactive and Functional Compounds*

The bioactive and functional compounds in other species' milk and equine milk share several similarities, but there are distinct differences in their composition and effects on human health [84].

Milk proteins primarily consist of casein proteins (alpha-casein, beta-casein, and kappa-casein), which account for approximately 80% of the total protein in cow milk and around 60–80% in the milk of other mammals, such as goats and sheep. These proteins form gels or clots in the stomach, promoting slow digestion and enhancing nutrient absorption. Caseins are also responsible for milk's white color and are a key source of essential amino acids and minerals, including calcium and phosphorus [85]. Whey proteins, including beta-lactoglobulin, alpha-lactalbumin, serum albumin, and immunoglobulins (IgA, IgG, and IgM), which are crucial for immune function, make up about 20% of the protein in cow milk and a higher percentage in goat and sheep milk. These proteins are quickly digested and are of high biological value, as they contain all the essential amino acids. Milk also contains small amounts of enzymes such as lipase and lacto-peroxidase, as well as proteins like lactoferrin and lysozyme, which have antimicrobial properties [86]. The comparison between species regarding these aspects is revealed in Table 4.

**Table 4.** The comparison between the bioactive and functional compounds of different breeds *.

| Item | | Species | | | | |
|---|---|---|---|---|---|---|
| | | **Horses** [a] | **Donkeys** [b] | **Cows** [c] | **Sheep** [d] | **Goat** [e] |
| Casein (%) | | 1.1–1.4 | 0.7–1.0 | 2.6–2.8 | 3.8–4.2 | 2.4–2.7 |
| Bioactive peptides (mg/mL) | | 1.5–2.5 | 1.0–1.8 | 2.0–4.0 | 3.5–5.0 | 2.5–3.5 |
| Conjugated linoleic acid (CLA)—mg/g fat | | 1.0–1.5 | 1.0–1.2 | 3.5–7.0 | 6.0–11.0 | 4.0–8.0 |
| Omega-3 fatty acids (mg/100 g fat) | | 70–100 | 60–90 | 30–50 | 40–60 | 20–40 |
| Unsaturated fats (%) | | 60–70 | 55–65 | 25–35 | 30–40 | 27–35 |
| Digestibility level | | High | Very high | Moderate | Low to moderate | high |
| Immunological properties | | Rich in lactoperoxidase and cytokines; anti-inflammatory peptides | Contains growth factors and cytokines; hypoallergenic | Contains beta-lactoglobulin and casein-derived peptides with immune effects | Rich in antioxidant peptides and lactoperoxidase | Contains bioactive peptides promoting gut health |
| Vitamins | A (µg/100 mL) | 18–25 | 20–30 | 28–40 | 50–75 | 40–50 |
| | D (µg/100 mL) | 0.1–0.2 | 0.1–0.3 | 0.1–0.2 | 0.3–0.5 | 0.2–0.4 |
| | E (µg/100 mL) | 40–60 | 50–70 | 60–80 | 80–100 | 50–70 |
| | C (mg/100 mL) | 1.2–1.8 | 1.5–2.0 | 1.0–1.5 | 1.0–1.8 | 1.2–1.8 |
| | B complex | Lower levels of B2 vitamin than cow and sheep milk, low but comparable to cow and goat milk levels of B9, and lower levels of B1 than cow milk | Moderate levels compared to goat milk | Rich in B2 vitamin: 0.16 mg/100 mL and B12 vitamin: 0.4 µg/100 mL | Higher levels of B2 and B9 vitamins compared to cow milk | High levels of B2 and B3 vitamins |
| Minerals (mg/100 mL) | Ca | 110–130 | 35–45 | 120–130 | 160–180 | 120–140 |
| | P | 85–100 | 20–30 | 90–100 | 110–130 | 90–100 |
| | Mg | 10–15 | 6–10 | 10–15 | 20–25 | 10–15 |
| | K | 150–200 | 150–200 | 140–150 | 150–200 | 180–200 |
| | Na | 35–40 | 15–20 | 40–50 | 45–55 | 40–50 |

* Sources: [a] [68]; [b] [86]; [c] [70]; [d] [69]; [e] [72].

Bioactive peptides are protein fragments released during digestion or fermentation that exert biological effects beyond basic nutrition. One such compound is lactoferrin-derived peptides. Lactoferrin, a multifunctional glycoprotein found in both the whey and casein fractions of milk, possesses antibacterial, antiviral, and anti-inflammatory properties. Additionally, peptides derived from lactoferrin may enhance iron absorption and help modulate immune responses [86–88].

Casein-derived peptides are another group of bioactive peptides. When casein is hydrolyzed, it releases peptides with a range of health benefits, including opioid-like peptides such as casomorphins, which have pain-relieving and relaxing effects. Some casein peptides also act as antihypertensive agents by inhibiting ACE (angiotensin-converting enzyme), potentially helping to lower blood pressure. Additionally, certain casein peptides have antioxidant properties, which help protect against oxidative stress and promote overall cellular health. Whey proteins are also a source of bioactive peptides with diverse health benefits. Among these are lactokinins, which have antihypertensive properties and contribute to lowering blood pressure. Peptides derived from alpha-lactalbumin are known to reduce inflammation and improve gut health. Additionally, immunomodulatory peptides, including those derived from immunoglobulins like IgA, help strengthen the immune system's defense against infections. Although not proteins, milk oligosaccharides—particularly abundant in human milk—serve as prebiotics, fostering the growth of beneficial gut bacteria and indirectly promoting digestive and immune health [85]. When comparing the protein content of milk from various species to donkey and horse milk, it is notable that cow's milk is predominantly composed of two types of proteins: casein (around 80%) and whey proteins (approximately 20%). Caseins are recognized for their gel-forming ability in the stomach, which supports slower digestion. Additionally, bioactive peptides derived from cow's milk proteins, such as lactoferrin and α-lactalbumin, have been shown to possess immunomodulatory, antibacterial, and anticancer properties. Casein-derived peptides exhibit antihypertensive effects and may support gut health. Sheep milk, with its higher protein content compared to cow milk—particularly its abundance of casein—offers greater nutritional value in terms of protein. It is also rich in bioactive peptides with antioxidant properties that enhance immune function and promote cardiovascular health. Additionally, sheep milk contains elevated levels of lactoferrin, a compound with potent antimicrobial properties [87].

Goat milk contains both casein and whey proteins, with a higher proportion of whey proteins than cow milk, which enhances its digestibility. Bioactive peptides derived from goat milk whey proteins offer anti-inflammatory and antimicrobial properties. Additionally, goat milk is rich in oligosaccharides, which support gut health by encouraging the growth of beneficial bacteria [73]. Donkey milk closely resembles human milk in composition, with a higher proportion of whey proteins and lower casein levels, making it potentially more suitable for individuals with milk allergies. Its bioactive peptides, including lysozyme, exhibit immune-modulating and antimicrobial properties. Additionally, donkey milk is rich in immunoglobulins, which play a crucial role in enhancing immune function [17,18,83]. Mare's milk exhibits a higher proportion of whey proteins and lower casein levels, akin to human milk, which enhances its digestibility. Additionally, it contains lysozyme and lactoferrin, which offer immune-modulating and antimicrobial benefits. Due to its lower allergenicity compared to cow milk, mare's milk serves as a suitable alternative for individuals with sensitivities [84].

When comparing the fat content across different species' milk, several key aspects emerge: cow's milk contains a mix of saturated and unsaturated fatty acids, which support energy needs and the absorption of fat-soluble vitamins. Additionally, the lipids in cow's milk include bioactive compounds such as conjugated linoleic acid (CLA), known for its potential anti-inflammatory and anti-cancer properties [86]. The sheep milk is richer in fat compared to cow milk, with a higher percentage of unsaturated fatty acids, including omega-3 fatty acids. These contribute to cardiovascular health and are beneficial for reducing inflammation and improving brain function [70]. Goat milk contains a higher proportion of short- and medium-chain fatty acids compared to cow milk, making it easier to digest and absorb. These smaller fat globules are particularly beneficial for individuals with digestive issues or absorption disorders. Additionally, goat milk contains conjugated linoleic acid (CLA), which is associated with potential health benefits, including anti-inflammatory effects [83].

When comparing the lactose content and digestibility of milk from different species, including donkey milk, several differences become apparent. While lactose-free alternatives are available, they may still contain other components that could cause issues for sensitive individuals [24]. Sheep milk has a higher lactose content than cow milk, but it is still tolerated by some people who have mild lactose intolerance due to its composition of proteins and fats that aid digestion [74]. Goat milk contains less lactose than cow milk, and its proteins are easier to digest. As a result, it is a better choice for individuals with mild lactose intolerance or those with sensitive digestive systems [73]. Although donkey milk contains similar levels of lactose to cow milk, its higher whey protein content makes it easier to digest for some individuals, including those with mild lactose intolerance. The mare milk contains similar lactose levels to cow milk but is more digestible due to its higher whey protein content. Considered easier to tolerate for those with mild lactose intolerance [75].

When comparing the immunological properties of different species' milk, several key points emerge. Cow milk contains immunoglobulins and other bioactive compounds, but its immune-modulating effects are generally less powerful than those found in donkey and sheep milk. While cow milk does support gut health and immune function to some extent, it is often regarded as less effective for individuals with immune system issues or allergies [89]. Sheep milk is rich in immunoglobulins and lactoferrin, which play a key role in enhancing immune responses. The milk's elevated protein content, particularly its immunoglobulins, aids in combating pathogens and promoting overall health [90]. Goat milk contains moderate levels of immunoglobulins and lysozyme, which provide immune support, though these levels are lower than those found in sheep and donkey milk [91]. On the other hand, donkey milk is rich in immunoglobulins, particularly IgG, making it an excellent option for enhancing the immune system. This has led to its use in helping individuals with weakened immune systems or digestive issues. Similarly, mare milk also contains high levels of immunoglobulins (especially IgG), which support immune health and are used to boost immunity in individuals with compromised immune systems [91].

When comparing the vitamins and minerals in different species' milk, including donkey milk, the following conclusions can be made. Cow milk is abundant in vitamins such as B2 (riboflavin), B12, and D, as well as essential minerals like calcium and phosphorus, all of which are vital for bone health and overall development. Sheep milk contains higher amounts of vitamins A, D, and E compared to cow milk, making it more nutrient-dense. It is also richer in calcium, iron, and zinc, which are crucial for bone health, immune function, and blood cell production [92]. Goat milk offers higher levels of calcium, potassium, and magnesium than cow milk, supporting bone health, nerve function, and muscle function [93–96]. Donkey milk stands out for its higher vitamin C content, which supports the immune system and acts as an antioxidant. It also contains moderate levels of calcium and magnesium, which aid in bone health. Mare milk also has elevated vitamin C levels, benefiting immune function and acting as an antioxidant, along with moderate amounts of calcium and magnesium to support bone health [97].

### 3.4. *The Sensory Analysis of Equine Milk*

Compared to cow's milk, equine milk is sweeter due to the higher content of lactose and smoother texture, and compared to goat's milk, it has a smoother texture, a more subtle flavor and it is less creamy [8]. The sensory analysis involves the fat content, flavor, sweetness, texture, color, smell, and astringency (Table 5).

The factors that influence the sensory attributes are as follows: the fat content is influenced by the breed (Percheron, Belgian Draft, Clydesdale), the diet (high-energy levels and season changes increase the fat content), the mare's health (healthy mares produce milk with more fat content), and the stage of lactation (at the beginning, the milk is lower in fat), the flavor is influenced by the diet, freshness, fat content (hay, grass, and oat can influence the sweetness and flavor, if the milk is fresh it has a milder flavor, while if it is older tends to have more pronounced aftertaste, and a higher fat content makes it creamier and flavorful), the sweetness is influenced by the stage of lactation (the colostrum is sweeter), the diet (good pastures lead to sweeter milk), and the health's mare (if the mare is healthy, produces sweeter milk); a lower fat content, and also freshness contributes to a lighter texture; some breeds as Purebred Arabian and high-carotene contents (found in pasture and carrots) produce a yellow color. The smell is influenced by grass and oats, which bring a pleasant scent, while improper storage can lead to specific and unwanted odors. The astringency is influenced by the fat levels (low fat content leads to less astringency) and by the mare's health (healthy mares have reduced astringency) [92].

**Table 5.** Sensory analysis of equine milk [98] *.

| Sensorial Attribute | Exposure |
|---|---|
| Fat content | - has a fat content that ranges between 1.5 and 2.5%, which makes it light but creamy as texture |
| Flavor | - sweet, mildly creamy flavor, slightly bitter |
| Sweetness | - naturally sweet due to high lactose content |
| Texture | - light, watery, smooth |
| Color | - white to pale yellow, sometimes slightly opalescent due to low fat content and smaller fat globules |
| Smell | - mild, pleasant, fresh odor |
| Astringency | - low, which means the aftertaste is now dry |

* Source [98].

## 3.5. *Dairy Equine Raising*

### 3.5.1. Dairy Equine Diets

Equines raised for milk, such as horses and donkeys, have unique nutritional needs. Their diets must be carefully balanced to support optimal milk production while maintaining their overall health [93–97,99]. First and foremost, their diet needs to be well-rounded, including high-quality forage like grass or hay (Table 6).

**Table 6.** Nutritional profile of high-quality forages for equine *.

| Item | Protein (%) | Digestibility (%) | Fiber (%) | Fat (%) | Energy (ME) (MJ/kg) | Minerals (%) | Vitamins |
|---|---|---|---|---|---|---|---|
| Grass/pasture | 10–15 | 60–80 | 20–30 | 1–2 | 8–12 | 0.3–0.6 Ca<br>0.2–0.3 P<br>0.1–0.2 Mg<br>2–3 K | 500–1000 IU/kg Vit. A<br>15–30 mg/kg vit. E<br>traces of Vit. K and D |
| Hay (timothy and meadow) | 6–14 | 50–60 | 30–40 | 1–2 | 6–9 | 0.4–0.6 Ca<br>0.2–0.4 P<br>0.2 Mg<br>1–2 K | 1000–1500 IU/kg<br>10–20 mg/kg vit. E<br>traces of vit. K |
| Alfalfa hay | 18–22 | 55–65 | 25–30 | 2–3 | 9–12 | 1.2–1.8 Ca<br>0.2–0.4 P<br>0.2 Mg<br>1–2 K | 1500–2500 IU/kg<br>30–50 mg/kg vit. E<br>traces of vit. K and D |
| Meadow hay | 8–15 | 50–60 | 30–40 | 1–2 | 6–10 | 0.3–0.6 Ca<br>0.2–0.4 P<br>0.1–0.2 Mg<br>1–2 K | 500–1000 IU/kg vit. A<br>15–25 mg/kg vit. E<br>traces of vit. K |

* Source [100].

In addition, they require supplemental grains and minerals to ensure sufficient intake of protein, energy, vitamins, and minerals. A fiber-rich diet is essential for promoting healthy digestion and supporting milk production. The digestive process in horses is specific, with the following characteristics [101–104]:

- Forage forms the cornerstone of equine milk diets, playing a vital role in supporting healthy digestion, especially since horses and donkeys are non-ruminant herbivores (hindgut fermenters). Providing high-quality hay in adequate quantities is essential to ensure they consume enough fiber to maintain digestive health. For lactating equines, hay intake may need to be adjusted to meet the higher energy demands. Pasture also plays a significant role in their feeding, as fresh pasture can supplement their diet when available and during favorable seasons. However, proper pasture management is crucial to prevent overgrazing and maintain nutrient balance, with rotational grazing being an effective strategy.
- Concentrates and grains are essential for meeting the high energy demands of lactating equines. Oats and barley are commonly used for this purpose, with oats being preferred for lactating mares due to their easy digestibility and steady energy release. It is recommended to grind and soak the feed in sufficient water before feeding to help prevent colic—a common equine digestive issue caused by insufficient water intake, sudden dietary changes, or the consumption of dry, hard-to-digest feed. Soaking or grinding feed improves its digestibility, reduces the risk of intestinal blockages, and promotes better hydration, which is critical for maintaining gastrointestinal function and reducing the risk of dehydration. Existing studies also underscore the importance of proper feeding practices, such as soaking hay and ensuring continuous access to clean water. Horses that receive dry hay or limited water access have a higher risk of developing colic (this type of forage leads to forming compacted masses in the digestive tract, causing pain and even obstructions), as the sudden changes in their diets could cause the same reaction. Equines are vulnerable to colic because of their unique digestive system (they have small stomachs compared to their body size—10–15 L or about 10% of their digestive systems—and a large cecum—approximately 30–40 L, which represents 15–20% of their digestive capacity, that functions properly with enough fiber and water intake; since they are non-ruminant and cannot vomit due to a specific gastroesophageal sphincter at the entrance in their stomach, it is necessary to consume small amounts of feed, throughout the entire day—this is a proper way to maintain the gastrointestinal motility). All these measures reduce the risk of impactions, gas buildup, and twists of the intestines, which could lead to colic [105]. Consulting an equine nutritionist or veterinarian is highly recommended to customize diets and feeding regimens to meet specific needs and conditions. Additionally, supplements are advisable for lactating equines to meet their specific nutritional needs. Depending on their condition and milk production levels, incorporating a small amount of vegetable oil, such as soybean or sunflower oil, into their diet can provide extra energy. This strategy helps increase calorie intake without disrupting the protein balance, ensuring that the animals maintain optimal health and sustain adequate milk production.
- Protein should come from both forage and concentrates to meet the nutritional needs of lactating equines. High-quality grass hay serves as an excellent protein source, but lactating mares and donkeys may require additional supplementation. Alfalfa hay, known for its higher protein content, is a suitable option. Alternatively, protein supplements such as soybean meal can be included in the diet. Amino acids are also essential for optimal milk production, with lysine and methionine being particularly crucial for dairy animals. For equines, the levels of lysine and methionine are also crucial. For a 500 kg horse, a quantity of 15–20 g of lysine/day and 10–15 g of methionine/day. Lysine contributes to collagen formation, calcium absorption, enzyme function, and milk production (for maintenance, the minimum daily intake is 0.45–0.5% of the total diet), and for lactating equines, the baseline could be increased

by 0.6–0.8% of the total diet (optimum sources are alfalfa hay, soybean, and canola meals, oats or lysine powder or pellet). Methionine is important in keratin formation, liver detoxification, fat metabolism, and lactation; for maintenance, 0.2–0.25% of the total diet is needed, and for lactation, a surplus of 0.3–0.35% of the total diet is recommended (good sources of methionine are grains as oats and barley, hay, soybean and canola meals and methionine powder or pellet) [106].

- Equines require adequate intake of essential minerals and vitamins to maintain strong bones and proper metabolic function. High-quality forage combined with targeted supplementation helps meet these nutritional demands—calcium and phosphorus are crucial with the calcium-to-phosphorus ratio balanced at approximately 2:1 and lactating equines requiring increased calcium levels to support milk production; magnesium is vital for proper muscle function while potassium is essential for maintaining electrolyte balance, particularly during lactation. Vitamins A, D, and E play critical roles in equine health. Vitamin D supports calcium absorption and bone strength, vitamin E acts as a powerful antioxidant and supports immune function, and vitamin A is essential for vision growth and overall health. Calcium plays an important role in various functions of the body, particularly in bone and muscle function, as in nerve transmission; the deficiency leads to parturient hypocalcemia (this occurs in the first days after foaling, causing weakness, lack of coordination, and in severe situations the inability to stand), muscle tremors (low calcium levels cause muscle spasms, tremor, and in severe situations even seizures). Magnesium is another crucial mineral that plays an important part in the same direction as calcium. The deficiency reveals specific signs as follows: muscle tremors and even stiffness (it is important in muscle relaxation, and low levels could create spasms, twitching, and stiffness), low milk production (magnesium deficiency negatively impacts lactation), and even behavioral changes (irritability and hyper-excitability are common reactions to magnesium deficiency). During lactation, inadequate levels of calcium, magnesium, and phosphorus can lead to metabolic bone disease that causes fragile bones that are susceptible to fractures. Also, vitamin deficiencies are dangerous. A lack of vitamin A can lead to vision problems, dry coat and hair loss, skin problems, impaired immune function, reproductive issues, and delayed growth in foals. A lack of vitamin E leads to muscle atrophy, neurological issues (wobbly gait, loss of reflexes or head tilting), excessive sweating, reproductive issues, immune system conditions, equine motor neuron disease (progressive muscle weakness, loss of muscle mass, and difficulty standing), dry hair and skin problems. The recommended supplements in this case are calcium carbonate/phosphate, dicalcium phosphate, calcium and phosphorus balancer, magnesium oxide, magnesium sulfate, or chloride. When speaking about vitamin deficiency, retinol available in oil or powder form, carotene from grass and carrots, or cod liver oil are recommended to supply with vitamin A. Natural vitamin E, vitamin E in oil or powder, and pasture or good quality hay are added to supply the lack of this vitamin; Biotin and methionine are indicated to satisfy vitamin B7 levels and to support the coat characteristics. Zinc and copper chelates and selenium supplements are precious for overall health, as are omega-3 fatty acids found in flaxseed, linseed, algal, or fish oil [107–110].
- Since milk is primarily composed of water, lactating equines have significantly increased water demands to support milk production. It is essential to provide them with ample fresh and clean water at all times to ensure optimal hydration and milk yield.

A model of diet used in dairy equines (mares and donkeys) is presented in Table 7.

It is also important to calculate the needs regarding the level of production, knowing that the peak of lactation in mares is in the first 2–3 months when the foal needs the most nutrients for growth. After this stage, milk production gradually declines as the foal starts consuming solid feed, so the milk is not his exclusive feed. By the age of 6 months, foals generally eat substantial amounts of grass, hay, and grain as they approach weaning (as a consequence, milk production starts to significantly reduce [112]) (Table 8). The quantity is also influenced by the breed (heavy ones produce more than the light breeds), nutrition (proper diets that meet the needs of supporting the production are crucial), age of the mare (old females may produce less milk than young ones), health of the mare including their mammary gland (health issues may influence the milk production), and also the foal's size (larger products demand more milk). Therefore, it is vital to monitor the mare's milk production to ensure proper development of the foal to adjust the diet based on the quantity produced (high-quality forage and concentrates are crucial, especially during the first stages of lactation period), and respect the weaning moment (generally when the foal is around 6–9 months; it is recommended to apply gradual weaning when the foal also consumes solid food) [113].

**Table 7.** Model of diet for 100 kg live weight of lactating equines *.

| Item | Quantity/Day | Protein (g) | Energy (MJ) | Fiber (g) |
|---|---|---|---|---|
| Forage | 2 kg | 300 | 16 | 6000 |
| Concentrates | 0.75 kg of oats or barley | 90 | 7.5 | 56.25 |
| Alfalfa hay (optional for protein intake) | 0.5 kg | 100 | 3 | 500 |
| Vegetable oil (optional) | 25 g | - | 0.925 | - |
| Mineral supplement | 15 g | - | - | - |

* Source: [111].

**Table 8.** Diet for lactating mare (live weight 500 kg; milk production 2–4 L/day) *.

| Item | Quantity/Day | Protein (g) | Energy (MJ) | Fiber (g) |
|---|---|---|---|---|
| Forage | 7–8 kg | 1050–1200 | 56–64 | 21,000–24,000 |
| Concentrates | 2.5–3 kg | 250–300 | 20–25 | 150–210 |
| Alfalfa hay (optional for protein intake) | 1.5–2 kg | 300–400 | 9–12 | 1500–2000 |
| Vegetable oil (optional) | 50–75 g | - | 1.85–2.775 | - |
| Mineral supplement | 50 g | - | - | - |

* Source: [111].

The milk production registered in donkeys is approximately 0.5–1 L/day in the first period of lactation; when the foal is 4–6 months, the quantity drops as he starts grazing more and consuming solid foods. The colostrum phase, when the female produces less than 1 L/day, lasts for 24–48 h after foaling, then the peak lactation comes for 1–3 months, when 0.5–1 L of milk is produced daily. The next stage is represented by the declining lactation phase (3–6 months/after foaling) when only 0.3–0.5 L of milk are produced per day; in the late phase of lactation (6–9 months/after foaling), only 0.2–0.3 L/daily are produced. A model of diet for a 250 kg donkey, with 0.5–1 L of milk/daily, is presented in Table 9 [114].

**Table 9.** Diet for lactating donkey (live weight 250 kg; milk production 0.5–1 L/day) *.

| Item | Quantity/Day | Protein (g) | Energy (MJ) | Fiber (g) |
|---|---|---|---|---|
| Forage | 6–7 kg | 800–1000 | 25–30 | 5000–6500 |
| Concentrates | 0.5–1 kg | 50–150 | 8–16 | 100–200 |
| Alfalfa hay (optional for protein intake) | 1–1.5 kg | 250–375 | 9–12 | 1000–1500 |
| Vegetable oil (optional) | 25–50 g | - | 1–2 | - |
| Mineral supplement | 20–30 g | - | - | - |

* Source: [111].

### 3.5.2. Dairy Equine Water Requirements

Lactating equines have higher water needs than those that are non-lactating, up to 50% higher due to the high water content of milk (each liter has 87–88% water). This means that 1.5–2 L of water is required for each liter of milk. Normally, horses consume approximately 50–60 L of water daily, and adding the water requirements per each liter for a production of 2–4 L of milk/day leads to a daily need of approximately 54–64 L of water (it also depends on the season—hot water or high humidity increases this need, diet—hay or grain also increases the water intake, level of effort—due to sweating some increments could appear, etc.). Equines need to have free access to fresh water, the hay should be soaked, and electrolytes are recommended to be added for better hydration and a really important aspect in water management for equines is to be monitored for any level of dehydration (when the milk production decreases, lethargy occurs, dry gums and sunken eyes are identified) [114].

### 3.5.3. Pasture Management for Dairy Equines

Research highlights the advantages of pasture-based management for equine herds, showing that grazing systems enhance both the nutritional quality of mare's milk and the socio-ecological benefits of sustainable agriculture. Particularly in arid and semi-arid regions, these systems provide a practical approach to equine milk production while promoting biodiversity and preserving natural landscapes [110]. Equine milk stands out for its exceptional nutritional value and therapeutic potential, positioning it as a valuable component of human diets and a contributor to sustainable agricultural practices. However, challenges remain in raising awareness about its benefits and production techniques, which may limit its widespread adoption and hinder the growth of the industry [36]. However, the management of broodmares and foals involved in milk production demands meticulous planning, particularly regarding their nutritional needs and overall care. Prioritizing the health and well-being of both the mares and their offspring is crucial for ensuring the sustainability of the operation. Providing mares and foals with appropriate nutrition is essential not only for maintaining their health but also for producing high-quality milk. Additionally, addressing knowledge gaps related to optimal nutrition in equine dairy systems is vital for the long-term success of such enterprises [101].

Another key consideration is the ecological impact of equine farming. To reduce environmental harm, sustainable farming practices must be adopted, particularly in marginal areas where pastures and natural meadows are essential for maintaining ecological balance [108–111].

Lastly, it is crucial to prioritize food security and animal welfare. As with all forms of animal husbandry, ensuring humane treatment of animals is essential, along with ensuring that the production system contributes to local food security. While equine milk can provide valuable nutrition, its production must be managed in a way that supports long-term sustainability [107,108,113]. Thus, the successful growth of the equine milk industry relies on bridging communication gaps, adopting effective management practices for mares and foals, optimizing animal nutrition, addressing environmental challenges, and safeguarding food security and animal welfare, all while encouraging sustainable practices for long-term success.

In conclusion, sustainable pasture management plays a crucial role in maintaining the health of dairy equines and fostering environmental sustainability. Implementing rotational grazing is a primary strategy in this regard. By moving equines between different pasture areas, overgrazing is minimized, soil health is preserved, and pastures are given time to regenerate. This approach enhances the long-term productivity of the land while reducing the risk of soil erosion [113].

Another crucial element is maintaining pasture diversity. Promoting a variety of grasses and legumes in pastures supports healthy ecosystems and provides the diverse nutrients required by dairy equines. A mix of plant species ensures sustained pasture productivity throughout the year, even in changing weather conditions [112].

Also, soil health is equally vital—implementing sustainable practices, such as composting and applying organic fertilizers, enhances soil fertility, fostering the growth of nutrient-rich pasture grasses. Limiting the use of chemical fertilizers and pesticides is essential to reduce environmental impact and maintain ecological balance [113].

Lastly, optimizing water management is crucial—ensuring reliable access to clean water is essential for dairy equines, particularly during periods of high milk production when their water needs significantly increase. Pasture management strategies should prioritize the availability of fresh water to support both animal health and productivity [114].

## 4. Equine Meat—Benefits in Human Consumption

Equine meat consumption is common in several parts of the world, such as France, Italy, Belgium, and the Netherlands, where it is considered as traditional and even premium dish, and also in Argentina and Brazil, countries where small markets for horse meat exist; the United States of America and Canada do not have a tradition in this sense, the horse consumption being something rare (although it was more common in the past, particularly during the early 20th century, the practice has decreased significantly due to cultural norms, animal rights concerns, and legal restrictions; however, horse meat is still available in some specialty markets or through imports). Also, in Japan, Mongolia, and Kazakhstan, horse meat is an important part of the diet; in Mongolia, it is a staple food because of the main role of this species in their culture [115]. However, the topic remains controversial and less widespread in others.

Equines used for meat production are represented by horses and donkeys, and in some cultures, they are also represented by mules. Regarding the breeds, there is no such thing as meat breeds as in cattle or sheep; however, certain horses raised for work, racing, or horseback riding can be slaughtered after being removed from their activity and being fattened for a while (some countries also raise horses, especially for this purpose). Donkeys are preferred in certain regions of Asia, such as China, Japan, and Africa, due to the quality of their meat (in these cases, it is preferred in stews, dried meat, and sausages); donkey meat is appreciated for its tenderness and nutritional values because it is leaner than horse meat (some cultures think that it has medicinal properties). Lastly, mules are occasionally raised for this purpose, although this situation is not that common. As they are the offspring of male donkeys and mares, they have peculiarities from both species; generally, they are used for strength and endurance, but there are countries that also prefer to consume their meat as it is culturally accepted [116].

### *4.1. The Chemical Composition of Equine Meat*

Horse meat is a lean meat, which offers a couple of health benefits. The chemical composition in this case can vary depending on factors like breed, age, diet, type of effort, and processing methods. However, on average, it is considered leaner than beef, sheep, and pork, with a slightly higher protein content and lower fat content. The water percent is around 70–75% (fresh meat), and the protein is around 20–24%, which makes this species a reference diet rich in high-quality protein that includes all essential amino acids (close to red meat); regarding the amino acids, horse meat also contains higher amounts of lysine and methionine than beef [117].

The total fat in horse meat is relatively low, ranging from 2–5%, which is lower than beef and pork; the fat content also varies depending on the part of the animal, with muscle cuts being leaner than fatty cuts. This aspect makes horse meat a leaner option, preferred by individuals who need a low-fat diet. Regarding saturated fat, horse meat tends to be lower in saturated fat compared to beef, with about 1–2% of saturated fat, which makes it a relatively healthier option (in beef is generally 5–7%); the monounsaturated (oleic acid—it is associated with reduced low-density lipoprotein cholesterol level, which improves the heart health) and the polyunsaturated fat (omega-3 and omega-6, which maintain the cellular function, sustain the cardiovascular health, reduce the inflammation in body, and supports the brain functions) are in higher proportion compared to beef [26]. Horse meat is also a healthier choice when compared with pork, meat with 9–20% total fat content, with high saturated fat (3–12% or higher if we include parts such as belly and ribs), and higher content of monounsaturated fat, especially in fatty cuts, compared to horse meat; pork is also poorer in omega-3, richer in saturated fat which contributes to elevated LDL cholesterol levels and it has more calories per serving as a consequence to higher fat content (245–290 kcal/100 g of meat, on average). Chicken breast is known as being low-fat, with a total fat content of 1.5–3.5% and only 0.5–1% saturated fat, while the chicken legs, which also contain the skin, have a total fat of 8–10% and 2–3% saturated fat [118].

Furthermore, horse meat has a moderate cholesterol content, similar to other lean meats (50–55 mg/100 g of meat). For beef and pork these limits range between 60 and 70 mg [119], and for chicken, between 60 and 75 g mg [47].

Horse meat is rich in protein [120] (20–24 g/100 g of meat) but lower than other species, such as beef, which has 20–26 g/100 g of meat (cuts like sirloin are higher in protein than horse meat); however, the beef keeps the disadvantage of having a higher fat content. The same problem also appears in pork, even if the protein is high (20–24 g/100 g of meat) [118]. When compared to skinless chicken breast (30 g of protein/100 g of meat), horse meat does not exceed the protein level in this case (other cuts, like legs or wings, have less protein per 100 g, and they also contain more fat, especially if the skin is on) [121].

As for the minerals, horse meat is an excellent source of hem iron, the type that is found in animal-based products (it is easily absorbed by the human body and it typically contains 2.5–4 mg/100 g of meat); horse meat also contains about 4–5 mg of zinc/100 g of meat, which is very important for the immune system, protein synthesis, and wound healing. There is also a content of 180–200 mg/100 g of meat phosphorus, which is essential for bone health and energy metabolism [122]. When comparing horse meat with beef, pork, and chicken, the iron level is the highest among the four (beef: 2–3 mg/100 g of lean cuts—sirloin and flanks; pork and chicken: 0.9–1.2 mg/100 g, lower than horse and beef meat). When comparing the zinc level, beef has 405 mg/100 g, pork has moderate levels of 2.5–3 mg/100 g, and chicken has 1–2 mg/100 g. Beef is a good source of phosphorus (180–210 mg/100 g), but with lower levels than horse meat; pork and chicken have similar levels of 200–220 mg/100 g [123].

Regarding the vitamin content, horse meat contains vitamin B12, which is essential for blood cell production and neurological function (10–20 mcg/100 g of horse meat, which covers the daily recommended intake a couple of times). Vitamin B6 is also present in horse meat in the proportion of 0.3–0.5 mg/100 mg of horse meat, which helps with amino acid metabolism and supports brain function. This product is also rich in vitamin B3 (5–7 mg/100 g), which is important for energy production and skin protection. Beef also has excellent levels of B12 (2–2.5 μg/100 g), good levels of B6, similar to horse meat (0.5–0.7 mg/100 g), and also good levels of B3 (4–5 mg/100 g). Pork, on the other hand, is a moderate source of B12 (0.8–1.0 μg/100 g), a good source of B6 (0.7–0.9 mg), and a very good source of B3 (6–8 mg/100 g). Lastly, chicken meat is lower in vitamin B12, compared to the other mentioned species (0.3–0.4 μg/100 g), but still a good source of vitamin B6 (0.6–0.8 mg/100 g) and a very good source of B3 (7–8 mg/100 g) [124]. The comparison between species is revealed in Table 10.

When it comes to carbohydrates, horses, like many other species, contain virtually no carbohydrates, making horse meat a protein-rich and also fat-rich product.

Donkey meat is similar to horse and mule meat in many aspects (Table 11), but it has its own unique nutritional profile. When describing donkey meat (per 100 g), it can be stated that it is a lean, high-protein meat (1–24 g, providing all the essential amino acids, supporting muscle growth and tissue repair), with moderate levels of iron, zinc, and phosphorus; it is also a rich source of B12, a moderate source of vitamin B6, and a moderate source of B3. It is also low in fat and cholesterol (50–60 mg), making it a heart-healthy option compared to some other meats. The vitamin content, particularly B12 and niacin, makes it beneficial for overall health, especially for energy metabolism and immune support. Lastly, mule meat, similar to donkey and horse meat, is considered lean and a nutritious source of protein, as well. When it comes to chemical composition (per 100 g), the mule meat has around 2–4% of fat, with 1–1.5% of saturated fat, and cholesterol of 50–60 mg, 20–24 g of protein (which, like donkey meat, provides all essential amino acids for muscle growth and other tissue protection) [123,124].

**Table 10.** The comparison between different types of meat *.

| Nutrient ** | Horse Meat [a] | Donkey [b] | Beef [a] | Pork [a] |
|---|---|---|---|---|
| Calories (kcal) | 120–140 | 130–150 | 250–300 | 245–290 |
| Protein (g) | 20–24 | 20–22 | 26–30 | 25–30 |
| Total fat (g) | 2–5 | 2–4 | 15–20 | 20–25 |
| Saturated fat (g) | 1–2 | 0.8–1.5 | 5–7 | 7–10 |
| Monounsaturated fat (g) | 1.5–2.5 | 1.0–2.0 | 6–10 | 7–10 |
| Polyunsaturated fat (g) | 0.5–1.5 | 0.5–1.0 | 0.5–1.5 | 1–3 |

* Sources: [a] [120]; [b] [121] ** per 100 g, on average.

**Table 11.** Comparative profile of different types of meat *.

| Nutrient (per 100 g) | Horse Meat | Donkey Meat | Mule Meat |
|---|---|---|---|
| Calories (kcal) | 120–140 | 130–150 | 130–150 |
| Protein (g) | 20–24 | 20–24 | 20–24 |
| Total fat (g) | 2–5 | 2–4 | 2–4 |
| Saturated fat (g) | 1–2 | 1–1.5 | 1–1.5 |
| Cholesterol (g) | 50–60 | 50–60 | 50–60 |
| Monounsaturated fat (g) | 1.5–2.5 | 1.0–2.0 | 1.0–2.0 |
| Polyunsaturated fat (g) | 0.5–1.5 | 0.5–1.0 | 0.5–1.0 |
| Iron (mg) | 3–4 | 3–4 | 3–4 |
| Zinc (mg) | 4–6 | 4–6 | 4–6 |
| Phosphorus (mg) | 200–250 | 200–250 | 200–250 |
| Vitamin B3 (mg) | 5–8 | 5–8 | 5–8 |
| Vitamin B6 (mg) | 0.3–0.5 | 0.3–0.5 | 0.3–0.5 |
| Vitamin B12 (μg) | 2.5–3.5 | 2.5–3.5 | 2.5–3.5 |

* Source [124].

### 4.2. *The Bioactive and Functional Compounds in Equine Meat*

Equine meat contains bioactive and functional compounds that contribute to several health benefits beyond basic nutrition (they support various physiological functions, such as antioxidant activity, immune modulation, and metabolic health) [125].

The bioactive peptides contained in equine meat are released during digestion and protein breakdown; these components have many health benefits, such as the following [126]:

- Antioxidant effect: Several peptides have the ability to destroy free radicals, reduce oxidative stress, and offer protection from cell damage (this aspect is very important for reducing the risk of heart disease and other chronic problems, even cancer).
- Immune-modulating activity: The bioactive peptides can modulate the immune function of the body by enhancing the body's defense mechanisms against viruses and bacteria and improving overall immune responses.
- Antihypertensive effects: Some peptides from equine meat may act as angiotensin-converting enzyme (ACE) inhibitors, supporting cardiovascular health and reducing blood pressure.

The iron [126]:

- Heme iron is a type of iron found in animal-based foods, very easy to be absorbed by the human body (compared to non-heme iron from plant sources), and equine meat is a very good source in this sense;
- Adequate iron intake can prevent iron deficiency anemia and support the overall metabolic processes (immune system and energy production).

Creatine [125]:

- Equine meat contains creatine, a substance that helps provide energy to muscles during intense physical activity (it is particularly important for muscle function and repair).
- The functional role of this compound is to support athletic performance, muscle recovery, and muscle health.

Taurine [125]:

- Equine meat is a good source of taurine, an amino acid that is important for heart health, nervous system, and bile salt formation.
- It has been proved that taurine has antioxidant properties and supports cardiovascular health because it improves blood vessel functioning while also improving the risk of heart diseases.

Vitamins [126]:

- Equine meat is rich in vitamin B12 (which helps prevent anemia and neurological problems) and B3 (which contributes to metabolic processes and energy production in human body).

Zinc levels [126]:

- Equine meat has high amounts of zinc, which is crucial for DNA and protein synthesis, immune system, skin protection, etc.

Phospholipids [126]:

- Phospholipids are in a small amount in equine meat (1–2% in muscle and 3–6% in organs like liver).
- Phospholipids are important in the formation of cell membranes and support the brain and nervous system.

Polyunsaturated fatty acids (PUFA) [125]:

- It is a well-known fact that equine meat is very rich in omega-3 and omega-6 fatty acids, which are crucial for brain and heart health.
- They have anti-inflammatory effects while also contributing to healthy cholesterol level maintenance (cardiovascular health support).

Conjugated linoleic acid (CLA) [125]:

- Equine meat contains modest amounts of conjugated linoleic acid, which in ruminant meats is in higher proportion.
- It has multiple purposes in human body from anti-cancer properties to weight control, since it improves lipid metabolism; it also has anti-inflammatory and antioxidant activity.

### *4.3. The Sensory Analysis of Horse Meat*

The sensory analysis comprises the juiciness, flavor, texture, color, fat content (marbling), tenderness, smell, and overall acceptability (Table 12). The factors influencing these aspects are the cultural influences (in many countries, horse meat is viewed as a delicacy and it has a positive sensory response; on the other hand, in some cultures it is considered a taboo, which makes the perception over these sensory qualities quite negative) and the type of cooking (grilling, slow-cooking or including the meat in minced dishes or sausages has an impact on the tenderness and juiciness of horse meat; also the overcooking leads to dryness, slow-cooking and marinating being indicated to obtain juicy and tender portions); on the other hand, the ethical perception and the consumer education are important in this sense: in several markets, ethical problems regarding this type of meat (where horses are seen as companions or when the animal welfare is a concern) can influence in a negative way the consumer choices, despite its qualities; also knowing the benefits of consuming horse meat such as for its high levels of protein and iron and low levels of cholesterol can lead to its acceptance) [127].

**Table 12.** Sensory analysis of horse meat *.

| Sensorial Attribute | Exposure |
|---|---|
| Juiciness | - Less juicy than other types of meat (due to its low fat content); to obtain juicy dishes is indicated to be marinated or slow-cooked) |
| Flavor | - Distinct, slightly sweet, savory, gamey (milder than venison, more intense than beef) |
| Texture | - Tender, often less fibrous than beef |
| Color | - Darker than beef (from deep red to dark purple) |
| Fat content | - Leaner than beef, less marbled (intramuscular fat) |
| Tenderness | - Can vary depending on the cut |
| Smell | - Slightly sweet, gamey flavor |
| Overall acceptability | - Often appreciated for its unique flavor and tenderness when horse meat consumption is accepted |

* Source [127].

The factors that influence the sensorial aspects are various; the flavor is influenced by breed, diet, and processing (the breed has slight influence over meat's flavor; the horses which are fed with hay or grass have milder, sweeter meat than those fed with grains, and aging and marination influence the tenderness and the flavor of the meat). The juiciness and the texture depend on the cut (tenderloin and ribs are juices, the shoulder and flanks are drier), on the type of cooking (marinating and slow-cooking are recommended for juicy dishes), and on the age of meat (when the horse is younger, the meat has a higher tenderness and juiciness), while the color is influenced by the freshness (fresh meat is generally darker), diet (grass-fed horses have darker colors—from intense red to purple—due to vitamin E and beta-carotene which contribute to myoglobin production), and post-mortem handling (how the meat is processed and stored can impact its color). The fat content depends on the breed and diet (some breeds such as Belgian Draft, Percheron, Clydesdale, etc., and those grain-fed may have slightly more fat content than the horses that receive only hay or grass), the age (older horses may have slightly higher fat content) [128,129].

### *4.4. Meat Equine Raising*

#### 4.4.1. Meat Equine Diets

The equines reared for meat production need proper nutrition to provide good meat quality (tenderness, flavor, and balanced fat content), as well as to keep their health and growth in optimal ranges [126,127].

The energy requirements: Equines raised for this purpose need adequate energy to support their muscle development and to have a good body condition; the necessary depends on age, body weight, activity level, and growth moment. The energy sources include high-quality forages such as grass and hay, grains such as barley, oat, and corn, and other concentrated feeds (which provide carbohydrates and fats needed in the diet). The fat intake should be controlled to maintain lean meat; high-energy grains are recommended in moderate quantities to prevent excessive fat deposits.

The fat: High amounts need to be avoided in administration for equine raised for meat because they negatively impact the meat quality; however, in colder months or high growth stages, they need moderate amounts of healthy fats to support their energy needs and muscle growth (as sources it would be recommended to include in their diet corn and oats as grains, sunflower, and soybean oil as vegetable oils and also high-quality forages).

The protein requirements: Since protein is vital for muscle development and growth, equines raised for meat production need a high-quality protein source and also an accurate amino acid profile, which is very important for young equines to develop correctly (lysine and methionine are particularly important for muscle growth and protein synthesis). For equines, the levels of lysine and methionine are also crucial. For a 500 kg horse, a quantity of 15–20 g of lysine/day and 10–15 g of methionine/day. The minimum daily intake of lysine is 0.45–0.5% of the total diet, and for muscle growth, 0.5–0.6% of the total diet should be added (optimum sources are alfalfa hay, soybean, and canola meals, oats or lysine powder or pellet). For maintenance, 0.2–0.25% of the total diet of methionine is needed, and for muscle growth, a surplus of 0.3–0.35% of the total diet is recommended (good sources of methionine are grains such as oats and barley, hay, soybean, and canola meals and methionine powder or pellet) [130].

The mineral and vitamin requirements: Refer first of all to calcium and phosphorus and the ratio between them (2:1), as the optimal balance between these minerals is crucial for bone health and muscle function; then, magnesium and potassium are important for muscle growth, nerve protection and function, and also electrolyte balance. Last but not least, vitamin A is important for the immune system, growth, and site. Vitamin D is crucial for calcium absorption and bone health. Vitamin E is a great antioxidant that supports the immune system and muscle growth. The B complex is very important for metabolic processes and energy production.

The fiber intake: The equines need a high-fiber diet to maintain healthy digestion and gastrointestinal function; the forage is the base of their diet as it contributes to gut health, reduces colic incidence, helps to assimilate nutrients, and, as a consequence, contributes to producing high-quality meat by supporting a good metabolism.

A type of diet used in equines raised for meat is presented in Table 13.

**Table 13.** Model of diet for meat equines *.

| Feeding Time | Type of Feed | Quantity | Effect |
|---|---|---|---|
| Morning | Forage | 2–3 kg of high-quality grass or alfalfa hay | Offers essential fiber, vitamins, and minerals for digestion and growth |
| | Concentrates | 1 kg of barley/oats | Offers energy and protein for growth and muscle development |
| | Protein | 0.5 kg of soybean meal | Offers enough protein for muscle development |
| | Salt | Mineral block/loose salt | Offers proper mineral balance (Ca, P, Mg) |
| Noon | Forage | 2–3 kg of hay | Offers essential fiber, vitamins, and minerals for digestion and growth |
| | Concentrates | 0.5–1 kg of corn | Offers energy and protein for growth and muscle development |
| | Vegetable oil | 50–100 mL of sunflower/soybean oil | Used to provide healthy fats and additional energy |
| Evening | Forage | 2–3 kg of hay/pasture | Helps the health and digestion process |
| | Protein supplement | 0.5 kg of alfalfa/legume-based protein supplement | Helps muscle growth |

**Table 13.** *Cont.*

| Feeding Time | Type of Feed | Quantity | Effect |
|---|---|---|---|
| Observation | Water | Fresh and clean water ad libitum | Provides proper hydration and digestive health |
| | Supplements | As required | Additional minerals and vitamins may be provided |
| | Pasture grazing | If available | Grazing during the day with supplemental feeding during cooler months |
| | Adjustments | As required | Some adjustments may be needed according to age, size, weight, sex, growth stage |

* Source: [131].

#### 4.4.2. Water Requirements for Meat Equines

The water intake: Water is essential for hydration and digestion, so equines need free access to clean, fresh water ad libitum, especially in summer or during periods of high feed intake. It contributes to nutrient absorption, protein metabolism, and muscle development, and very importantly, sustains overall health, also preventing digestive disorders (colic and impaction). Water maintains a normal body temperature (which is crucial and harder to regulate in hot climates or during effort), it contributes to cellular functions (nutrient absorption, metabolic processes, toxin elimination) and digestion (as their diets are based on high-fiber intake, they need proper quantities of water, for the microbial fermentation in the cecum and colon to function well); water is important also in saliva production (it contains enzymes that contribute to carbohydrate digestion), being also a nutrient carrier (it dissolves and carry nutrients as amino acids, vitamins, and minerals, which are important for muscle development of horses raised for meat). It is also crucial in protein metabolism, which is involved in muscle growth: water is involved in the dissolving and transportation of the amino acids, leading to high protein synthesis, and also in urea excretion (excessive nitrogen from protein metabolism is converted into urea in the liver, and has to be eliminated through the urine). Adequate water intake also improves feed consumption (equines consume high amounts of feed when they have water ad libitum), which leads to muscle growth through the following scheme: they require an adequate balance of electrolytes like so sodium, calcium, and potassium, and the water transports and balance these substances, essential for muscle development (these also help in muscle cramp occurrence). The water intake ranges between 8 and 12 L/day/100 kg live weight (a surplus of 20–30% could be needed for high-fiber diets); also, if the weather is very hot, consumption increases by 50% [132].

#### 4.4.3. Pasture Management for Meat Equines

Pasture management is very important for equines raised for meat production, as it helps support their health, ensures efficient feeding, and helps maintain environmental sustainability. This includes the following [110]:

- Grazing intensity: The management should be directed to ensure pastures are not overgrazed (pastures should be grazed down to a height of 4–7 cm; if the graze is under the lowest limit, there is a sign of overgrazing).
- Rotational grazing: The pastures should be divided into portions, and equines should be rotated through different sections; also, the pastures need to regenerate, and through rotational grazing the parasite buildup is reduced, which is crucial for health of the animals and also of the land (70% of the ground has to be covered by vegetation; the rotational grazing involves dividing the pasture into paddocks that can be grazed for 7–14 days and then left to regenerate for 30–60 days).
- Pasture rotation in winter: During wintertime, the grazing should be limited to prevent damage to grass and land; also, it is recommended to supplement with hay and other supplements during the cold weather.

- Pasture rest time: Generally, the pastures need to rest for 3–6 weeks, depending on season and grazing intensity (if the grass regrowth is slow or it does not exist, the pasture is considered as being overgrazed).
- Pasture diversity: This refers to encouraging a mix of grass species and legumes (for example, fescue and clover); also, diverse pastures are more resilient, and they provide nutrients while maintaining quality over time.
- Vegetation control: If the pastures are regularly monitored, the risk of toxic grass growing would be lower, which contributes to high-quality pastures.
- Fertilization and soil health: Involves using organic fertilizers to promote soil fertility; overuse of chemical fertilizers has to be limited to prevent long-term damage to soil health.
- Water access: Fresh and clean water has to be ad libitum, especially during grazing.
- Fencing: This helps prevent accidents and equine escapes.
- Pasture utilization efficiency: Sometimes, controlled grazing with proper feeding can maximize pasture productivity; also, grazing periods should be long enough to provide adequate nutrition while also allowing the pasture to rest.

These pasture management rules assure rational grazing, where equines raised for meat production benefit from healthy maintaining conditions, optimal growth rates, and more sustainable farming practices. For example, for a 500 kg horse, the recommendation is to beneficiate 1.5–2 ha, and for a 250 kg donkey, the recommendation is approximately 1.0–1.5 ha [111].

## 5. Sustainability in Equine Farming

These practices involve balancing environmental, economic, and social factors as follows:

- Reduced environmental impact: by applying farm management practices such as solar energy resources for barn lightning, manure management practices (composting or biogas stations), and agroforestry, equines raised for milk and meat production can reduce their carbon footprint. When compared to other species, equines are between the impact of swine and cattle in terms of sustainability and with benefits such as moderate land use and lower methane emissions; the disadvantage is they require careful pasture management. The cattle have the highest environmental impact as a consequence of high land and water use, low feed efficiency, and, more than this, significant greenhouse gas (GHG) emissions (one cow needs around 2.5–10 ha of land/year, including pasture and land used for crops). Swine has a lower impact than beef but is less sustainable than poultry and equine due to moderate GHG emissions and water dispense (0.1–0.3 ha/pig is recommended). Chicken is considered the most environmentally friendly option, with low resource requirements and efficient feed conversion (1 kg of chicken meat needs 1.3 m$^2$ of land for production). Equines need 1.5–2 ha/capita for grazing [133].
- Minimizing feed miles by sourcing feed locally is a practical strategy to reduce the environmental impact of equine dairy or meat production. By relying on nearby feed producers, farms can significantly lower greenhouse gas emissions from transportation, cut costs associated with fuel and logistics, and support the local agricultural economy. Local feed sourcing ensures fresher, higher-quality feed with less risk of spoilage, benefiting animal nutrition and productivity. Strategies such as partnering with local farmers, growing forage on-site, using seasonal feeds, and integrating crop-animal systems further enhance sustainability. This approach not only reduces environmental degradation but also strengthens community ties and promotes efficient, eco-friendly farming practices [133].
- Water is another significant environmental input for livestock; for 1 kg of beef, approximately 15,000 L are needed (dairy cows require around 100 L of water/day), for 1 kg of pork, around 6000–10,000 L of water are consumed, and equines requests range from 30–100 L/daily (the water footprint of horse meat per kg is approximately 4000–5000 L) [134].

- The methane emissions represent another concern these days; cattle emit up to 200–500 L of methane/daily or 80–120 kg/year/cow (methane emissions from cattle account for about 60% of the total methane emissions from livestock globally); pigs produce 0.4–0.6 kg/year, and equines emit 0.1–0.3 kg/capita/year [135].
- Ethical animal care is a cornerstone of sustainable equine farming, whether for milk or meat production, ensuring the animals' well-being and fostering long-term farm productivity. Providing spacious, well-ventilated barns with comfortable bedding helps reduce stress and prevent common health issues such as respiratory problems or joint strain. Access to high-quality pasture allows equines to graze, exercise, and exhibit natural behaviors like socialization and herd dynamics, which are essential for their mental and physical health. Lactating mares benefit from stress-free environments, as this directly influences milk yield and quality, while equines raised for meat develop better muscle tone and health through regular activity. Tailored nutrition, including balanced diets rich in fiber, protein, and essential nutrients, combined with humane handling practices, further ensures ethical care. Routine health monitoring and veterinary attention help prevent disease and promote welfare. Such practices not only enhance animal well-being but also improve the quality of milk and meat, contributing to sustainable and ethical farming operations [107].
- Climate change resilience is crucial for the sustainability of equine farming, particularly as changing weather patterns and extreme conditions impact forage availability and water resources. Implementing adaptive practices such as planting drought-resistant grasses and legumes in pastures ensures that equines have access to consistent, nutrient-rich forage even during dry spells. Improved water storage systems, such as rainwater harvesting or pond construction, can provide a reliable water supply to meet the high hydration demands of equines, especially those raised for milk or meat production. Rotational grazing strategies can help maintain pasture health by preventing overgrazing and promoting regrowth during periods of environmental stress. Additionally, incorporating climate-resilient shelter designs, such as barns with enhanced ventilation and insulation, helps protect equines from extreme heat or cold. By adopting these measures, equine farms can mitigate the adverse effects of climate change, ensuring both animal welfare and the continuity of production systems [134].

## 6. Conclusions

Dairy and meat equine farming represents a sustainable and nutritious alternative for milk and meat production. The milk is suitable for dietary restrictions, is low-fat, and high in vitamin C—similar to human breast milk. Equine meat is lean, with high levels of omega-3 fatty acids, iron, and zinc, which makes it a healthier option compared to other types of meat. On the other hand, equine farming also supports sustainable agriculture by utilizing pastures, sustaining biodiversity, and promoting food security. Overall, equine raising is a key practice in long-term environmental and economic resilience since sustainability in equine farming hinges on a holistic approach that balances environmental, economic, and social considerations.

**Author Contributions:** Conceptualization, C.P. (Claudia Pânzaru); methodology, M.G.D.; software, R.-M.R.-R.; validation, C.P. (Constantin Pascal) and V.M.; formal analysis, R.-M.R.-R.; investigation, C.P. (Claudia Pânzaru); resources, M.G.D. and M.-A.D.; data curation, M.-A.D.; writing—original draft preparation, C.P. (Claudia Pânzaru); writing—review and editing, C.P. (Claudia Pânzaru); visualization, M.-A.D. and C.P. (Constantin Pascal); supervision, V.M.; project administration, C.P. (Claudia Pânzaru). All authors have read and agreed to the published version of the manuscript.

**Funding:** This research received no external funding.

**Conflicts of Interest:** The authors declare no conflicts of interest.

## References


1. Nielsen, S.S.; Joosten, L. Sustainability in livestock production: Impacts and opportunities. *Sustainability* **2022**, *14*, 82–100.
2. Yachi, S.; Loreau, M. Biodiversity and ecosystem productivity in a fluctuating environment: The insurance hypothesis. *Proc. Natl. Acad. Sci. USA* **1999**, *96*, 1463–1468. [CrossRef] [PubMed]
3. Cosentino, C.; Paolino, R.; Musto, M.; Freschi, P. Innovative use of jenny milk from sustainable rearing. In *The Sustainability of Agro-Food and Natural Resource Systems in the Mediterranean Basin*; Vastola, A., Ed.; Springer International Publishing: Cham, Switzerland, 2015; pp. 113–132.
4. Jost, L.; Smulders, R.; Brem, G. *Historical and Current Significance of Horses as Food-Producing Animals and Production of Horse Meat in Central Europe*; Wiener Tierärztliche Monatsschrift: Purkersdorf, Austria, 2017; pp. 15–20.
5. Miraglia, N.; Saastamoinen, M.; Martin-Rosset, W. Role of pastures in mares and foals management in Europe. In *Nutrition and Feeding of the Broodmare*; Academic Publisher: Wageningen, The Netherlands, 2006; Volume 120, pp. 279–297.
6. Kaya, Y.; Saka, S.; Tuncer, D. Effect of hippotherapy on balance, functional mobility, and functional independence in children with Down syndrome: A randomized controlled trial. *Eur. J. Pediatr.* **2023**, *182*, 3147–3155. [CrossRef] [PubMed]
7. Tseng, S.H.; Chen, H.C.; Tam, K.W. Systematic review and meta-analysis of the effect of equine-assisted activities and therapies on gross motor outcomes in children with cerebral palsy. *Disabil. Rehabil.* **2013**, *35*, 89–99. [CrossRef]
8. Verhulst, L.; Kerre, S.; Goossens, A. The unsuspected power of mare's milk. *Contact Dermat.* **2016**, *74*, 376–377. [CrossRef]
9. Rivero, M.J.; Cooke, A.S.; Gandarillas, M.; Leon, R.; Merino, V.M.; Velásquez, A. Nutritional composition, fatty acids profile and immunoglobulin G concentrations of mare milk of the Chilean Corralero horse breed. *PLoS ONE* **2024**, *19*, e0310693. [CrossRef] [PubMed]
10. Karami, Z.; Akbari-adergani, B. Bioactive food derived peptides: A review on correlation between structure of bioactive peptides and their functional properties. *J. Food Sci. Technol.* **2019**, *56*, 535–547. [CrossRef] [PubMed]
11. Nagarajan, S.; Radhakrishnan, S.; Kalkura, S.N.; Balme, S.; Miele, P.; Bechelany, M. Overview of Protein-Based Biopolymers for Biomedical Application. *Macromol. Chem. Phys.* **2019**, *220*, 1900126. [CrossRef]
12. Li, M.; Kang, S.; Zheng, Y.; Shao, J.; Zhao, H.; An, Y.; Cao, G.; Li, Q.; Yue, X.; Yang, M. Comparative metabolomics analysis of donkey colostrum and mature milk using ultra-high-performance liquid tandem chromatography quadrupole time-of-flight mass spectrometry. *J. Dairy Sci.* **2020**, *103*, 992–1001. [CrossRef]
13. Jastrzębska, E.; Daszkiewicz, T.; Górecka-Bruzda, A.; Feliś, D. Current situation and prospects for the horse meat market in Poland and the world. *Med. Weter.* **2019**, *75*, 196–202. [CrossRef]
14. Reséndiz González, G.; Alarcón Zúñiga, B.; Villegas Velázquez, I.; Albores Moreno, S.; Aranda Osorio, G. Composición nutricional de la carne equina y grado de sustitución de la carne bovina por equina en expendios de la Ciudad de México. *Rev. Mex. Cienc. Pecu.* **2021**, *12*, 742–755. [CrossRef]
15. Sebbane, M.; Vial, C.; Lamy, A. A horse on your plate? A cluster analysis of French consumers hippophagy acceptance. *Meat Sci.* **2023**, *203*, 109220. [CrossRef] [PubMed]
16. Kaskous, S. Cow's milk consumption and risk of disease. *Emir. J. Food Agric.* **2021**, *33*, 1–11.
17. Bertino, E.; Cavallarin, L.; Cresi, F.; Tonetto, P.; Peila, C.; Ansaldi, G.; Raia, M.; Varalda, A.; Giribaldi, M.; Conti, A.; et al. A Novel Donkey Milk–derived Human Milk Fortifier in Feeding Preterm Infants: A Randomized Controlled Trial. *J. Pediatr. Gastroenterol. Nutr.* **2019**, *68*, 116–123. [CrossRef]
18. Cresi, F.; Maggiora, E.; Pirra, A.; Tonetto, P.; Rubino, C.; Cavallarin, L.; Giribaldi, M.; Moro, G.E.; Peila, C.; Coscia, A. Effects on Gastroesophageal Reflux of Donkey Milk-Derived Human Milk Fortifier Versus Standard Fortifier in Preterm Newborns: Additional Data from the FortiLat Study. *Nutrients* **2020**, *12*, 2142. [CrossRef]
19. Peila, C.; Spada, E.; DeAntoni, S.; Iuliano, E.; Moro, G.E.; Giribaldi, M.; Cavallarin, L.; Cresi, F.; Coscia, A. The "Fortilat" Randomized Clinical Trial Follow-Up: Neurodevelopmental Outcome at 18 Months of Age. *Nutrients* **2020**, *12*, 3807. [CrossRef] [PubMed]
20. Tidona, F.; Criscione, A.; Devold, T.G.; Bordonaro, S.; Marletta, D.; Vegarud, G.E. Protein composition and micelle size of donkey milk with different protein patterns: Effects on digestibility. *Int. Dairy J.* **2014**, *35*, 57–62. [CrossRef]
21. Simeanu, D.; Radu-Rusu, R.-M.; Macri, A.M.; Mierliță, D. Animal Nutrition and Productions: Series II. *Agriculture* **2024**, *14*, 448. [CrossRef]
22. Fotschki, J.; Szyc, A.M.; Laparra, J.M.; Markiewicz, L.H.; Wróblewska, B. Immune-modulating properties of horse milk administered to mice sensitized to cow milk. *J. Dairy Sci.* **2016**, *99*, 9395–9404. [CrossRef] [PubMed]
23. Trinchese, G.; Cavaliere, G.; De Filippo, C.; Aceto, S.; Prisco, M.; Chun, J.T.; Penna, E.; Negri, R.; Muredda, L.; Demurtas, A.; et al. Human Milk and Donkey Milk, Compared to Cow Milk, Reduce Inflammatory Mediators and Modulate Glucose and Lipid Metabolism, Acting on Mitochondrial Function and Oleylethanolamide Levels in Rat Skeletal Muscle. *Front. Physiol.* **2018**, *9*, 32. [CrossRef] [PubMed]
24. Ugidos-Rodríguez, S.; Matallana-González, M.C.; Sánchez-Mata, M.C. Lactose malabsorption and intolerance: A review. *Food Funct.* **2018**, *9*, 4056–4068. [CrossRef]
25. Rațu, R.N.; Cârlescu, P.M.; Usturoi, M.G.; Lipșa, F.D.; Veleșcu, I.D.; Arsenoaia, V.N.; Florea, A.M.; Ciobanu, M.M.; Radu-Rusu, R.-M.; Postolache, A.N.; et al. Effects of Dairy Cows Management Systems on the Physicochemical and Nutritional Quality of Milk and Yogurt, in a North-Eastern Romanian Farm. *Agriculture* **2023**, *13*, 1295. [CrossRef]

26. Ren, W.W.; Bekhit, A.E.D.A.; Li, F.; Yang, H.Y.; Jiang, X.F.; Zhang, W.; Kong, L.M. Physicochemical properties of pastirma from horse meat, beef, mutton and pork. *J. Food Qual.* **2015**, *38*, 369–376. [CrossRef]
27. Martin-Rosset, W.; Trillaud-Geyl, C. Horse meat production and characteristics: A review. *EAAP Sci. Ser.* **2015**, *136*, 197–222.
28. Frunză, G.; Ciobanu, M.-M.; Murariu, O.C.; Rațu, R.N.; Radu-Rusu, R.-M.; Simeanu, C.; Boișteanu, P.-C. Effect of Gender and Muscle Type on Fatty Acid Profile, Sanogenic Indices, and Instrumental and Sensory Analysis of Flemish Giant Rabbit Meat. *Agriculture* **2023**, *13*, 2265. [CrossRef]
29. Lorenzo, J.M.; Munekata, P.E.; Campagnol, P.C.B.; Zhu, Z.; Alpas, H.; Barba, F.J.; Tomasevic, I. Technological aspects of horse meat products—A review. *Food Res. Int.* **2017**, *102*, 176–183. [CrossRef]
30. Țenu, F.; Șerban, A.; Donosă, R.E.; Alecu, C.; Maciuc, V. Study Regarding the Beef Meat Production Evolution Worldwide and National Level. *Sci. Pap. Ser. D Anim. Sci.* **2023**, *66*, 22–26.
31. Curea, C.D.; Usturoi, M.G.; Custură, I.; Radu-Rusu, R.M.; Rațu, R.N.; Prisăcaru, M.; Usturoi, A. Efficiency of Growing of Chicken Broilers Under Conditions of Compliance with Eu Rules of Welfare. *Sci. Pap. Ser. D Anim. Sci.* **2023**, *66*, 352–359.
32. Singer, J.W.; Bamka, W.J.; Kluchinski, D.; Govindasamy, R. Using the recommended stocking density to predict equine pasture management. *J. Equine Vet. Sci.* **2002**, *22*, 73–76. [CrossRef]
33. Mădescu, B.M.; Lazăr, R.; Ciobanu, M.M.; Boişteanu, P.C. Research on the characterization of beef proteins. *Sci. Pap. Anim. Sci. Ser.* **2020**, *74*, 97–102.
34. Pascal, C.; Gîlcă, I.; Nacu, G. Researches concerning the improvement of carcasses issued from reformed and fattened adult sheep. In Proceedings of the 45th and 5th International Symposium on Agriculture, Opatija, Croatia, 15–19 February 2010; pp. 1052–1057.
35. Pascal, C.; Dărăban, S.V.; Pădeanu, I. Researches Regarding the Meat Aptitudes Of Sheep Youth From Romanian Indigenous Breeds. *Sci. Pap. Anim. Sci.* **2012**, *57*, 275–280.
36. Horse Meat. Available online: https://www.sciencedirect.com/topics/food-science/horse-meat#:~:text=Adult%20horse%20meat%20is%20very,the%20age%20of%20the%20animal (accessed on 7 December 2024).
37. Humane Society International. Horsemeat Production Worldwide. Available online: https://www.hsi.org/wp-content/uploads/assets/pdfs/Horses_meatproduction_worldwide.pdf (accessed on 23 November 2024).
38. Pregowski, M.P. An Appetite for Dogs: Consuming and Loving Them in Vietnam. In *Companion Animals in Everyday Life: Situating Human Animal Engagement Within Cultures*; Palgrave Macmillan: London, UK, 2016; pp. 111–128.
39. Stanciu, S. Horse Meat Consumption—Between Scandal and Reality. *Proc. Econ. Fin.* **2015**, *23*, 697–703. [CrossRef]
40. Miraglia, N.; Salimei, E.; Fantuz, F. Equine Milk Production and Valorization of Marginal Areas—A Review. *Animals* **2020**, *10*, 353. [CrossRef]
41. Working Equids in Numbers: Why Data Matters for Policy. Available online: https://www.fao.org/agroecology/database/detail/en/c/1473377/ (accessed on 23 November 2024).
42. Martini, M.; Altomonte, I.; Tricò, D.; Lapenta, R.; Salari, F. Current Knowledge on Functionality and Potential Therapeutic Uses of Donkey Milk. *Animals* **2021**, *11*, 1382. [CrossRef]
43. Livestock Counts, World, 1890 to 2014. Available online: https://ourworldindata.org/grapher/livestock-counts (accessed on 23 November 2024).
44. Gateway to Dairy Production and Products. Available online: https://www.fao.org/dairy-production-products/production/dairy-animals/en (accessed on 23 November 2024).
45. Gateway to Dairy Production and Products. Available online: https://www.fao.org/dairy-production-products/dairy/other-animals/en (accessed on 23 November 2024).
46. Camillo, F.; Rota, A.; Biagini, L.; Tesi, M.; Fanelli, D.; Panzani, D. The current situation and trend of donkey industry in Europe. *J. Equine Vet. Sci.* **2018**, *65*, 44–49. [CrossRef]
47. Musaev, A.; Sadykova, S.; Anambayeva, A.; Saizhanova, M.; Balkanay, G.; Kolbaev, M. Mare's Milk: Composition, Properties, and Application in Medicine. *Arch. Razi Inst.* **2021**, *76*, 1125–1135. [PubMed]
48. Rzekęć, A.; Vial, C.; Bigot, G. Green Assets of Equines in the European Context of the Ecological Transition of Agriculture. *Animals* **2020**, *10*, 106. [CrossRef] [PubMed]
49. Data on Working Equids. Available online: https://www.thebrooke.org/our-work/data-working-equids (accessed on 23 November 2024).
50. Crops and Livestock Products. Available online: https://www.fao.org/faostat/en/#data/QCL (accessed on 24 November 2024).
51. Kim, D.S.; Joo, N. Texture characteristics of horse meat for the elderly based on the enzyme treatment. *Food Sci. Anim. Res.* **2020**, *40*, 74. [CrossRef]
52. *Herodotus, The Histories*; Godley, A.D., Ed. Available online: https://www.perseus.tufts.edu/hopper/text?doc=Perseus:text:1999.01.0126:book=4&force=y (accessed on 11 December 2024).
53. Lloyd, G.E.R. The Hippocratic Writings. In *A New Translation*; Harvard University Press: Cambridge, MA, USA, 1978; pp. 78–81.
54. *Pliny the Elder, The Natural History*; John Bostock, M.D., F.R.S., H.T. Riley, Esq., B.A., Eds. Available online: https://www.perseus.tufts.edu/hopper/text?doc=plin.+nat.+toc (accessed on 11 December 2024).
55. Temkin, O. *Galenism: Rise and Decline of a Medical Philosophy*; Cornell University Press: Ithaca, NY, USA, 1956; pp. 23–37.
56. Yule, H.; Burnell, A.C. *The Travels of Marco Polo*; Penguin Classics: London, UK, 1993; pp. 54–55.
57. Dhanani, A. Avicenna's Canon of Medicine: An Introduction to its Use in Western Medicine. *J. Islamic Stud.* **2007**, *18*.

58. Studies on Equine Milk and Comparative Studies on Equine and Bovine Milk Systems. Available online: https://cora.ucc.ie/items/08b0f256-4294-4ee5-aaa7-862080cf6f26 (accessed on 11 December 2024).
59. Uniacke-Lowe, T.; Huppertz, T.; Fox, F.P. Equine milk proteins: Chemistry, structure and nutritional significance. *Int. Dairy J.* **2010**, *20*, 609–629. [CrossRef]
60. Doherty, A.M.; Lodge, C.J.; Dharmage, S.C.; Dai, X.; Bode, L.; Lowe, A.J. Human Milk Oligosaccharides and Associations with Immune-Mediated Disease and Infection in Childhood: A Systematic Review. *Front. Pediatr.* **2018**, *6*, 91. [CrossRef] [PubMed]
61. Williams, C.A.; Kenny, L.B.; Burk, A.O. Effects of grazing system, season, and forage carbohydrates on glucose and insulin dynamics of the grazing horse. *J. Anim. Sci.* **2019**, *26*, 2541–2554. [CrossRef]
62. Bott-Knutson, R.; Greene, E.A.; Martinson, K.L.; Siciliano, P.; Williams, C.; Trottier, N.L.; Burk, A.; Swinker, A.M. Production and Environmental Implications of Equine Grazing. *J. Equine Vet. Sci.* **2013**, *33*, 1031–1043. [CrossRef]
63. Effect of Grazing Cell Size on Horse Pasture Utilization. Available online: https://lpelc.org/effect-of-grazing-cell-size-on-horse-pasture-utilization/ (accessed on 11 December 2024).
64. Agroecological Transitions: A Systematic Review of Research Approaches and Prospects for Participatory Action Methods. Available online: https://www.pan-europe.info/sites/pan-europe.info/files/public/resources/Webinars/Agroecological%20Transitions_N_Shaban.pdf (accessed on 11 December 2024).
65. Lonnerdal, B. Human milk: Bioactive proteins/peptides, their role in infant health and development. *Front. Nutr.* **2016**, *3*, 26.
66. Maciuc, V.; Pânzaru, C.; Ciocan-Alupii, M.; Radu-Rusu, C.-G.; Radu-Rusu, R.-M. Comparative Assessment of the Nutritional and Sanogenic Features of Certain Cheese Sorts Originating in Conventional Dairy Farms and in "Mountainous" Quality System Farms. *Agriculture* **2024**, *14*, 172. [CrossRef]
67. Ugwu, C.P.; Abarshi, M.M.; Mada, S.B.; Sanusi, B.; Nzelibe, H.C. Camel and Horse Milk Casein Hydrolysates Exhibit Angiotensin Converting Enzyme Inhibitory and Antioxidative Effects In Vitro and In Silico. *Int. J. Pept. Res. Ther.* **2019**, *25*, 1595–1604. [CrossRef]
68. Li, Q.; Li, M.; Zhang, J.; Shi, X.; Yang, M.; Zheng, Y.; Cao, X.; Yue, X.; Ma, S. Donkey milk inhibits triple-negative breast tumor progression and is associated with increased cleaved-caspase-3 expression. *Food Funct.* **2020**, *11*, 3053–3065. [CrossRef]
69. Malacarne, M.; Criscione, A.; Franceschi, P.; Bordonaro, S.; Formaggioni, P.; Marletta, D.; Summer, A. New Insights into Chemical and Mineral Composition of Donkey Milk throughout Nine Months of Lactation. *Animals* **2019**, *9*, 1161. [CrossRef]
70. Marcone, S.; Belton, O.; Fitzgerald, D.J. Milk-derived bioactive peptides and their health promoting effects: A potential role in atherosclerosis. *Br. J. Clin. Pharmacol.* **2017**, *83*, 152–162. [CrossRef]
71. López, C. Chemical composition of sheep milk and its contribution to the development of dairy products. *J. Dairy Sci.* **2005**, *88*, 29–45.
72. Haenlein, G.F.W. Goat milk: Characteristics and nutritional aspects. *Small Rumin. Res.* **2004**, *51*, 1–17. [CrossRef]
73. Sofyan, D. Health benefits of sheep milk. *Intern. Dairy J.* **2020**, *109*, 104–112.
74. Papadopoulou, A. Functional properties of goat milk proteins. *Food Res. Internsh.* **2019**, *120*, 604–615.
75. Fratini, F.; Turchi, B.; Pedonese, F.; Pizzurro, F.; Ragaglini, P.; Torracca, B.; Tozzi, B.; Galiero, A.; Nuvoloni, R. Does the addition of donkey milk inhibit the replication of pathogen microorganisms in goat milk at refrigerated condition? *Dairy Sci. Technol.* **2016**, *96*, 243–250. [CrossRef]
76. Yuniati, H.; Sahara, E. Protein and fat bioactive components in wild horses milk. *Bul. Penelit. Kesehat.* **2012**, *40*, 66–74.
77. Zhu, L.; Zeng, C.; Yang, S.; Hou, Z.; Wang, Y.; Hu, X.; Senoo, K.; Wei, W. Diversity and specificity of the bacterial community in Chinese horse milk cheese. *MicrobiologyOpen* **2020**, *9*, e1066. [CrossRef] [PubMed]
78. Businco, L.; Giampietro, P.G.; Lucenti, P.; Lucaroni, F.; Pini, C.; Di Felice, G.; Iacovacci, P.; Curadi, C.; Orlandi, M. Allergenicity of mare's milk in children with cow's milk allergy. *J. Allergy Clin. Immunol.* **2000**, *105*, 1031–1034. [CrossRef] [PubMed]
79. Sarti, L.; Martini, M.; Brajon, G.; Barni, S.; Salari, F.; Altomonte, I.; Ragona, G.; Mori, F.; Pucci, N.; Muscas, G.; et al. Donkey's milk in the management of children with cow's milk protein allergy: Nutritional and hygienic aspects. *Ital. J. Pediatr.* **2019**, *45*, 102. [CrossRef]
80. Aroua, M.; Jemmali, B.; Said, S.B.; Kbaier, H.B.H.; Mahouachi, M. Physicochemical properties of north African donkey milk. *Agric. Res. Technol.* **2019**, *21*, 1–3. [CrossRef]
81. Perna, A.; Intaglietta, I.; Simonetti, A.; Gambacorta, E. Donkey Milk for Manufacture of Novel Functional Fermented Beverages. *J. Food Sci.* **2015**, *80*, S1352–S1359. [CrossRef]
82. Li, L.; Liu, X.; Guo, H. The nutritional ingredients and antioxidant activity of donkey milk and donkey milk powder. *Food Sci. Biotechnol.* **2017**, *27*, 393–400. [CrossRef] [PubMed]
83. Equine Milk and Its Potential Use in the Human Diet. Available online: https://www.scielo.br/j/cta/a/TpqyRP5b7krCgZTWwF5BtnD/?format=pdf&lang=en (accessed on 11 December 2024).
84. Kaskous, S.; Pfaffl, M.W. Milk Properties and Morphological Characteristics of the Donkey Mammary Gland for Development of an Adopted Milking Machine—A Review. *Dairy* **2022**, *3*, 233–247. [CrossRef]
85. Van der Burg, L.J.; Muller, I. Horse milking industry in The Netherlands and Flanders. *Tijdschr. Diergeneeskd.* **2011**, *136*, 257–261.
86. Barni, S.; Sarti, L.; Mori, F.; Muscas, G.; Belli, F.; Pucci, N.; Novembre, E. Tolerability and palatability of donkey's milk in children with cow's milk allergy. *Pediatr. Allergy Immunol.* **2018**, *29*, 329–331. [CrossRef] [PubMed]
87. Reiter, A.S.; Reed, S.A. Lactation in Horses. *Anim. Front.* **2023**, *13*, 103–107. [CrossRef]

88. Mori, F.; Sarti, L.; Barni, S.; Pucci, N.; Belli, F.; Stagi, S.; Novembre, E. Donkey's Milk Is Well Accepted and Tolerated by Infants with Cow's Milk Food Protein–Induced Enterocolitis Syndrome: A Preliminary Study. *J. Investig. Allergol. Clin. Immunol.* **2017**, *27*, 269–271. [CrossRef] [PubMed]
89. Liu, L.L.; Fang, C.; Ma, H.Y.; Yu, X.; Lv, S.P.; Liu, W.J. Development and validation of KASP markers for the milk traits genes in Kazakh horse. *J. Appl. Anim. Res.* **2020**, *48*, 293–299. [CrossRef]
90. Lv, K.; Yang, Y.; Li, Q.; Chen, R.; Deng, L.; Zhang, Y.; Jiang, N. Identification and comparison of milk fat globule membrane and whey proteins from Selle Français, Welsh pony, and Tieling Draft horse mare's milk. *Food Chem.* **2024**, *437*, 137915. [CrossRef]
91. Barłowska, J.; Polak, G.; Janczarek, I.; Próchniak, T. Chemical composition, whey protein profile, and fatty acid profile of milk from Sokólski horses in relation to Polish Halfbred horses. *Ann. Anim. Sci.* **2023**, *23*, 587–596. [CrossRef]
92. Massouras, T.; Triantaphyllopoulos, K.A.; Theodossiou, I. Chemical composition, protein fraction and fatty acid profile of donkey milk during lactation. *Int. Dairy J.* **2017**, *75*, 83–90. [CrossRef]
93. Salgado, M.J.G.; Silva, A.R.; de Souza, C.O.; Lemos, P.V.F.; Oliveira, R.L.; Oliveira, C.A.D.A.; Ribeiro, C.V.D.M. Days in milk alters the milk fatty acid profile of grazing donkeys: A preliminary study. *J. Anim. Physiol. Anim. Nutr.* **2021**, *105*, 1173–1178. [CrossRef] [PubMed]
94. Martini, M.; Salari, F.; Licitra, R.; La Motta, C.; Altomonte, I. Lysozyme activity in donkey milk. *Int. Dairy J.* **2019**, *96*, 98–101. [CrossRef]
95. Licitra, R.; Chessa, S.; Salari, F.; Gattolin, S.; Bulgari, O.; Altomonte, I.; Martini, M. Milk protein polymorphism in Amiata donkey. *Livest. Sci.* **2019**, *230*, 103845. [CrossRef]
96. Martini, M.; Altomonte, I.; Licitra, R.; Salari, F. Short communication: Technological and seasonal variations of vitamin D and other nutritional components in donkey milk. *J. Dairy Sci.* **2018**, *101*, 8721–8725. [CrossRef] [PubMed]
97. Elia, J.B.; Erb, H.N.; Houpt, K.A. Motivation for hay: Effects of a pelleted diet on behavior and physiology of horses. *Physiol. Behav.* **2010**, *101*, 623–627. [CrossRef]
98. Yvon, S.; Olier, M.; Leveque, M.; Jard, G.; Tormo, H.; Haimoud-Lekhal, D.A.; Peter, M.; Eutamene, H. Donkey milk consumption exerts anti-inflammatory properties by normalizing antimicrobial peptides levels in Paneth's cells in a model of ileitis in mice. *Eur. J. Nutr.* **2018**, *57*, 155–166. [CrossRef]
99. Holscher, H.D. Dietary fiber and prebiotics and the gastrointestinal microbiota. *Gut Microbes* **2017**, *8*, 172–184. [CrossRef]
100. Dumont, B.; Rook, A.J.; Coran, C.; Röver, K.U. Effects of livestock breed and grazing intensity on biodiversity and production in grazing systems. 2. Diet selection. *Grass Forage Sci.* **2007**, *62*, 159–171. [CrossRef]
101. Silva, F.L.; Silva, M.H.M.; Oliveira-Junior, E.; Lopez-Alonso, M.; Pierangeli, M.A.P. Fertility and carbon stock in pasture and forest environments in the Southern Amazon1. *Rev. Bras. Eng. Agríc. Ambient.* **2024**, *28*, e270888. [CrossRef]
102. Furtado, T.; King, M.; Perkins, E.; McGowan, C.; Chubbock, S.; Hannelly, E.; Rogers, J.; Pinchbeck, G. An Exploration of Environmentally Sustainable Practices Associated with Alternative Grazing Management System Use for Horses, Ponies, Donkeys and Mules in the UK. *Animals* **2022**, *12*, 151. [CrossRef] [PubMed]
103. The Agricultural Transition: Building a Sustainable Future. Available online: https://www.mckinsey.com/industries/agriculture/our-insights/the-agricultural-transition-building-a-sustainable-future (accessed on 11 December 2024).
104. Polidori, P.; Vincenzetti, S. Farm management and feeding strategies for donkey milk production. *Agric. Res. Updates* **2017**, *14*, 93–111.
105. Guide to Good Animal Welfare Practice for the Keeping, Care, Training and Use of Horses. Available online: https://food.ec.europa.eu/system/files/2022-07/aw_platform_plat-conc_guide_equidae_en.pdf (accessed on 5 December 2024).
106. Dai, F.; Dalla Costa, E.; Burden, F.; Judge, A.; Minero, M. The development of guidelines to improve dairy donkey management and welfare. *Ital. J. Anim. Sci.* **2019**, *18*, 189–193. [CrossRef]
107. Awin Welfare Assessment Protocol for Donkeys, University of Milan (Italy) and Donkey Sanctuary (UK). Available online: https://air.unimi.it/retrieve/handle/2434/269100/384805/AWINProtocolDonkeys.pdf (accessed on 18 October 2024).
108. McLean, A.K.; Navas Gonzalez, F.J. Can scientists influence donkey welfare? Historical perspective and a contemporary view. *J. Equine Vet. Sci.* **2018**, *65*, 25–32. [CrossRef]
109. Raspa, F.; Cavallarin, L.; McLean, A.K.; Bergero, D.; Valle, E. A review of the appropriate nutrition welfare criteria of dairy donkeys: Nutritional requirements, farm management requirements and animal-based indicators. *Animals* **2019**, *9*, 315. [CrossRef]
110. Best Management Practices for Equine Disease Prevention. Brett D. Scott, Extension Horse Specialist. The Texas A&M System. Available online: https://navarro.agrilife.org/files/2024/02/Best-Management-Practices-for-Equine-Disease-Prevention.pdf (accessed on 5 December 2024).
111. Moore-Colyer, M.; Westacott, A.; Rousson, L.; Harris, P.; Daniels, S. Where Are We Now? Feeds, Feeding Systems and Current Knowledge of UK Horse Owners When Feeding Haylage to Their Horses. *Animals* **2023**, *13*, 1280. [CrossRef]
112. Soares, D.; Rolim, J.; Fradinho, M.J.; do Paço, T.A. Production of Preserved Forage for Horses under Water Scarcity Conditions: A Case Study. *Water* **2022**, *14*, 388. [CrossRef]
113. Lindinger, M.I. Oral Electrolyte and Water Supplementation in Horses. *Vet. Sci.* **2022**, *9*, 626. [CrossRef]
114. Gill, C.O. Safety and storage stability of horse meat for human consumption. *Meat Sci.* **2005**, *71*, 506–513. [CrossRef]
115. Jiang, L.; Lv, J.; Liu, J.; Hao, X.; Ren, F.; Guo, H. Donkey milk lysozyme ameliorates dextran sulfate sodium-induced colitis by improving intestinal barrier function and gut microbiota composition. *J. Funct. Foods* **2018**, *48*, 144–152. [CrossRef]

116. Demine, S.; Renard, P.; Arnould, T. Mitochondrial Uncoupling: A Key Controller of Biological Processes in Physiology and Diseases. *Cells* **2019**, *8*, 795. [CrossRef] [PubMed]
117. Lee, C.E.; Seong, P.N.; Oh, W.Y.; Ko, M.S.; Kim, K.I.; Jeong, J.H. Nutritional characteristics of horsemeat in comparison with those of beef and pork. *Nutr. Res. Pract.* **2007**, *1*, 70–73. [CrossRef]
118. Popoola, I.O.; Bruce, H.L.; McMullen, L.M.; Wismer, W.V. Consumer sensory comparisons among beef, horse, elk, and bison using preferred attributes elicitation and check-all-that-apply methods. *J. Food Sci.* **2019**, *84*, 3009–3017. [CrossRef]
119. Hedenborg, S.; Kronborg, M.; Sätre, A.; Radmann, A.; Torell Palmquist, G.; Andersson, P. Pro-Environmental Transformation of the Equine Sector—Facilitators and Challenges. *Animals* **2024**, *14*, 915. [CrossRef]
120. Radu-Rusu, R.M.; Teuşan, V.; Vacaru-Opriş, I. Aspects concerning the histological structure of the biceps brachialis muscles in chicken broilers. *Sci. Pap. Anim. Sci.* **2009**, *52*, 266–270.
121. Ludwig, E.K.; Hobbs, K.J.; McKinney-Aguirre, C.A.; Gonzalez, L.M. Biomarkers of Intestinal Injury in Colic. *Animals* **2023**, *13*, 227. [CrossRef]
122. Di Giuseppe, A.M.; Giarretta, N.; Lippert, M.; Severino, V.; Di Maro, A. An improved UPLC method for the detection of undeclared horse meat addition by using myoglobin as molecular marker. *Food Chem.* **2015**, *169*, 241–245. [CrossRef]
123. Marino, R.; Della Malva, A.; Maggiolino, A.; De Palo, P.; d'Angelo, F.; Lorenzo, J.M.; Sevi, A.; Albenzio, M. Nutritional profile of donkey and horse meat: Effect of muscle and aging time. *Animals* **2022**, *12*, 746. [CrossRef]
124. Belaunzaran, X.; Bessa, R.J.; Lavín, P.; Mantecón, A.R.; Kramer, J.K.; Aldai, N. Horse-meat for human consumption—Current research and future opportunities. *Meat Sci.* **2015**, *108*, 74–81. [CrossRef]
125. Muirhead, T.L.; Wichtel, J.J.; Stryhn, H.; McClure, J.T. The selenium and vitamin E status of horses in Prince Edward Island. *Can. Vet. J.* **2010**, *51*, 979–985. [PubMed]
126. Swirsley, N.; Spooner, H.S.; Hoffman, R.M. Supplement Use and Perceptions: A Study of US Horse Owners. *J. Equine Vet. Sci.* **2017**, *59*, 34–39. [CrossRef]
127. Bò, C.D.; Simonetti, P.; Gardana, C.; Riso, P.; Lucchini, G.; Ciappellano, S. Horse meat consumption affects iron status, lipid profile and fatty acid composition of red blood cells in healthy volunteers. *Int. J. Food Sci. Nutr.* **2012**, *64*, 147–154. [CrossRef]
128. Della Malva, A.; Gagaoua, M.; Albenzio, M.; Sevi, A. Proteomics as an Emerging Tool in Equine Meat Research: An Overview. *Meat Technol.* **2023**, *64*, 401–406. [CrossRef]
129. Edouard, N.; Fleurance, G.; Duncan, P.; Baumont, R. Determinants of the use of pastures by grazing horses. *Prod. Anim. Paris* **2009**, *22*, 14–17.
130. Badgery, W.B.; Simmons, A.T.; Murphy, B.W.; Rawson, A.; Andersson, K.O.; Lonergan, V.E. The influence of land use and management on soil carbon levels for crop-pasture systems in Central New South Wales, Australia. *Agric. Ecosyst. Environ.* **2014**, *196*, 147–157. [CrossRef]
131. Meyer, H.; Radicke, S.; Kienzle, E.; Wilke, S.; Kleffken, D.; Illenseer, M. Investigations on preileal digestion of starch from grain, potato and manioc in horses. *Zentralblatt Vet. A* **1995**, *42*, 371–381. [CrossRef]
132. Managing Equine Grazing for Pasture Productivity. Available online: https://edisciplinas.usp.br/pluginfile.php/8029267/mod_resource/content/1/3-s2.0-B9780128129197000093-main.pdf (accessed on 11 December 2024).
133. Tackling Climate Change Through Livestock—A Global Assessment of Emissions and Mitigation Opportunities. Available online: https://www.fao.org/4/i3437e/i3437e.pdf (accessed on 8 December 2024).
134. Herrero, M.; Gerber, P.; Vellinga, T.; Garnett, T.; Leip, A.; Opio, C.; Westhoek, H.; Thornton, P.; Olesen, J.; Hutchings, N.; et al. Livestock and Greenhouse Gas Emissions: The Importance of Getting the Numbers Right. *Anim. Prod. Sci.* **2015**, *55*, 9–13. [CrossRef]
135. Shi, R.; Irfan, M.; Liu, G.; Yang, X.; Su, X. Analysis of the Impact of Livestock Structure on Carbon Emissions of Animal Hus-bandry: A Sustainable Way to Improving Public Health and Green Environment. *Front. Public Health* **2022**, *10*, 835210. [CrossRef]

*agriculture*

MDPI

*Review*

# The Potential of Grape Polyphenols Additive in Pig Nutrition: Chemical Structure, Bioavailability and Their Effect on Intestinal Health of Pigs

**Andrei Claudiu Proca [1,†], Loredana Horodincu [1,†], Carmen Solcan [1] and Gheorghe Solcan [2,*]**

1 Department of Preclinics, Faculty of Veterinary Medicine, Iasi University of Life Sciences Ion Ionescu de la Brad, 700490 Iași, Romania; andrei.proca@iuls.ro (A.C.P.); loredana.horodincu@iuls.com (L.H.); csolcan@uaiasi.ro (C.S.)

2 Internal Medicine Unit, Clinics Department, Faculty of Veterinary Medicine, Iasi University of Life Sciences, Ion Ionescu de la Brad, 700490 Iași, Romania

* Correspondence: gsolcan@uaiasi.ro; Tel.: +40-745302684

† These authors contributed equally to this work.

**Abstract:** The recognition of the necessity for employing natural additives in animal feed has grown alongside the ban on antibiotics in the animal feed sector. Grapes, as well as by-products of the wine-making industry (grape marc and seed extracts), possess biologically active chemical constituents that can be used to improve animal production by incorporating them into animal feed. Grapes are a valuable resource of polyphenols, especially flavonoids, stilbenes and phenolic acids, most of them showing therapeutic or health-promoting properties. The purpose of this review is to elucidate the impact of polyphenols on animal gut health. The first section of the review discusses the chemical structure of the major polyphenols in grapes and the polyphenols' bioavailability and metabolism in pigs. The second and major part of the review reviews the results of investigations into the antioxidant, antimicrobial and prebiotic effects of grape polyphenols in pig diets, as well as their regulation of intestinal barrier functions through signalling pathways and intestinal responses. All of this is supported by previous research, findings and conclusions. There are fewer recorded pig studies, but the inclusion of up to 9% grape by-products resulted in improved performance with an increased mean daily gain. Ultimately, this analysis concluded that supplementation of pigs with grape phenolic compounds as natural feed additives enhanced their antioxidant capacity, improved humoral and cellular immune responses, and promoted gut ecosystem biodiversity and the overall production performance in pigs.

**Keywords:** grape by-products; polyphenols; antioxidant; growth performance; pig

## 1. Introduction

The continuous growth of the human population in recent decades has led to an increase in the global demand for animal products, one of which is pork [1]. In order to meet this demand, it is essential to improve the efficiency of pig production [2,3]. Feed accounts for the largest proportion of the total cost of livestock production [4]. Therefore, the identification of economically and sustainable viable alternatives to conventional feed is of paramount importance. Agro-industrial by-products are a notable option, as they are generated in large quantities annually, most of which are discarded or landfilled as waste [4].

Numerous studies have shown that the reuse of agricultural by-products can not only reduce costs but also improve environmental conditions [5,6]. These by-products are abundant and inexpensive and are abundant sources of nutritional constituents such as fibres, protein, minerals, antioxidants and vitamins. Due to their content of bioactive principles, they are well suited for use as supplementary ingredients in pig diets [7]. In addition, these co-products may be utilised as antibacterial agents, thus limiting the use of antibiotics [4]. A fundamental characteristic of phenolic compounds is their considerable antioxidant activity. In fact, certain phenolics can have multiple benefits, such as boosting immune function, reducing inflammation, promoting gut health and having antimicrobial activity.

The powerful antioxidant properties of procyanidins, a compound found in grapes, have been of great interest to scientists around the world. Scientific research has shown that the antioxidant capacity of procyanidins is about 20 times stronger than that of vitamin E and 50 times stronger than that of vitamin C [8]. Grapeseed procyanidins extracted from grapeseeds have been widely recognized as having potentially beneficial properties, including an antioxidant effect [9], an anti-inflammatory action [10,11] and an immunomodulating capacity [12]. Procyanidins' biological effects are well-studied in animal models and in vitro [13–15]. Their potential as feed additives or feed ingredients in animal farming, however, is still mostly unexplored.

The aim of this review is to elucidate the biological activity mechanisms of these bioactive compounds with the goal of discovering their potential applications and highlighting recent scientific advancements that support their useful properties, their bioavailability and their use as additives or ingredients in pigs' diets.

## 2. Grape Polyphenols: Their Chemical Structure and Bioavailability in the Gut

Grapes (*Vitis* spp.) are one of the most agronomically, zootechnically and economically important plant species due to their various uses in wine production and other food by-products [16]. The use of grapes has a long history, dating back to ancient times, spreading throughout the modern world, especially through the wine industry.

Grapes are one of the most cultivated fruit fruits in the world, with about 74 million of tones produced annually. Of this, 37.5% is cultivated in Europe, 36.5% in Asian countries and 17.2% in the USA [17]. Grapes grow on all continents in temperate regions characterised by abundant rainfall, hot, dry summers and mild winters [18]. About 50% of grape cultivation is used to produce wine; a third is consumed as raw fruit, and the remaining grapes are transformed into food items such as raisins, juice, grapeseed extract and oil and vinegar [17]. For this reason, there are numerous literature studies reviewing and characterising grapes, wine and grape derivatives such as grape pomace [19,20].

The production of *Vitis* spp. is of major importance due to the nutritional value and pharmaceutical properties of grapes, both raw and dried, as well as their derivatives, such as extracts and grapeseed oil [21,22]. Grapes are one of the richest fruits in carbohydrates (17 g/100 g), are high in calories (65 kcal/100 g) and have a relatively low glycemic index. Grape berries represent approximately 20 to 25 percent of the total mass of grapes used in the production of wine [23]. Grapes contain vitamin B6, thiamine and vitamin C and are an excellent source of manganese and potassium. They represent one of the most abundant sources of polyphenols, which are mainly found in the grape skin [24]. Grape berries contain fibres, protein, lipids and minerals. Essential amino acids such as lysine, arginine, glutamic acid, aspartic acid, glycine methionine and threonine are found in the protein of dry weight [25–27].

By-products derived from fruit are manufactured on a vast scale throughout the world, and the industrial grape segment is of great importance to the economy [4] and generates several thousand tonnes residues every year, creating a major waste management challenge [28]. These solid residues include a variety of by-products, such as vine shoots, stems, skins, seeds, lees, filter cakes and grape pomace. Until recently, grape marc was considered to be industrial waste and was left to decompose in the fields of the nearby winery or in huge landfill sites [29]. About 20% of the total weight of grapes used in winemaking is estimated to be marc. After the grapes have been processed for wine, they can be divided into two fractions: seedless grape marc (the remaining pulp, stalks and skins) and seeded grape marc (Figure 1). The first fraction is rich in fibres, while the seed is mainly used for its oil, containing unsaturated fatty acids. On a dry matter basis, the seeds account for 38–52% and the hulls for 5–10% of the grape pomace [30]. Grapeseeds are composed of fibres (47%, of which 60–70% is indigestible), complex carbohydrates (29%), fat (13%) rich in essential fatty acids, protein (11%), minerals and extractable phenolic compounds such as phenolic acids and flavonoids [23,31]. The protein from grape seeds contains all the essential amino acids, with an average of 3.6% methionine and 4.5% lysine. The main amino acids are those found in the grape, with the exception of threonine, which is present at 4.0% [32]. Grapeseed extract and grapeseed oil are two by-products obtained from grape seeds after processing wine or grape juice. The seeds are extracted, dried and purified to obtain these co-products, like grapeseed extract, which contains a high concentration of polyphenols, and grapeseed oil, high in essential fatty acids [13]. These substances are known for their powerful antioxidant action, preventing harmful oxidative reactions and removing free radicals from the body [33,34], as well as their antimicrobial properties [31].

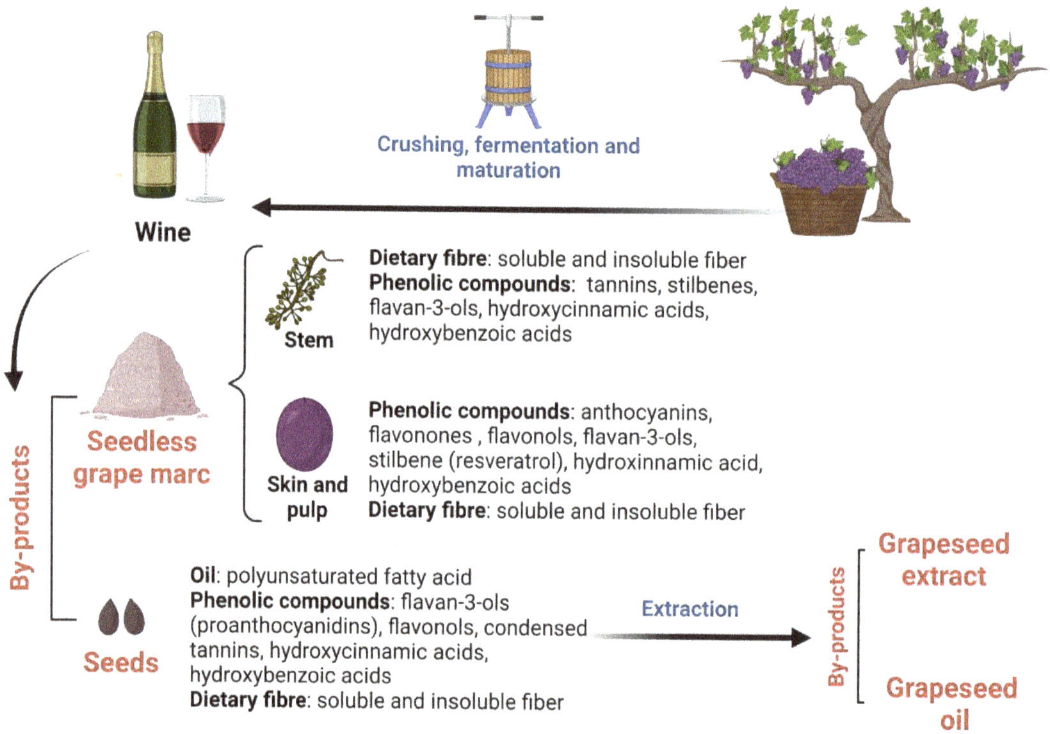


**Figure 1.** The process of obtaining grape by-products and the main components of each part. The figure was created with www.BioRender.com (accessed on 10 May 2024).

The nutrient values of grape by-products, such as fibre and organic matter digestibility and energy content and bioactivity, are the main determinants of their potential value in animal feed. Grape by-products are especially high in a variety of polyphenols. Previously called tannins, polyphenols have been seen as antinutritional agents because their occurrence in some legumes has negative impacts on the nutrition of animals. The main restrictions on the use of tescovine in monogastric diets are the presence of a lignified cell structure and the high tannin level. Grape skins and seeds contain the highest levels of dietary fibres (74% by weight), mainly hemicelluloses, covered with a whitish film [35] and tannins [36]. However, the stalk of the grape is lignified and consists entirely of tannins, which make up over half of the overall polysaccharides, making it an economical source of fibre [37].

In the last few years, in vivo and in vitro studies have demonstrated the positive effects of administering these bioactive substances. In fact, fibre enhances intestinal peristaltic action and works as a bacterial buffer and prebiotic, promoting the growth of friendly bacteria in the gut [4], while polyphenols may serve as an antioxidant, antimicrobial and immunomodulator [4,15,24,31,38–41]. Dietary fibre and polyphenolic substances can help to maintain or increase the growth performance and health of pigs according to the dosage and processing of the grape by-products included in the diet [15,38,39].

The chemistry of grape by-products is dependent on a number of factors, including ripeness, grape variety, soil type, weather conditions, geographical location [42,43] and the wine-making technology used [4,44,45]. All these aspects can alter the physical qualities of the grapes, including their flavour, aroma, texture and appearance [42]. Polyphenols are compounds with one or several phenolic hydroxyl groups linked to one or several benzene rings [46]. They are divided into classes based on the number of phenolic groups contained and the structural elements attached to these benzene rings [47]. The biosynthesis of these components in plants has been the subject of much research [48–50]. An understanding of the biosynthesis of phenolic components is essential for the effective management of their production in plants and, hence, of the by-products that can be used in pig feed.

Polyphenols are a group of chemicals biosynthesised in plants via the shikimate- or acetate-pathway of photosynthesis. These secondary metabolites are beneficial to the plant's lifespan and are formed as naturally occurring phytoalexins to confer pathogen resistance, protection against damage caused by the sun's ultraviolet rays and to deter predators due to their strong astringency when ingested [51,52]. Polyphenols are ubiquitous in plant life and are often part of our daily diet, occurring in a wide variety of fruits, some greens and even drinks [51,53,54]. The identification of phenolic compounds in grapes began in the late 19th century and continues today. To date, more than 8000 different phenolic structures have been discovered [51,55,56]. In general, polyphenols can be divided into about 10 classes with different basic structures, according to Bravo et al. [51]. Within grape pips, polyphenols can be divided into two major categories: flavonoids and non-flavonoids. Flavonoids in grape pips are divided into flavanols, flavonols, proanthocyanidins, anthocyanins and anthocyanidins, while non-flavonoids include phenolic acids and stilbenes [4,30,57] (Table 1).

**Table 1.** Chemical structures of major grape non-flavonoids.

| Typ | Name | Chemical Structure | Radicals | | | | Compound |
|---|---|---|---|---|---|---|---|
| | | | R1 | R2 | R3 | R4 | |
| Flavonoid | Flavanol | 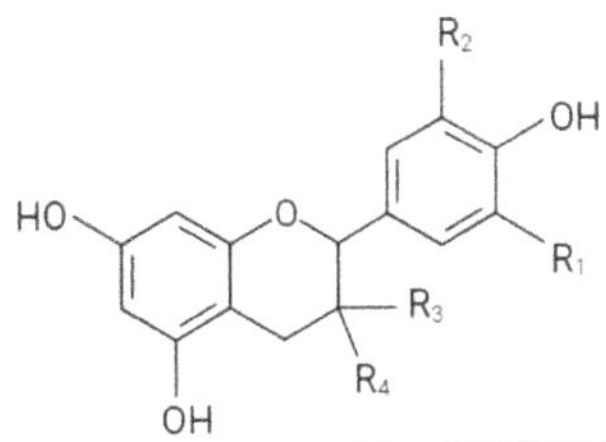  | H | OH | H | OH | Catechin |
| | | | H | OH | OH | H | Epicatechin |
| | | | H | OH | G | H | Epicatechingallate |
| | | | PH | OH | OH | H | Epigallocatechina |
| | | | OH | OH | H | OH | Gallocatechin |
| | Anthocyanin | 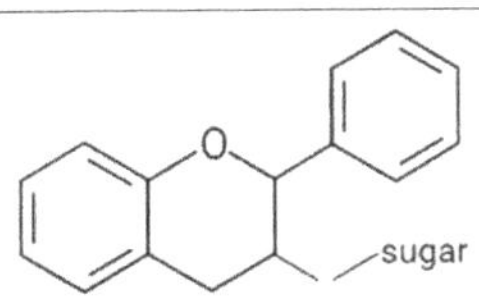  | | | | | |

**Table 1.** *Cont.*

| Typ | Name | Chemical Structure | Radicals | | Compound |
|---|---|---|---|---|---|
| Flavonoid | Proanthocyanidin B-type link | 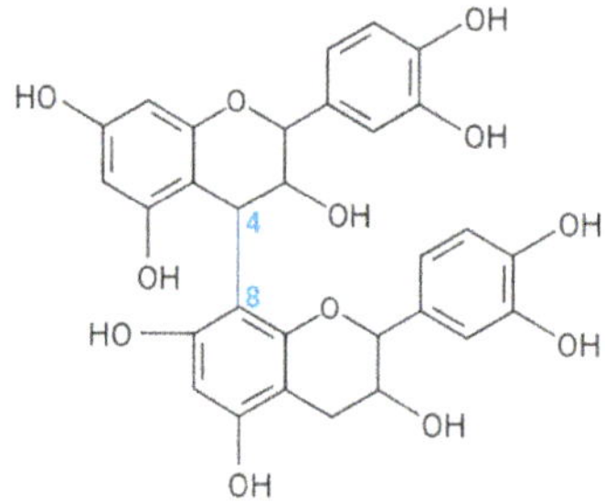 | | | |
| | Proanthocyanidin B-type link | 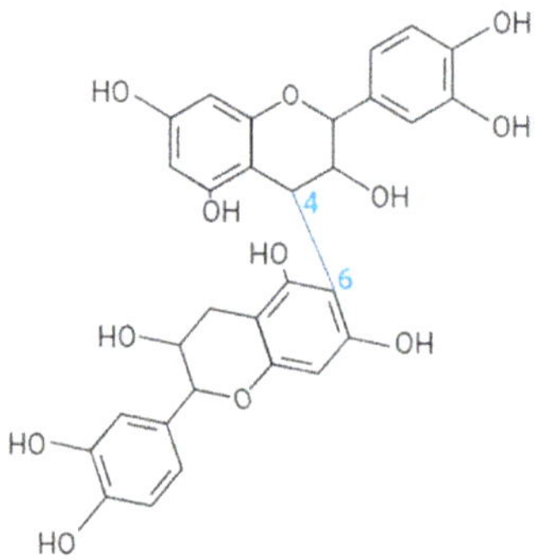 | | | |
| | Proanthocyanidin A-type link | 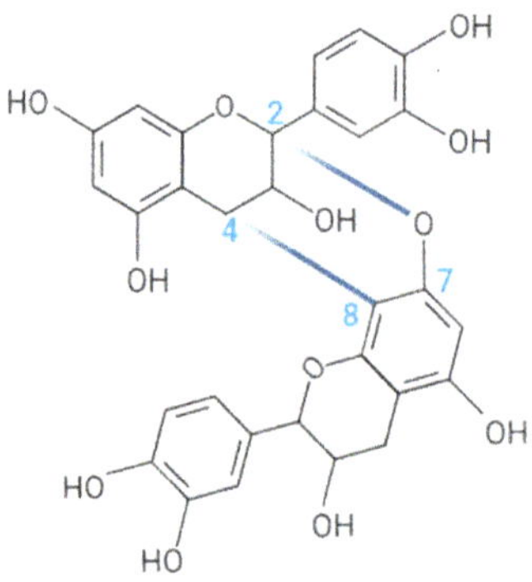 | | | |
| | Flavonol | 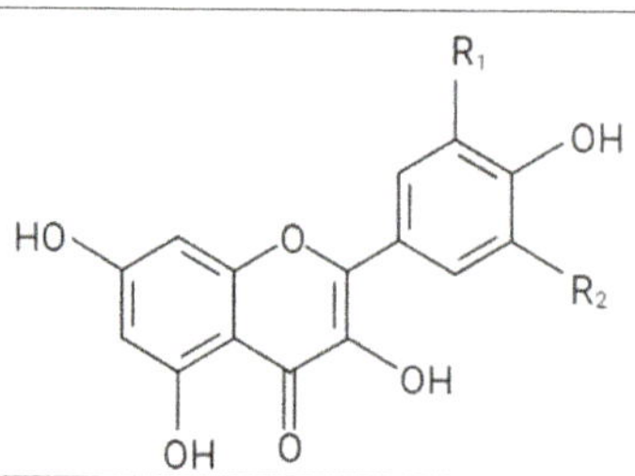 | R1 | R2 | |
| | | | OH | OH | Myricetin |
| | | | OH | H | Quercetin |
| | | | H | H | Kaempferol |
| | | | OCH3 | H | Isorhamnetin |

**Table 1.** *Cont.*

| Typ | Name | Chemical Structure | Radicals | | | | Compound |
|---|---|---|---|---|---|---|---|
| | | | R1 | R2 | R3 | R4 | |
| Non-flavonoid | Hydroxybenzoic acid | 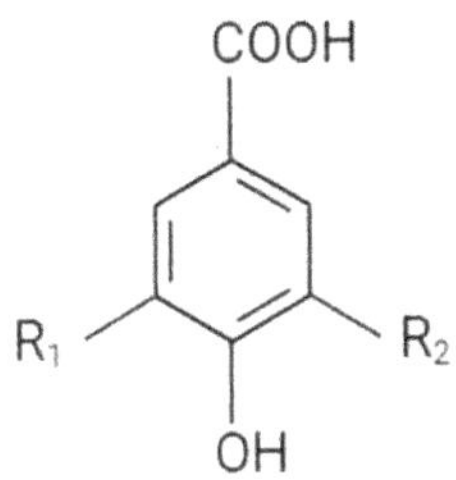 | OH | OH | | | Gallic acid |
| | | | OH | H | | | Protocatechuic acid |
| | | | OCH3 | OCH3 | | | Syringic acid |
| | | | H | H | | | P-Hydroxybenzoic acid |
| | Hydroxycinnamic acid | 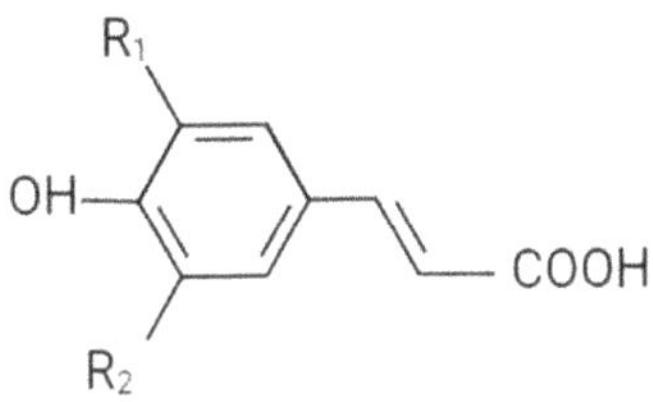 | OH | H | | | Cafeic acid |
| | | | OCH3 | H | | | Ferulic acid |
| | | | OCH3 | OCH3 | | | Sinaptic acid |
| | | | H | H | | | p-Coumaric acid |
| | Stilbene | 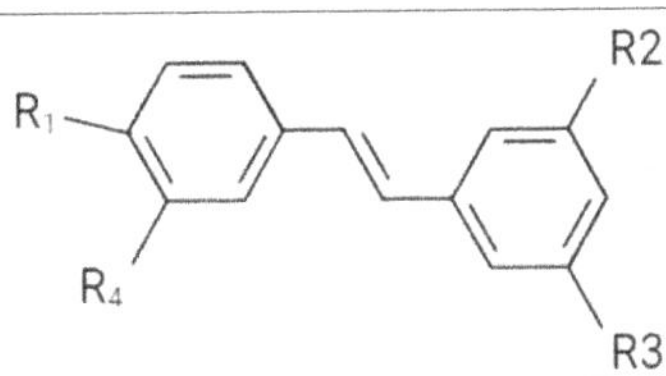 | OH | OH | OH | H | Resveratrol |
| | | | OH | OH | OH | OH | Piceatannol |

Flavonoids are biologically the most potent phytonutrients within the grape polyphenols. They can be divided into more than 13 subclasses with more than 6000 different chemical structures [58–60]. According to studies by Bravo et al. [51], Motohashi et al. [61] and Aron and Kennedy [62], the basic structure of flavonoids is two phenyl radicals (rings A and B) joined by three carbon atoms to create an oxygen-containing heterocycle (ring C). The flavonoids are further classified according to their oxidation state and distribution of hydroxyl radicals on the heterocyclic ring [47,63] (Table 1). The diversity of the chemical structures of flavonoids is responsible for their wide range of physiological and biological activities. The majority of the flavonoids are contained in the grape skin's epidermal cells, while around 60–70% of all of the polyphenols are found in the seeds [16,64,65]. Flavonoids are the major class of soluble phenolic compounds found in grapes and are the major contributors to the biological activity of grape products [66]. The bioactivities of flavonoids are significantly influenced by the degree of glycosylation, the type of sugar radical present and the subsequent acid esterification [67]. Therefore, selecting different varieties with distinct flavonoid profiles can have effects on pig health. Flavonoids are renowned for their cardioprotective, neuroprotective, antimicrobial, anti-ageing [68–72], antidiarrheal [73], antibiotic [74], anti-inflammatory [75,76] and antioxidant properties [55,77]. They also aid in improving vision [78] and cardio-protection [79], among other benefits [8]. Additionally, they provide UV protection, define flower colour, help attract pollinators and protect tissues from pathogen invasion or damage from oxidative stress [80].

The most common flavonoids in grapes are flavanols, including monomers, such as catechin and epicatechin, oligomeric proanthocyanidins (2 to 5 units) and polymeric polyphenols (more than five units), which are known as condensed tannins [81,82]. The structure of flavanols includes a hydroxy radical in the C3 atom and a B-ring linked to the C2 atom [63,83,84] (Table 1). Flavanols are synthesised before flowering, and their concentration increases until veraison—the time of grape ripening [85]. The main flavan-

3-ol monomers in grapes and wine include (+)-catechin, (−)-epicatechin, (−)-epicatechin-3-gallate [86] and (−)-epigallocatechin, and traces of (+)-gallocatechin. In the skins and stems of grapes, the main monomers are (+)-catechin, (−)-epicatechin and (−)-epicatechin 3-O-gallate, while the seeds contain catechin, epicatechin 3-O-gallate and epicatechin [87, 88]. The contents of (−)-epicatechin and (+)-catechin are higher in red than in white varieties [89]. Catechin is the most abundant flavanol found in seeds and grape skins, with traces also in grape pulp [30]. These compounds contribute to the bitterness in wine and may cause some astringency. Initial studies on these compounds began in the 1920s [90] and were further explored in grape seeds and during wine production. A significant proportion of flavan 3-ol monomers originate from grape seeds [91–95], with higher temperatures, higher alcohol concentrations and longer extraction times increasing their concentration in wine. Flavanols are not glycosylated in food, unlike other flavonoid classes [30].

Proanthocyanidins (condensed tannins) are made up of two to five flavan-3-ol subunits. They are called oligomeric proanthocyanidins because the acid-catalysed cleavage of polymer chains produces anthocyanidins. Flavan-3-ols are characterised by the presence of the hydroxyl group (-OH) at position 3 of the flavonoid basic structure. The chemistry of tannins varies according to their origin, containing up to twenty hydroxyl radicals and a molecular mass of 500 Da to 3000 Da [96]. Structural variations between proanthocyanidins depend mainly on the interflavanic bonds, subclassified into type A (C2-O-C5 or C2-O-C7 bonds) and type B (C4-C6 or C4-C8 bonds) [62]. Procyanidin C1 is a C4-C8 bonded trimer [75,97,98] (Table 1).

The difference in interflavonoid bond positions and constituent units confers structural diversity to the higher oligomers, increasing the number of isomers with a polymerization degree [92,99–101]. Limited knowledge about proanthocyanidin chemistry is due to the analytical methods focusing on each oligomer as a class without identifying proanthocyanidins within each class [102]. Galloylated oligomeric proanthocyanidins are characteristic of condensed grapeseed polyphenols [103]. Tannins' ability to bind proteins underlies their protective characteristics [104] and nutritional benefits [105]. They have a distinctive odour and astringent taste and appear as a loose or shiny white or yellow powder [106,107]. Grapes and wine have one of the highest phenolic levels among fruits, vegetables and beverages. Depending on the variety, the proanthocyanidin content at the point of harvest can range from 0.5 to approximately 6.4 mg/g of fresh berry fruit weight [108]. Proanthocyanidins contribute to the astringency of red wines, extracted from the seeds, steams and skin [23]. About 30% of proanthocyanidins are in the seeds, and 15% are in the peel [109]. Extraction from seeds requires breaking the cell walls [110]. The final proanthocyanidin content in seeds is obtained at a later stage than in the peel, a couple of weeks after the onset of ripening. These components were among the most recent important phenolic compounds to be structurally identified [111,112]. Proanthocyanidins with a low molecular weight are found in very low concentrations and are easily hydrolysed [113], while higher molecular weight ones are associated with astringent and UV protective properties. These compounds are popular in functional food formulations for their health benefits [114–116]. The proanthocyanidin content is influenced by climatic and geographical conditions, grape variety, fertilisation, cultivation practices and soil [89].

Anthocyanins, with the aromatic B-ring attached to the C2 position, have two bonds in the heterocyclic ring. They constitute the glycosylated version of the anthocyanidins (aglycone), resulting from the linkage of the C3 hydroxyl group with the sugar moiety. Anthocyanins are the most prevalent polyphenols in the skins of red grapes and act as natural dyes [44,57,63,83,84,117]. They are mostly found in the skin, but in certain 'teinturier' (or coloured) varieties, anthocyanin pigments also accumulate in the berry pulp [118,119]. There is an intimate relationship between the levels of anthocyanin biosynthesis and the development of the berry, starting at 'veraison' when the biosynthesis of proanthocyanidins is complete and peaking at 'ripeness' [120]. There is a unique set of anthocyanins in each grape species and variety [121]. For instance, European grapes produce mainly 3-O-monoglucoside anthocyanidins, while muscadine grapes produce mainly 3,5-O-diglucoside anthocyanidins. The identification of the anthocyanin structure was preceded by scientific research into red wine's colour [122]. The general structure of anthocyanins was determined in the early 20th century [123,124]. The structures of the major anthocyanins in *Vitis vinifera* grapes were identified in 1959, with malvidin-3-O-glucoside as the main anthocyanin present, together with its acylated forms [125]. The work of Ribéreau-Gayon also demonstrated that anthocyanins in *Vitis vinifera* are structurally distinct from those found in other species, being exclusively monoglucosides, while the non-vinifera species also contain 3,5-diglucosides. Subsequent studies on the distribution and structure of anthocyanins in grape species have furthered our understanding of grape anthocyanins [126–128].

Found mainly as 3-O-glycosides in grape skins, flavonols are the next most abundant flavonoid in grapes. In wines and juices, they can also be found as aglycones, such as kaempferol, quercetin, isorhamnetin and myricetin, due to hydrolytic attack by acids in processing and during storage [129]. The flavonol structure includes a C2-C3 double linkage, where the hydroxyl radical is at the C3 atom and the B-ring is attached to the C2 atom of the keto group [130] (Table 1). The flavonol profile varies between grape varieties, but generally, quercetin 3-O-glucoronide and quercetin-3-O-glucuside predominate in most varieties [121]. Quercetin derivatives, isorhamnetin and kaempferol, are found in both white and red varieties, while the derivatives of myricetin are found only in red grapes [121,129].

Unlike flavonoids, non-flavonoid polyphenols have a single ring as their main structure. The non-flavonoid molecules found in grapes include stilbenes and phenolic acids [47,63,131]. Phenolic acids in grapes are derivatives of hydroxycinnamic acids, such as caffeic, p-coumaric, ferulic and sinapic acids, and hydroxybenzoic acids, such as gentisic, gallic, protocatechuic, p-hydroxybenzoic and syringic acids [16,47,63,131,132] (Table 1). These phenolic acids, especially hydroxycinnamic acids in the tartaric acid ester form, are the main phenolic components of white wine and are essential for its colour. They are mainly derived from grape pulp but are also found in similar quantities in red wines. In the mid-20th century, hydroxycinnamic acids were identified in grapes [125]. These compounds were earlier observed to be in the form of free acids. Later, it was discovered that grapes contain no free hydroxycinnamic acids but do have esterified tartaric acid [125]. Other compounds have been characterised [133,134] and shown in berries before ripening [93]. Stilbenes consist of two aromatic rings linked by an ethylene radical. The best-known stilbene is resveratrol. Stilbenes are found in grapes, wine and their derivatives [47,63,131] (Table 1).

Bioavailability is a measure of how much of a food's natural structure is available to its destination after it has been ingested through the gastrointestinal tract. The amount of a compound that is absorbed, metabolised and circulates throughout the body is called bioavailable [135]. Digestive metabolism, bioactivity, tissue partition, hepatic and intestinal metabolism, and absorption by intestinal epithelial cells are termed bioavailability [136]. Thus, bioavailability is rigorously based on bioavailability activities [137]. The bioavailability and efficient delivery of polyphenols to target tissues is required to explain the biological effects of polyphenols. It is, therefore, important to have an understanding of how they are absorbed, metabolised and excreted out of the body (Figure 2).

Absorption studies are complicated by the molecular complexity of polyphenol-rich extracts or foods due to factors such as the degree of conjugation and polymerization with other phenols. The majority of polyphenols are available in plants in the form of esters, polymers or glycosides, which cannot be absorbed in their original structure. Most polyphenols are linked to cell wall components like proteins, arabinoxylans or other organic compounds such as lipids and acids [138,139]. Before they can be absorbed, they must be hydrolysed by microbiota or endogenous enzymes. After absorption, polyphenols are recognised by the body as foreign substances, resulting in relatively low bioavailability compared to macro- and micronutrients.

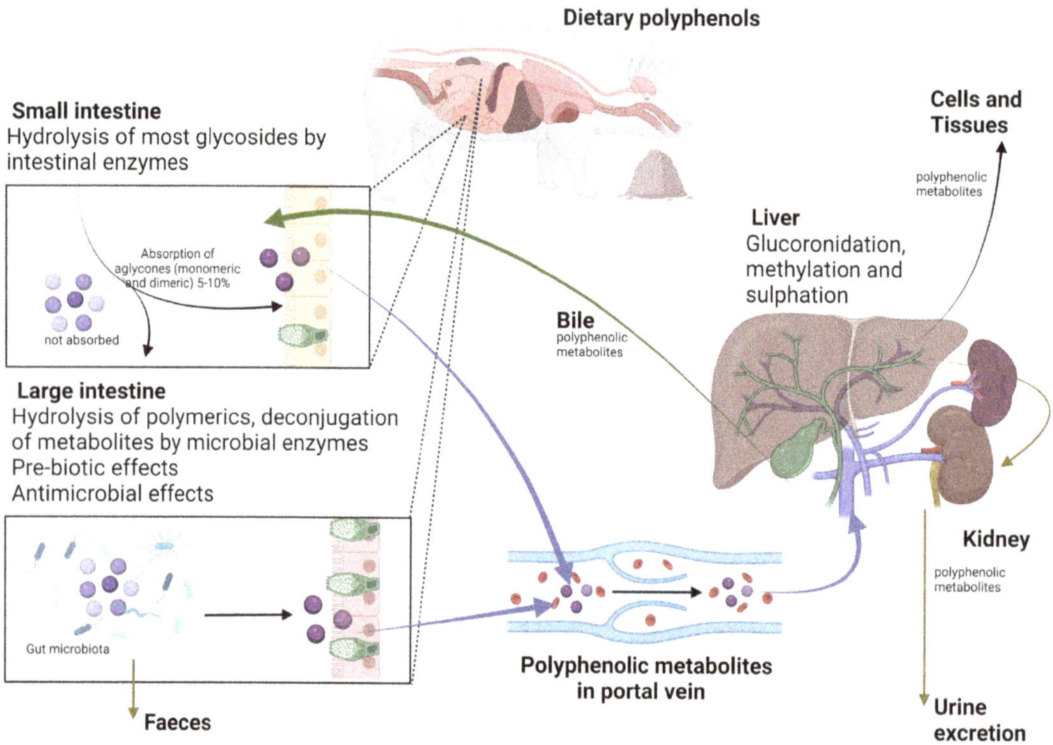


**Figure 2.** Bioavailability of polyphenols in the digestive tract of pigs. The figure was created with www.BioRender.com (accessed on 3 June 2024).

The metabolism of polyphenols involves a number of common reactions aimed at reducing their cytotoxic potential by increasing hydrophilicity and favouring biliary or urinary excretion [54]. It is the chemical structure of the polyphenols, rather than their concentrations, that defines the speed and magnitude of absorption, as well as the characteristics of the circulating metabolites in the plasma. According to estimates by Faria et al. [140,141] and Corrêa et al. [140,141], the absorption of polyphenols in plant substances in the small intestine is less than 5–10%.

Depending on their complexity and degree of polymerisation, the absorption of polyphenols may vary. Polyphenols with a simple structure (monomeric and dimeric) are easily absorbed in the small intestine, whereas oligomeric and polymeric polyphenols reach the colon almost unchanged [54,142–144]. Previous studies have shown that procyanidin trimers and dimers are very stable under gastric and duodenal digestion conditions, and the consumption of dimers is considered to be approximately 100 times lower than that of monomers [145]. Furthermore, it has been observed that proanthocyanidins from peels tend to have a greater degree of polymerisation than those obtained from seeds [146]. However, recent research suggests that only polymers with polymerisation degrees lower than 5 are absorbed [147] or are subsequently degraded to their flavan-3-ol monomers in the intestinal lumen.

After their absorption, less complex polyphenols, known as aglycones, can be hydrolysed and bio-transformed in enterocytes and then in hepatocytes [148,149]. A variety of water-soluble metabolites, such as methyl derivatives, glucuronides and sulphates, are formed during this process. These metabolites rapidly enter the circulation, are distributed to various tissues and are subsequently eliminated in the urine [54,150]. Some of these metabolites are excreted in the bile into the colon, where they are hydrolysed by bacterial enzymes, particularly beta-glucuronidase. This enterohepatic recycling may extend the duration of polyphenols in the organism. However, most polyphenols reach the colon in an intact state, where they preserve the intestinal barrier integrity and exert their bioactive properties, among them anti-inflammatory and antioxidant activities. While polyphenols are distributed to various tissues, the majority end up in the colon, where they are subject to the enzymatic activity of the gut microbiota along with metabolites eliminated in the lumen of the intestine. The result of these microbial activities is the production of a wide range of metabolites that can be incorporated into the blood stream or be excreted in the faeces. Metabolites taken up by the body are transported via the portal veins to the liver, where some can be glucuronidated, methylated or sulphated, and then dispersed to various tissues or eliminated in the urine.

The binding affinity of polyphenolic metabolites to albumin, a factor influenced by the chemical composition of the polyphenolic components, determines their bioavailability in target tissues [151]. Moreover, certain metabolites may exhibit prebiotic effects, promoting an increase in beneficial gut bacteria and inhibiting the growth of pathogenic strains [152–156]. Therefore, the microbiota of the gut has a pivotal function in both the degradation of polyphenols and the formation of metabolites responsible for the biological effects of polyphenols.

Bioavailable nutrients, as defined by Prada et al. [157], represent the most crucial fraction of nutrients reaching the bloodstream. The focus is on releasing these nutrients into systemic circulation rather than solely considering the total excreted amount. Moreover, many beneficial bioactive substances are often present in forms that are not readily available, emphasising the need to modify dietary matrices for enhanced accessibility. Variations in flavonoid bioavailability, including plasma and urinary concentrations and compound availability, may be due to many factors, including the source, chemical characteristics, food matrix, dosage, individual differences, analytical methodology and detection limits [158,159]. Research gaps exist in understanding the grape by-product polyphenols' digestibility in livestock and how it affects nutrient digestibility. Significant differences based on sex and species affect xenobiotic metabolism, leading to variations in polyphenol metabolite spectra, tissue distribution and blood concentrations [160–167].

## 3. Antioxidant and Anti-Inflammatory Effects of Grape Polyphenols in Pig Feed

Oxidants consist of both radical and non-radical molecules containing oxygen, chlorine or nitrogen and are called reactive oxygen species (ROS), chlorine or reactive nitrogen species (RNS). Oxidants can be produced in the course of metabolism (superoxide radicals formed in the mitochondrial chain of respiration) during a response to inflammation (superoxide radicals produced by the oxidase NADPH in the activated immune cells) and as a result of exogenous insults (free metals such as iron and copper, which lead to the production of OH radicals form $H_2O_2$) [168].

Animals can be exposed to oxidative stress due to a number of factors, including diets containing fungal toxins, elevated ambient temperatures, a number of pathological situations in which the immunological system is heightened (vaccinations, infections), ascites, pulmonary hypertension and coccidiosis [169–171].

The antioxidant system works to avoid oxidative stress through the reduction and/or scavenging of oxidants and is made up of the following: (1) Antioxidant enzymes (catalase—CAT, glutathione peroxidase—GPX and superoxide dismutase—SOD); (2) antioxidants of low molecular weight (vitamin C, vitamin E, uric acid, carotenoids, glutathione and polyphenols); and (3) proteins sequestering free-transition metals (ferritin, metallothionein and ceruloplasmin) [168].

Phenolic compounds, which are natural metabolites, are recognised for their ability to counteract oxidative stress, which is associated with serious disorders of the metabolism by damaging the extracellular and cellular macromolecules [172,173]. These components are highly antioxidant and are essential in providing health benefits [174]. Flavonoids, a wide class of lower molecular weight compounds, have significant antioxidant activity.

Due to their particular chemical structure, they are able to decrease oxidative stress by various pathways [70,175]. In vitro, for example, flavonoids can serve as chain-breaking and protective antioxidants, scavenging alkoxyl, peroxyl, hydroxyl and superoxide radicals and blocking low-density lipoprotein oxidation. (LDL) [176,177]. Furthermore, flavonoids may function as metal chelators and blocking enzymes that are involved in ROS production, such as protein kinase C, xanthine oxidase, lipoxygenase, glutathione S-transferase, cyclooxygenase, mitochondrial NADH oxidase, mitochondrial succinoxidase and microsomal monooxygenase [173,177,178].

The antioxidant role of phenols depends on the position and number of hydroxyl groups and their relationship to carboxyl functional groups [173,179,180]. Among phenolic compounds, the monomer forms are not as efficient as hydrogen scavengers compared to phenols in the polymer form [181]. The structure–function relationship [182] and glycosylation [183] influence the antioxidant properties of phenolic compounds. In other words, the ability to donate hydrogen or electrons and delocalise unpaired electrons in the ring of the phenol is the primary mechanism to protect the molecules from oxidation [184,185]. ROS may damage the intestinal mucosal barrier and interfere with nutrient absorption, and antioxidants play a critical role in neutralising these radicals and maintaining an optimal environment at the intestinal surface [186].

Oxidants activate NF-κB, a key regulator of inflammation, and oxidative stress is closely linked to inflammation [187]. In its inactive state, NF-κB is bound to inhibitory proteins in the intracellular cytosol and is found in approximately all animal cell types. Following oxidative stimulation and factors, for example, bacterial stimuli, viruses, UV radiation and cytokines, the repressor proteins are removed from NF-κB, allowing the translocation of active NF-κB. This allows active NF-κB to relocate to the nucleus and enable the transcription of a wide range of inflammation genes [188]. The genes targeted by NF-κB encode typical proteins like adhesion molecules, inflammatory enzymes, chemokines, inflammatory cytokines and a variety of receptors [188–194]. Many of the proteins regulated by NF-κB, including chemokines and cytokines, promote the formation of oxidants from activated neutrophils and mitochondria, increasing oxidative stress and creating a 'vicious circle' [188,192,194–196]. If no intervention can be made to stop this vicious circle and the production of oxidants becomes overproduced, the process of inflammation will become chronic [188,192,197–201], and the cells and tissues of the body will be damaged, and in the case of pigs as farm animals, the following diseases can occur: lung inflammation, intestinal inflammation and septicemia [202].

By stimulating the production of immunoglobulins and decreasing the release of pro-inflammatory cytokines, polyphenols can improve gut health and immunity in the diet of monogastric animals [203]. These phenolic components may also enhance the action of antioxidant enzymes, thereby limiting inflammation [204]. Studies have shown that grape polyphenols may decrease inflammation by regulating inflammatory pathways and reducing levels of reactive oxygen species (ROS). Being natural substances, the flavonoids and proanthocyanidins in grapes may act in multiple ways against chronic inflammation, which may make them better than single-target synthetic chemical anti-inflammatory medications [205,206].

The anti-inflammatory activity of the polyphenols is achieved through complex cellular pathways. The majority of these mechanisms involve an inhibition of NF-κB, which is the key regulatory molecule in inflammation. Polyphenols can inhibit NF-κB activation through the inhibition of phosphorylation and proteasomal degradation of IκB, and this activity may be, at last, partly attributed to the polyphenols' antioxidant properties [207]. Polyphenols can directly scavenge free oxygen radicals and cause transcription factor Nrf2 to be activated. This causes a number of antioxidant enzymes to be activated [208]. Both direct ROS scavenging and Nrf2 activation help prevent oxidative stress, which initiates the pro-inflammatory responses through the activation of NF-κB, mitogen-activated protein kinases (MAPK) and activator protein 1 (AP-1) [209]. Furthermore, polyphenols can engage transcription proteins such as peroxisome proliferator-activated receptor gamma (PPAR-γ), thereby counteracting inflammation through the inhibition of NF-κB activation [209].

The activation of Nrf2 by polyphenols is an example of a common hormonal pathway activated by polyphenols and other phytochemicals. The idea of hormesis suggests that while higher doses of some polyphenols may be harmful, sub-toxic levels consumed by herbivorous animals may cause minor cellular stress responses such as Nrf2 activity. This leads to the initiation of vitagenesis, including genes encoding antioxidants, biotransforming enzymes and heat shock proteins, which maintain cell stability under stress conditions and confer tolerance to greater stress [210]. Stress-related phytochemicals not only offer protection against increased doses of the same compound but against other less specific compounds or stress factors, such as metabolic, oxidative and inflammatory stresses [211], which are relevant to farm animals. In response to these non-specific stressors, such as ROS and reactive nitrogen species, Nrf2 cytoprotective pathway activation stimulates autophagy [212]. Autophagy is a well-conserved lysosomal 'self-digestion' process. This process leads to the breakdown of long-lasting proteins, as well as cell organelles and the generation of fatty and amino acids and nucleotides that may be reused for protein synthesis and ATP production in times of cellular stress [213]. Activation of autophagy by hormetic phytochemicals and caloric restriction reduces ER, inflammatory and oxidative stresses, thereby contributing to an increased cellular capacity and organismal health [213].

Grapeseed proanthocyanidins (PACs) were shown to have strong anti-inflammatory activity by scavenging radicals, preventing the peroxidation of lipids and inhibiting the production of pro-inflammatory cytokines [214]. In the in vitro studies, PACs showed anti-inflammatory activity on enteric cells and macrophages [215,216]. Reduced inflammatory cytokine production, oxygen free radicals (ROS) and NF-κB translocation were observed when the macrophages or dendritic cells were exposed to pro-inflammatory stimuli in the presence of PAC [114,217]. PACs also effectively attenuated inflammation-induced mitochondrial dysfunction and oxidative stress in epithelial cells. The precise mode of action of PACs is still not fully understood, but it appears that they modulate the signalling pathways related to lysosomal activity and secondary messengers [217,218]. These results are in agreement with the immunological modifications seen in vivo, suggesting that at least part of the anti-inflammatory effects of PAC are due to the direct modulation of immune cells in the mucosa [219–221]. In addition, PAC can activate innate immune cells, such as γδ-T cells [222,223], which have a key role in enhancing the immune response of the mucosa against pathogens and in signalling the activation of other immune cells, including neutrophils [224]. Stimulating intestinal organelles with CAP causes a marked increase in the regulated antimicrobial defence, suggesting that the gut of mammals has developed to perceive CAP as a sign to enhance the innate immune response to avoid and reduce inflammation [225].

The gastrointestinal luminal surface is protected by a mucus gel layer, providing an essential physicochemical barrier against chemical, enzymatic, mechanical and microbial damage. Mucin is a major glycoprotein of this layer. It forms an effective barrier that prevents microbial adhesion and subsequent invasion [226]. It is thought that polyphenols may influence the properties of this protective layer, as well as the absorption of nutrients and the viscoelastic environment of the bacteria in the gut [227]. Some research has suggested that certain polyphenols, such as resveratrol and ellagic acid, can increase the production of mucus by goblet cells in the intestinal lining, indicating a potential benefit in inflammatory bowel disease [228,229]. Polyphenols have also been found to interact with the gastrointestinal mucus layer, acting as crosslinkers for purified gastric and duodenal mucin, which could affect mucus layer elasticity [227].

Another important mucosal defence mechanism is the production of antimicrobial peptides. These are known as host defence peptides or antibiotic peptides. Due to their antimicrobial and immunomodulatory activities, these molecules play a crucial function in innate immunity. To date, more than 400 peptides with antimicrobial activity have been reported in animals and plants, which are produced by both immune cells and mucosal epithelia [230]. In pigs, information on the activity and function of defensins is limited. However, several types of defensins have been studied, including porcine β-defensin 1 (pBD-1) [231]. Recently, new potential variants of β-defensins have been described on the basis of sequence homology [232]. It is believed that these peptides, such as pBD-2, may play an essential role in promoting gut health. More recently, Wan et al. [233] have reported that the EGCG black tea extract decreased the translocation of bacteria in IPEC-J2 cell monolayers by inducing the production of the antimicrobial biallelic peptides, pBD-1 (porcine b-defensins 1) and pBD-2, having greater antimicrobial efficacy towards *E. coli*. A subsequent mechanical investigation showed that the EGCG extract from black tea increased pBD-2 but not pBD-1 through the p38 mitogen-activated protein kinase (MAPK)-dependent pathway. This research indicates the possibility that some polyphenols, such as EGCG, may modulate epithelial immune barrier function by inducing defensin production. Additional research is needed to gain a greater perspective on how polyphenols such as EGCG affect intestinal mucosal barrier function and to assess the potential of other polyphenols in this regard.

Few investigations have evaluated the impact of polyphenol-rich grape by-products on pro-inflammatory gene expression in the gut in relation to the possible influence of polyphenols on the inflammation of pigs. A nutrition study by Gessner et al. [234] found that a diet containing grapeseed extract and grape pomace meal reduced the expression of a variety of pro-inflammatory markers in the duodenum of pigs during growth. The grape pomace and seed extracts supplementation also resulted in an improvement in the ratio of villus height to crypt depth, indicating that plant-derived polyphenols may have a beneficial effect on gut microarchitecture. In a different study, two plant extracts rich in polyphenols were used, namely either an extract of grapeseed and grape pomace meal or an extract of hops, both at a dietary level of 10 g/kg of feed. In different areas of the gut (duodenum, ileum or colon), both extracts reduced the expression of several inflammatory genes (CCL2, IL-1B, ICAM-1, TNF, IL-8) [235] (Table 2.). In particular, the genes are controlled by nuclear factor κB (NF-κB), the key inflammatory regulator [190,236,237] (Figure 3). Chemokines and cytokines are a class of small proteins that are essential for the modulation of a wide range of biological events, such as adaptative and innate immunity and the regulation of inflammatory responses. They may be generated by a different cell type, for example, by immune cells (like lymphocytes, macrophages and dendritic cells) or other cell types (like intestinal epithelial cells (IECs)). Some cytokines are expressed constitutively by the intestinal epithelium to maintain the homeostasis and growth of epithelial cells; these include granulocyte-macrophage colony-stimulating factor (GM-CSF), tumour growth factor alpha (TGF)-a and interleukins IL-18, IL-15, IL-10, IL-6 and (IL)-1 [238]. However, when the gut becomes inflamed, there is a substantial upregulation of cytokines and chemokines, including IL-8, IL-6, IL-1b and TNF-a [239]. Excessive secretion of these pro-inflammatory cytokines may play a central role in the pathogenesis of inflammatory bowel diseases (IBD) such as ulcerative colitis and Crohn's disease. Properly controlling the secretion of these cytokines is critical to maintaining intestinal homeostasis [240]. Significant reductions in the inflammatory mediators Nrf2 and NF-κB in the mucosa of the duodenum were observed in pigs fed diets rich in polyphenols, including seed extracts and grapeseed extract, which reduced the risk of intestinal disease. The potent anti-inflammatory and antioxidant properties of polyphenols, by reducing the production of reactive oxygen species locally in the small intestine, may be responsible for this inhibition of Nrf2. In contrast, no effects on NF-κB and Nrf2 gene expression were observed in pig liver. In pig liver, however, no effects were observed for NF-κB and Nrf2 gene expression. For example, dietary supplementation with grapeseed procyanidins at doses of 100–150 mg/kg resulted in improved serum IgM and IgG concentrations, considered as indicators of the humoral immune responses, in a study by [15] in young piglets. According to research by Ramiro-Puig and colleagues (2007) [241], the action of polyphenols on the humoral defence response is based on their effect on B-cells and on their differentiation in the direction of immunoglobulin-secreting cells. In the same context, Hao and colleagues (2015) [15] found that grapeseed teskovine-supplemented diets enhanced serum levels of IL-2, a key cytokine in the differentiation of T- and B-cells. The use of the grapeseed cake diet in the above experiment resulted in a statistically significant increase in IgA levels in plasma following 24 days of grapeseed polyphenol supplementation in the Taranu 2018 diet [242]. After the consumption of curcumin-derived polyphenolic compounds, a comparable elevation in intestinal IgA concentration has been observed in rats [243]. Secretory immunoglobulin A is the major form of immunoglobulin in the lumen of the intestine and has multiple properties essential for mucosal immunity and homeostasis. The secretory component of sIgA confers protection against degradation by resisting digestive and proteolytic enzymes found in the intestinal tract [244]. sIgA is localised in the intestinal lumen, an environment rich in microbes, unlike other types of antibodies, such as IgG, which are found in an almost sterile systemic compartment. As a result, the function of sIgA is different from other antibodies, exerting its effect through steric inhibition, receptor blocking or immune elimination, resulting in a lower inflammatory response [245].

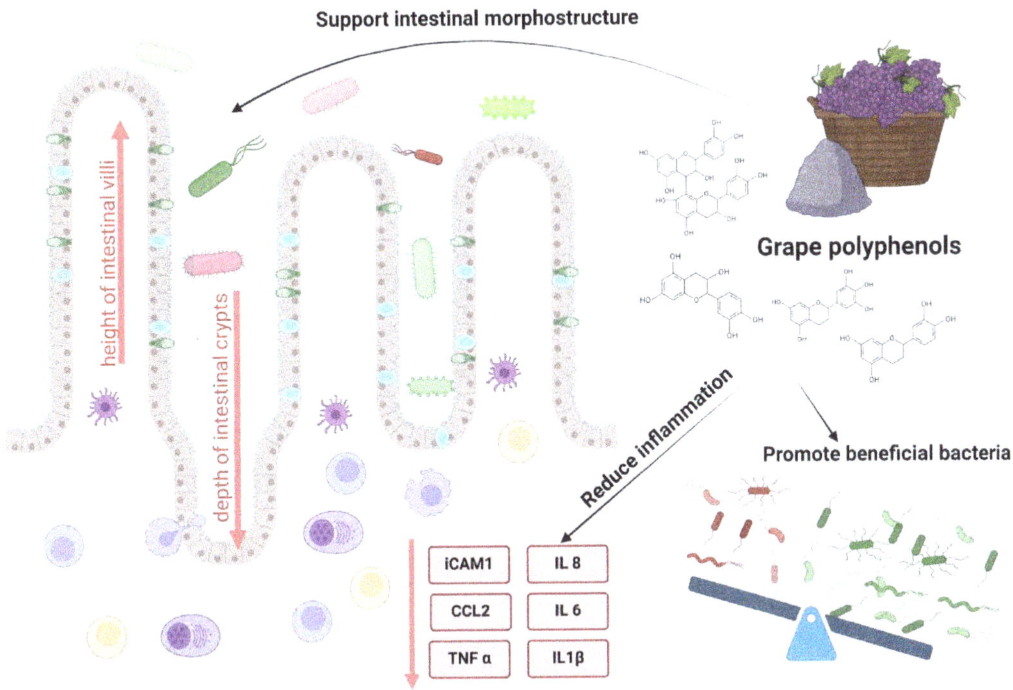


**Figure 3.** Effect of polyphenols on intestinal health. The figure was created with www.BioRender.com (accessed on 27 May 2024).

In pigs, adding a mixture of vegetable extracts, including grape seeds (2%), to the diet reduced plasmatic TBARS production in the weaned piglets without affecting the total antioxidant capacity, according to research by Zhang et al. [246] involving the modulation of plasma antioxidants. In another study by Gessner et al. [234], grapeseed and grape tescovine did not affect the plasma and liver TBARS concentrations in piglets or plasma antioxidant defences. Furthermore, the TBARS-MDA levels and liver antioxidant capacity in pigs on the high grapeseed inclusion diet were not significantly different from those in the control diet in a report by Taranu et al. [242]. Based on several studies to date, Gessner et al. [247] found that polyphenols from plants had a lesser impact on antioxidant capacity in healthy subjects, partly due to their lower bioavailability. However, due to their systemic anti-inflammatory effects, these compounds may improve the overall antioxidant balance and reduce free radical formation in challenged animals.

**Table 2.** Major physiological effects of grape by-product polyphenols in pigs.

| Grape By-Product | Dose | Effect | References |
| --- | --- | --- | --- |
| Fermented grape pomace | 3% | ADFI, final bodyweight and ADG were not affected. FCR decreased | [248] |
| Grapeseed cake | 5% | ADG and ADFI were not affected<br>Elevated plasma IgA levels and TBARS were significantly reduced | [242] |
| Complex polyphenpol extracts, including grape seeds | 1% | Reduced the level of plasma MDA | [246] |
| Grapeseed extract (procyanidins) | 0.04% | ADG increased and FCR decreased<br>Increased expression of CAT, SOD and GSH-Px genes associated with antioxidant activity in the liver and could reduce MDA levels in muscle tissue, liver and serum | [39] |
| Grape pomace | 5% | Higher jejunal villus height and villus height/crypt depth ratio<br>ADG, ADFI and FCR were not affected | [39] |
| Grapeseed extract (procyanidins) | 250 mg/kg | Improving the barrier function and morphology of the intestinal mucosa<br>Enhanced the biodiversity of the gut ecosystem | [249] |
| Grapeseed and grape marc extract | 1% | Increased small intestine villus height/crypt depth ratio<br>Gain/feed ratio improved<br>Duodenal mucosal inflammation inhibition | [234] |
| Grape pomace | 9% | Increased ADG and final body weight<br>Enhancement of antioxidant mechanisms and prevention of oxidative stress damage to lipids and proteins<br>Enhances intestinal barrier function and health | [250] |
| Resveratrol | 0,2% | Antimicrobial effect: *E. coli* and *Salmonella*<br>Bacteria growth promoting activity: *Lactobacillus* spp. | [251] |
| Grapeseed extract | 1% | Reducing *E. coli*-induced diarrhoea in weaned pigs | [252] |
| Grape seeds | 8% | There has been an increase in *Bacteroidetes* phylum and a significant decrease in *Firmicutes* phylum | [253] |
| Grapeseed extract | 1% | Microbiome ecological shift | [254] |
| Grapeseed procyanidins | 0.5, 1, and 1.5% | No significant effect on growth performance, increased antioxidant capacity, improved humoral and cellular immune responses, reduced incidence of diarrhoea | [15] |
| Grape seeds and grape marc | 1% | Modifies intestinal microbiota and reduces inflammation | [235] |

## 4. The Antimicrobial and Prebiotic Effects of Grape Polyphenols in the Intestine of Pigs

An important tool for improving growth performance and feed efficiency is the manipulation of gut function and microbial habitat in livestock with feed additives. Research into alternative options for increasing the antimicrobial efficacy in animal production has become a priority due to the increasing antimicrobial resistance of pathogens isolated from humans and animals and the ban on the use of antibiotics as feed additives [255].

The concept of a "healthy gut" has gained popularity in recent years, as gut dysfunction has been associated with a variety of diseases, both at the local and systemic levels [256]. A disturbance in the immune homeostasis of the gastrointestinal tract (GIT) results in a weakened intestinal barrier, increasing susceptibility to infection with opportunistic pathogens and facilitating the translocation of intestinal bacteria to the basal part of the mucosa, which can lead to systemic inflammation [257,258]. It is well established that the modulation of gut microbiota benefits gut function. This includes improvements to the integrity of the gut barrier, a reduction in bacterial components entering the circulation and stimulation of the immune system to adapt [259]. In line with this, modulation of gut microbiota by plant polyphenols has been reported to be associated with reduced levels of systemic inflammatory markers, such as C-reactive protein, as well as reduced tissue inflammatory gene expression [152,154,156].

There is an increasing body of evidence that indicates the potential importance of gut microbiota as an underlying factor in the observed benefits of polyphenols on health [260]. The intestinal microbiota has been demonstrated to metabolise dietary polyphenols into bioactive compounds with diverse physiological significance. Furthermore, it has been shown to influence the gut bacterial population's composition and activity [261]. It can be observed that dietary phenolic compounds are frequently modified by gut microbiota and that gut microbial populations are influenced by dietary polyphenols in a bidirectional phenol–microbiota interaction. It is evident that polyphenols and their metabolites exert a profound influence on gut ecology. This is because a considerable proportion of these compounds are not fully absorbed, but are metabolised in the liver, excreted in the bile as glucuronides and accumulate in the ileum and colonic lumen [262]. The presence of significant quantities of unabsorbed phenols in feed can have a substantial impact on the gut environment, exerting a regulatory influence on the growth of specific components of gut microbiota.

The precise mechanism by which polyphenol-rich foods affect the composition of the gut microbiota remains unclear. This is an active area of research. A considerable field of study has been devoted to the metabolism of PAC molecules by the gut microbiota, with a particular focus on the colon [263]. It has been demonstrated, at least in vitro, that the bacterial metabolism of PAC is influenced by the degree of polymerisation (DP). It has been demonstrated that polymers are more resistant to degradation in comparison to catechin monomers [264]. The stimulation of gut microbiota-mediated polyphenol anthocyanin digestion in the in vitro models resulted in the active depolymerisation of polyphenols, followed by the appearance of phenolic metabolites. These metabolites exhibited similarities to those observed in the in vivo systemic circulation of animals fed with polyphenols [264]. It can be concluded that polyphenols may act as a direct prebiotic substrate in a manner similar to dietary fibres. Furthermore, the direct antibacterial effects of polyphenols have been extensively investigated. It is established that PAC induces the growth inhibition of certain bacteria, either by inhibiting enzymes, depriving them of growth substrates or exerting a direct effect on bacteria cell membranes [265]. The in vitro studies indicate that polyphenols have the potential to interact directly with host intestinal cells, thereby stimulating the production of mucins and other proteins that may serve as a selective nutrient source for bacteria such as *Akkermansia* [225]. *Akkermansia muciniphila* abundance has been observed to increase significantly in several species, including mice, pigs and humans, in response to diets rich in prebiotics and probiotics [266–268]. This bacterium has been identified as a potential biomarker for gut health, given its association with mucosal barrier integrity and mucin production. Furthermore, *Akkermansia* is capable of producing metabolites that directly suppress inflammatory responses in the intestinal epithelium. This suggests that its enhanced response to certain dietary components may contribute to its purported health benefits [269]. Other metabolites with documented anti-inflammatory properties, such as short-chain fatty acids (particularly propionate), have also been observed to be elevated in the digesta of animals fed diets rich in polyphenols [267,270]. It is evident that bacterial growth may be indirectly influenced by the effects of PAC on host cells. It is evident that the consumption of PAC alters the composition of the gut microbiota and produces soluble metabolites with anti-inflammatory or immunomodulatory activity. This prebiotic capacity of PAC may be an important mechanism for the observed health benefits in various disease models.

The term "prebiotics" has been defined as non-viable dietary components that confer health benefits on the host by modifying the microbiota [271]. Polyphenols have the potential to fill the definition described above due to their capacity to combat pathogens. In contrast, beneficial bacteria remain unharmed or are even stimulated by polyphenols. In their study, Han et al. [249] demonstrated that GSP increased the abundance and diversity of bacteria in both the ileum and colon. This indicates that GSP has a considerable positive impact on the biodiversity of the gut ecosystem. The principal mechanism by which phenolic compounds exert their effect is their lipophilic nature, which permits their accumulation in the lipid layer of the bacterial membrane and mitochondria. This can have an impact on the normal function of these membranes and cellular organelles [272]. Furthermore, phenolic compounds have the capacity to enhance the permeability of the inner bacterial membrane, thereby reducing ATP production and inhibiting DNA gyrase. These processes are essential for the production of DNA and RNA in bacteria. Phenols can disrupt cellular homeostasis and induce bacterial cell death by denaturing proteins, which results in ion loss [273,274]. Another noteworthy aspect is the active role of phenolic compounds in combating microorganisms, which is attributed to their structural characteristics. The hydroxyl (-OH) groups present in phenolic compounds exhibit bactericidal activities, which facilitate the killing of bacteria and the inhibition of their growth [275].

The gut microbiota is frequently conceptualised as a 'metabolic organ' with the capacity to influence nutrient absorption and interact with the immune system [276–278]. A healthy microbiota composition can act as a physical barrier against infection. Conversely, a perturbed microbiota balance can increase susceptibility to pathogens, contributing to the development of a range of diseases, including obesity, inflammatory bowel disease and cancer [279–282]. Consequently, an understanding of the interactions between polyphenols and the gut microbiota is crucial for elucidating their impact on gut health. It is important to note that the bioavailability and bioactivity of polyphenols are significantly influenced by bacterial metabolism. Polyphenols can also influence the composition of gut bacteria, with one of the most studied aspects being their ability to inhibit the growth of certain bacteria. The antimicrobial activity of foods rich in polyphenols has been the subject of extensive investigation, with studies ranging from simple in vitro experiments to complex in vivo studies. For instance, in vitro studies have demonstrated that certain pathogenic bacteria, including *Escherichia coli*, *Staphylococcus aureus*, and *Candida albicans*, are susceptible to phenolic acids [283].

It can be demonstrated that the inhibitory effect of polyphenolic compounds on bacteria can be attributed to a number of complex mechanisms, the most significant of which are their ability to adhere to the cell membranes of bacteria, interact with the enzymes of bacteria and bind to the metal ions present in the environment [68,265,284,285]. For example, the extract of grapeseed, derived from the 'Bangalore Blue Grapes' variety of *Vitis vinifera*, was found to possess potent antibacterial activity against Gram-positive bacteria [286]. A series of experiments conducted with extracts of juice, skin and seeds from 'Ribier' black table grapes indicated that they possess significant inhibitory effects against the proliferation of *Listeria monocytogenes* [287]. A number of studies have demonstrated that red grape pigments, including anthocyanin pigments, grape juice and grape skin extract, display pH-dependent anti-Listeria activity, whereas the seed extract exhibits pH-independent anti-Listeria activity [287]. Extracts derived from red and white grape berries, seeds, pomace and stems also demonstrated significant antimicrobial activity against *L. monocytogenes*. Catechin, epicatechin and epicatechin gallate were identified as the major active compounds [288]. A substantial body of in vitro research [119,289–292] has demonstrated that flavonoids present in grape by-products possess the capacity to inhibit the growth of a range of organisms, including *Staphylococcus aureus*, *Escherichia coli*, *Candida albicans* and *Campylobacter* spp. Additionally, certain polyphenolic compounds, including resveratrol, hydroxytyrosol, quercetin and phenolic acids, have been demonstrated to possess antimicrobial activity against intestinal pathogens such as *Salmonella* spp. and *Helicobacter pylori* [293,294]. In addition, grape polyphenols have been demonstrated to inhibit the growth of a range of pathogens, with polymeric flavonoids (procyanidins) exhibiting greater efficacy than their monomeric counterparts [74].

The existing literature on the interaction of polyphenolic compounds with the gut microbiota in pig nutrition is limited. Nevertheless, the in vivo studies have demonstrated that resveratrol has the potential to serve as an alternative to antibiotics in counteracting the adverse effects of weaning stress on growth performance, immunity and the microbial environment in piglets exposed to *E. coli* and *Salmonella* spp. challenges [251]. The addition of grapeseed extract (GSE) (10 g/kg) to the diet of weaned pigs also resulted in a reduction in Escherichia coli-induced diarrhoea, as reported in reference [252].

A study by Fiesel et al. [235] observed that the consumption of grapeseed and grapeseed meal extract resulted in alterations in the microbial composition, which led to a reduction in the number of *Streptococcus* spp. and *Clostridium* spp. bacteria present in the faecal microbiota. Nevertheless, no differences in the number of microorganisms in the faeces and caecum were observed in weaned pigs fed polyphenol-containing extracts (including GSE), according to a separate study by Zhang et al. [295]. Furthermore, administration of grapeseed proanthocyanidins in pigs has been found to reduce *Campylobacter jejuni* infection. The mechanisms underlying the effects of grapeseed proanthocyanidins in pigs appear to involve improvements in the mucosal barrier function, which may result from reduced oxidative damage and, thus, less disruption to the epithelial tight junctions [250]. Grapeseed extracts have demonstrated the capacity to regulate the composition of the intestinal microbiota. In both in vivo [24,296] and in vitro studies, they have been shown to inhibit the growth of *Clostridium* spp. species. However, in a recent study [253], it was demonstrated that grapeseed administration resulted in a significant decrease in bacteria, such as *Lactobacillus*, *Lachnospiraceae*, *Bacteroidales* and *Campylobacter*, while having a positive effect on other species, such as *Megasphaera*, *Clostridiales*, *Anaerovibrio* and *Prevotella*. The higher relative abundance of *Megasphaera*, *Prevotella* and bacteria of the order *Clostridiales* observed in pigs fed the high grapeseed diet may have beneficial effects on the host due to their involvement in carbohydrate metabolism and synthesis of short-chain fatty acids [297]. *Prevotella*, *Megasphaera* and *Anaerovibrio* are strictly anaerobic bacteria that have been identified as the dominant species in the large intestine of pigs and are also abundant in the ileum [298–300]. These bacteria have been identified as playing a pivotal role in the breakdown of complex carbohydrates in the lower gastrointestinal tract [301]. It is possible that the growth of these bacteria in the colon of piglets fed a grapeseed-rich diet is a response to the increased fibre content of this diet.

The in vivo studies utilising grape borage have demonstrated an increase in the abundance of *Clostridiales*, which is consistent with the aforementioned results. An ecological change in the microbiome was observed in sows fed a diet containing grapeseed extract (1%) [254], characterised by an increase in *Lachnospiraceae*, *Clostridiales*, *Lactobacillus* and *Ruminococcacceae*. The effects of tea polyphenols were also observed in pigs and calves. There was a significant increase in *Lactobacilli*, a decrease in the total number of bacteria and *Bacteroidaceae*, and a downward trend in the number of *Clostridium perfringens* [302,303]. The class *Firmicutes* includes the order *Clostridia*, which comprises obligate anaerobic bacteria such as *Clostridium* [300]. The effects of *Clostridia* on animal health are diverse, and they constitute a normal component of the intestinal flora. It has been demonstrated that certain members of the *Clostridiales* family, such as *Clostridium difficile*, are linked to the onset of adverse effects such as inflammatory bowel disease [304]. Others, such as *C. leptum* and *C. coccoides*, are significant components of the gut microbiome and contribute to healthy ageing [305]. The *Ruminococcaceae* are typically associated with an enhanced feed conversion in piglets as a consequence of their capacity to degrade cellulose [306]. This process contributes to the conversion of complex polysaccharides, otherwise resistant to digestive enzymes, into more readily metabolizable forms of energy [307].

Given the beneficial roles of phenolic compounds, it can be posited that they may serve as effective natural feed additives. Furthermore, agricultural by-products represent an excellent source of phenolic compounds and antioxidants that can be employed as functional ingredients in animal feed [308]. The hypothesis put forth is that the positive effects on antioxidant status and a reduction in ROS levels observed in some porcine studies [309–311] cannot be attributed to the direct antioxidant effects of the diet. It is possible that these effects are secondary to improved gut health, thereby reducing the translocation of pro-inflammatory and prooxidative stimuli. Nevertheless, further studies are required to provide more evidence of the potential beneficial effects of plant polyphenols in pigs. The beneficial roles of phenolic substances can be exploited as effective natural feed additives (Figure 3). Furthermore, agricultural by-products represent an excellent source of phenolic compounds and antioxidants, which can be employed as functional feed ingredients [308].

## 5. The Effects of Grape Polyphenols on the Production of Pigs

The incorporation of polyphenols into the diets of farm animals has the potential to enhance production performance and the oxidative stability of their feeds [312]. The literature on the influence of grape by-products on animal growth performance is inconclusive, with studies reporting either enhanced growth, depressed growth or no effect [248]. This apparent inconsistency appears to depend on the amount of polyphenols in grape products, which may interact with the digestive enzymes and gut proteins to affect nutrient digestibility and, thus, animal performance [248].

In growing pigs, the performance of the animals, as indicated by the body weight and gain/feed ratio, was enhanced by the administration of grapeseed and grape tannin meal containing 8.5% polyphenols at a dosage of 1%. This improvement was not due to an enhanced digestibility of nutrients but rather to alterations in the composition of the microbial population and the downregulation of several pro-inflammatory genes in different regions of the gut [234,235]. Similarly, another study [248] conducted over 15 weeks in pigs demonstrated that a diet supplemented with 30 g/kg of fermented grape seeds improved their nutrient digestibility without affecting their growth. Similar outcomes were observed by Taranu et al. [242], who noted that pigs fed 5% grapeseed cake exhibited no change in performance. Consequently, the total polyphenol content of the diet was probably insufficient to affect nutrient digestibility and, as a consequence, animal performance.

It is notable that the incorporation of grapeseed cake into the diet of GS pigs led to a notable reduction in plasma cholesterol levels. This finding is consistent with the previous observations from animal and human studies that have identified similar beneficial effects of grape by-products or whole grapes [313,314]. Polyphenols, including catechin, gallic acid, epicatechin and other active compounds in grape seeds, have been linked to reduced cholesterol absorption by inhibiting pancreatic enzyme cholesterol esterase, resulting in reduced cholesterol solubility in the mycelia and thus blood [315–317]. Furthermore, it has been demonstrated that certain polyphenols, including resveratrol and epigallocatechin, are capable of binding directly to miRNAs, namely miR-33a and miR-122, which are key regulators of the genes involved in hepatic lipid metabolism. This binding may result in reduced cholesterol concentrations [318]. Furthermore, another study reported that diets containing a mixture of plant extracts, including grapeseed, had no effect on the blood parameters in piglets. It has been demonstrated that a dietary intake of antioxidants can influence the humoral immune response [295]. However, the effect may vary depending on the source, the ratio and the duration of administration [15,241,319].

In the field of animal production, oxidative stress has the potential to exert considerable influence. For example, it can result in a reduction in body weight due to the disruption of optimal metabolic processes. Additionally, oxidative stress can impact meat quality by increasing plasma corticosterone accumulation, which is associated with a reduction in the pigmentation of breast meat in broilers [320]. Additionally, it can cause biological damage to the DNA, proteins and lipids, which can result in adverse health effects that may impact the productive capacity of farm animals [321]. The replacement of 50% of the vitamin E in the diet with polyphenols had no effect on the growth performance. However, it may improve the antioxidant status of sows or their offspring [322]. In piglets, the addition of polyphenols, a mixture of apples, grape seeds, green tea leaves and olive leaves, to the diet has been shown to reduce plasma MDA levels [246]. A recent study [38] demonstrated that diets supplemented with grapeseed procyanidins, a type of phenolic compound, exhibited enhanced resilience to weaning stress. This was evidenced by an increase in the expression of antioxidant-related genes, including GSH-Px, SOD and CAT, in the liver. The levels of malondialdehyde (MDA) in the serum, liver and muscle tissue were found to be reduced [38]. The bioactive compounds were also responsible for an increase in the ratio of intestinal villus height to crypt depth when grape pomace was administered to piglets at 5% [39], 1% GSGME [250] and grapeseed proanthocyanidins at 250 mg/kg [249]. These data suggest that polyphenols may improve intestinal mucosal barrier function, which may explain the increase in average daily gain (ADG) and decrease in the feed conversion ratio (FCR) [38] and the improved effects on growth performance in piglets fed grape by-products [15]. The impact of grape by-products on pig growth performance depends mainly on the age of the animal and the dosage of the by-products. In principle, the effect of low levels of grape by-products on growth performance was reduced, but this was dependent on the animal's age and the doses applied. Despite the variability of the results between studies, the increase in ADG demonstrated by Kafantaris [250] (9% grape pomace) deserves to be highlighted as a promising effect for pig production.

## 6. Conclusions

Grape by-products have many industrial applications, which include animal feed. They are a valuable feed due to their richness in polyphenols, which can modulate intestinal microbiota and morphology and boost anti-inflammatory and antioxidant capacities, thus maintaining intestinal health and production in pigs. It is difficult to establish the ideal dose of polyphenols in animal diets due to the variable composition of the phenolic components in these by-products. The best results, both in terms of health and high pig production, were obtained when 9% grape marc was added to the diet. For pig diets, future studies could focus on the optimisation of the dose and digestibility of grape by-products.

**Author Contributions:** Conceptualization, A.C.P., L.H., G.S. and C.S.; methodology, A.C.P., L.H., G.S. and C.S.; validation, G.S. and C.S.; formal analysis, A.C.P., L.H., G.S. and C.S.; investigation, A.C.P., L.H., G.S. and C.S.; writing—original draft preparation, A.C.P., L.H., G.S. and C.S.; writing—review and editing, C.S. and G.S; supervision, C.S and G.S.; project administration, C.S.; funding acquisition, C.S. and G.S. All authors have read and agreed to the published version of the manuscript.

**Funding:** This research received no external funding.

**Institutional Review Board Statement:** Not applicable.

**Data Availability Statement:** Not applicable.

**Conflicts of Interest:** The authors declare no conflicts of interest.

## References

1. Henchion, M.; Moloney, A.P.; Hyland, J.; Zimmermann, J.; McCarthy, S. Review: Trends for Meat, Milk and Egg Consumption for the next Decades and the Role Played by Livestock Systems in the Global Production of Proteins. *Animal* **2021**, *15*, 100287. [CrossRef]
2. Cheng, Y.; Song, M.; Zhu, Q.; Azad, M.A.K.; Gao, Q.; Kong, X. Dietary Betaine Addition Alters Carcass Traits, Meat Quality, and Nitrogen Metabolism of Bama Mini-Pigs. *Front. Nutr.* **2021**, *8*, 728477. [CrossRef]
3. Matarneh, S.K.; Silva, S.L.; Gerrard, D.E. New Insights in Muscle Biology That Alter Meat Quality. *Annu. Rev. Anim. Biosci.* **2021**, *9*, 355–377. [CrossRef] [PubMed]
4. Erinle, T.J.; Adewole, D.I. Fruit Pomaces—Their Nutrient and Bioactive Components, Effects on Growth and Health of Poultry Species, and Possible Optimization Techniques. *Anim. Nutr.* **2022**, *9*, 357–377. [CrossRef] [PubMed]
5. Reckmann, K.; Blank, R.; Traulsen, I.; Krieter, J. Comparative Life Cycle Assessment (LCA) of Pork Using Different Protein Sources in Pig Feed. *Arch. Anim. Breed.* **2016**, *59*, 27–36. [CrossRef]
6. Altmann, B.; Neumann, C.; Velten, S.; Liebert, F.; Mörlein, D. Meat Quality Derived from High Inclusion of a Micro-Alga or Insect Meal as an Alternative Protein Source in Poultry Diets: A Pilot Study. *Foods* **2018**, *7*, 34. [CrossRef]
7. Spigno, G.; Marinoni, L.; Garrido, G.D. State of the Art in Grape Processing By-Products. In *Handbook of Grape Processing By-Products*; Galanakis, C.M., Ed.; Academic Press: New York, NY, USA, 2017; pp. 1–27. ISBN 978-0-12-809870-7. [CrossRef]

8. Shi, J.; Yu, J.; Pohorly, J.E.; Kakuda, Y. Polyphenolics in grape seeds—Biochemistry and functionality. *J. Med. Food* **2003**, *6*, 291–299. [CrossRef]
9. Terra, X.; Pallarés, V.; Ardèvol, A.; Bladé, C.; Fernández-Larrea, J.; Pujadas, G.; Salvadó, J.; Arola, L.; Blay, M. Modulatory effect of grape-seed procyanidins on local and systemic inflammation in diet-induced obesity rats. *J. Nutr. Biochem.* **2011**, *22*, 380–387. [CrossRef]
10. Park, S.-S.; Kim, J.-M.; Kim, E.-J.; Kim, H.-S.; An, B.-K.; Kang, C.-W. Effects of Dietary Turmeric Powder on Laying Performance and Egg Qualities in Laying Hens. *Korean J. Poult. Sci.* **2012**, *39*, 27–32. [CrossRef]
11. Chu, H.; Tang, Q.; Huang, H.; Hao, W.; Wei, X. Grape-seed proanthocyanidins inhibit the lipopolysaccharide-induced inflammatory mediator expression in RAW264. 7 macrophages by suppressing MAPK and NF-κb signal pathways. *Environ. Toxicol. Pharmacol.* **2016**, *41*, 159–166.
12. Dinh, J.; Angeloni, J.T.; Pederson, D.B.; Wang, X.; Cao, M.; Dong, Y. Cranberry extract standardized for proanthocyanidins promotes the immune response of Caenorhabditis elegans to *Vibrio cholerae* through the p38 MAPK pathway and HSF-1. *PLoS ONE* **2014**, *9*, e103290. [CrossRef] [PubMed]
13. Lau, D.W.; King, A.J. Pre-and post-mortem use of grape seed extract in dark poultry meat to inhibit development of thiobarbituric acid reactive substances. *J. Agric. Food Chem.* **2003**, *51*, 1602–1607. [CrossRef] [PubMed]
14. Rababah, T.; Ereifej, K.; Al-Mahasneh, M.; Al-Rababah, M. Effect of plant extracts on physicochemical properties of chicken breast meat cooked using conventional electric oven or microwave. *Poult. Sci.* **2006**, *85*, 148–154.
15. Hao, R.; Li, Q.; Zhao, J.; Li, H.; Wang, W.; Gao, J. Effects of Grape Seed Procyanidins on Growth Performance, Immune Function and Antioxidant Capacity in Weaned Piglets. *Livest. Sci.* **2015**, *178*, 237–242. [CrossRef]
16. Ali, K.; Maltese, F.; Choi, Y.H.; Verpoorte, R. Metabolic Constituents of Grapevine and Grape-Derived Products. *Phytochem. Rev.* **2010**, *9*, 357–378. [CrossRef] [PubMed]
17. OIV. The OIV Provides Sound, Timely and Comparable Data and Statistics on the Vine and Wine Sector to Inform Policy Makers, Analysts, and Members of Civil Society around the World. Available online: https://www.oiv.int/what-we-do/statistics (accessed on 28 May 2024).
18. Vivier, M.; Pretorius, I.S. Genetic Improvement of Grapevine: Tailoring Grape Varieties for the Third Millennium—A Review. *S. Afr. J. Enol. Vitic.* **2000**, *21*, 5–26. [CrossRef]
19. Cotoras, M.; Vivanco, H.; Melo, R.; Aguirre, M.; Silva, E.; Mendoza, L. In Vitro and In Vivo Evaluation of the Antioxidant and Prooxidant Activity of Phenolic Compounds Obtained from Grape (*Vitis vinifera*) Pomace. *Molecules* **2014**, *19*, 21154–21167. [CrossRef]
20. Ky, I.; Lorrain, B.; Kolbas, N.; Crozier, A.; Teissedre, P.-L. Wine By-Products: Phenolic Characterization and Antioxidant Activity Evaluation of Grapes and Grape Pomaces from Six Different French Grape Varieties. *Molecules* **2014**, *19*, 482–506. [CrossRef]
21. Bail, S.; Stuebiger, G.; Krist, S.; Unterweger, H.; Buchbauer, G. Characterization of Various Grape Seed Oils by Volatile Compounds, Triacylglycerol Composition, Total Phenols and Antioxidant Capacity. *Food Chem.* **2008**, *108*, 1122–1132. [CrossRef]
22. Beres, C.; Costa, G.N.; Cabezudo, I.; da Silva-James, N.K.; Teles, A.S.; Cruz, A.P.; Mellinger-Silva, C.; Tonon, R.V.; Cabral, L.M.; Freitas, S.P. Towards Integral Utilization of Grape Pomace from Winemaking Process: A Review. *Waste Manag.* **2017**, *68*, 581–594. [CrossRef]
23. Yu, J.; Ahmedna, M. Functional Components of Grape Pomace: Their Composition, Biological Properties and Potential Applications. *Int. J. Food Sci. Tech.* **2013**, *48*, 221–237. [CrossRef]
24. Viveros, A.; Chamorro, S.; Pizarro, M.; Arija, I.; Centeno, C.; Brenes, A. Effects of Dietary Polyphenol-Rich Grape Products on Intestinal Microflora and Gut Morphology in Broiler Chicks. *Poult. Sci.* **2011**, *90*, 566–578. [CrossRef] [PubMed]
25. Llobera, A.; Canellas, J. Dietary fibres content and antioxidant activity of Manto Negro red grape (*Vitis vinifera*): Pomace and stem. *Food Chem.* **2007**, *101*, 659–666. [CrossRef]
26. Vlaicu, P.A.; Panaite, T.D.; Cornescu, M.G.; Ropota, M.; Olteanu, M.; Drăgotoiu, D. The influence of by-products on the production parameters and nutrient digestibility in fattening pig diet (60–100 kg). *AgroLife Sci. J.* **2019**, *8*, 261–269.
27. Erinle, T.J.; Oladokun, S.; MacIsaac, J.; Rathgeber, B.; Adewole, D. Dietary grape pomace–effects on growth performance, intestinal health, blood parameters, and breast muscle myopathies of broiler chickens. *Poult. Sci.* **2022**, *101*, 101519.
28. Skuras, D.; Psaltopoulos, D. A Broad Overview of the Main Problems Derived from Climate Change That Will Affect Agricultural Production in the Mediterranean Area. In *Building Resilience for Adaptation to Climate Change in the Agriculture Sector*; FAO: Rome, Italy; Volume 23, pp. 217–260. Available online: https://api.semanticscholar.org/CorpusID:55083905 (accessed on 10 May 2024).
29. Lu, Y.; Foo, L.Y. The Polyphenol Constituents of Grape Pomace. *Food Chem.* **1999**, *65*, 1–8. [CrossRef]
30. Brenes, A.; Viveros, A.; Chamorro, S.; Arija, I. Use of Polyphenol-Rich Grape by-products in Monogastric Nutrition. A Review. *Anim. Feed Sci. Tech.* **2016**, *211*, 1–17. [CrossRef]
31. Gungor, E.; Altop, A.; Erener, G. Effect of Raw and Fermented Grape Seed on Growth Performance, Antioxidant Capacity, and Cecal Microflora in Broiler Chickens. *Animal* **2021**, *15*, 100194. [CrossRef]
32. Costa, M.M.; Alfaia, C.M.; Lopes, P.A.; Pestana, J.M.; Prates, J.A. Grape By-Products as Feedstuff for Pig and Poultry Production. *Animals* **2022**, *12*, 2239. [CrossRef]
33. Yilmaz, Y.; Toledo, R.T. Major Flavonoids in Grape Seeds and Skins: Antioxidant Capacity of Catechin, Epicatechin, and Gallic Acid. *J. Agric. Food Chem.* **2004**, *52*, 255–260. [CrossRef]

34. Felhi, S.; Baccouch, N.; Ben Salah, H.; Smaoui, S.; Allouche, N.; Gharsallah, N.; Kadri, A. Nutritional Constituents, Phytochemical Profiles, in Vitro Antioxidant and Antimicrobial Properties, and Gas Chromatography–Mass Spectrometry Analysis of Various Solvent Extracts from Grape Seeds (*Vitis vinifera* L.). *Food Sci. Biotech.* **2016**, *25*, 1537–1544. [CrossRef] [PubMed]
35. Igartuburu, J.M.; Pando, E.; Rodríguez-Luis, F.; Gil-Serrano, A. Structure of a Hemicellulose A Fraction in Dietary Fiber from the Seed of Grape Variety Palomino (*Vitis vinifera* Cv. Palomino). *J. Nat. Prod.* **1998**, *61*, 876–880. [CrossRef] [PubMed]
36. Nardoia, M.; Romero, C.; Brenes, A.; Arija, I.; Viveros, A.; Ruiz-Capillas, C.; Chamorro, S. Addition of Fermented and Unfermented Grape Skin in Broilers' Diets: Effect on Digestion, Growth Performance, Intestinal Microbiota and Oxidative Stability of Meat. *Animal* **2020**, *14*, 1371–1381. [CrossRef] [PubMed]
37. Prozil, S.O.; Costa, E.V.; Evtuguin, D.V.; Lopes, L.P.C.; Domingues, M.R.M. Structural Characterization of Polysaccharides Isolated from Grape Stalks of *Vitis vinifera* L. *Carbohydr. Res.* **2012**, *356*, 252–259. [CrossRef] [PubMed]
38. Fang, L.; Li, M.; Zhao, L.; Han, S.; Li, Y.; Xiong, B.; Jiang, L. Dietary Grape Seed Procyanidins Suppressed Weaning Stress by Improving Antioxidant Enzyme Activity and MRNA Expression in Weanling Piglets. *J. Anim. Physiol. Anim. Nutr.* **2020**, *104*, 1178–1185. [CrossRef] [PubMed]
39. Wang, R.; Yu, H.; Fang, H.; Jin, Y.; Zhao, Y.; Shen, J.; Zhou, C.; Li, R.; Wang, J.; Fu, Y. Effects of Dietary Grape Pomace on the Intestinal Microbiota and Growth Performance of Weaned Piglets. *Arch. Anim. Nutr.* **2020**, *74*, 296–308. [CrossRef] [PubMed]
40. Alonso, Á.M.; Guillén, D.A.; Barroso, C.G.; Puertas, B.; García, A. Determination of Antioxidant Activity of Wine Byproducts and Its Correlation with Polyphenolic Content. *J. Agric. Food Chem.* **2002**, *50*, 5832–5836. [CrossRef] [PubMed]
41. Torres, J.L.; Varela, B.; García, M.T.; Carilla, J.; Matito, C.; Centelles, J.J.; Cascante, M.; Sort, X.; Bobet, R. Valorization of Grape (*Vitis vinifera*) Byproducts. Antioxidant and Biological Properties of Polyphenolic Fractions Differing in Procyanidin Composition and Flavonol Content. *J. Agric. Food Chem.* **2002**, *50*, 7548–7555. [CrossRef] [PubMed]
42. Chedea, V.S.; Drăgulinescu, A.-M.; Tomoiagă, L.L.; Bălăceanu, C.; Iliescu, M.L. Climate Change and Internet of Things Technologies—Sustainable Premises of Extending the Culture of the Amurg Cultivar in Transylvania—A Use Case for Târnave Vineyard. *Sustainability* **2021**, *13*, 8170. [CrossRef]
43. Iuga, M.; Mironeasa, S. Potential of Grape Byproducts as Functional Ingredients in Baked Goods and Pasta. *Compr. Rev. Food Sci. Food Saf.* **2020**, *19*, 2473–2505. [CrossRef]
44. de la Cerda-Carrasco, A.; López-Solís, R.; Nuñez-Kalasic, H.; Peña-Neira, Á.; Obreque-Slier, E. Phenolic Composition and Antioxidant Capacity of Pomaces from Four Grape Varieties (*Vitis vinifera* L.). *J. Sci. Food Agric.* **2015**, *95*, 1521–1527. [CrossRef]
45. Muncaciu, M.L.; Marin, F.Z.; Nastasia, P.; Babeș, A.C. Comparative Polyphenolic Content of Grape Pomace Flours from 'Fetească Neagră'and 'Italian Riesling'Cultivars. *Not. Bot. Horti Agrobo.* **2017**, *45*, 532–539. [CrossRef]
46. Vermerris, W.; Nicholson, R. The Role of Phenols in Plant Defense. In *Phenolic Compound Biochemistry*; Springer: Dordrecht, Germany, 2006; pp. 211–234. ISBN 978-1-4020-5164-7. [CrossRef]
47. Ozcan, T.; Akpinar-Bayizit, A.; Yilmaz-Ersan, L.; Delikanli, B. Phenolics in Human Health. *Int. J. Chem. Eng. Appl.* **2014**, *5*, 393–396. [CrossRef]
48. Hrazdina, G. Biosynthesis of Flavonoids. In *Plant Polyphenols: Synthesis, Properties, Significance*; Springer: Boston, MA, USA, 1992; pp. 61–72.
49. Winkel-Shirley, B. It Takes a Garden. How Work on Diverse Plant Species Has Contributed to an Understanding of Flavonoid Metabolism. *Plant Physiol.* **2001**, *127*, 1399–1404. [CrossRef] [PubMed]
50. Dixon, R.A.; Xie, D.; Sharma, S.B. Proanthocyanidins—A Final Frontier in Flavonoid Research? *New Phytol.* **2005**, *165*, 9–28. [CrossRef]
51. Bravo, L. Polyphenols: Chemistry, Dietary Sources, Metabolism, and Nutritional Significance. *Nutr. Rev.* **1998**, *56*, 317–333. [CrossRef]
52. Seigler, D.S. *Plant Secondary Metabolism*; Springer Science & Business Media: New York, NY, USA, 1998; ISBN 978-1-4613-7228-8. [CrossRef]
53. Scalbert, A.; Williamson, G. Dietary Intake and Bioavailability of Polyphenols. *J. Nutr.* **2000**, *130*, 2073S–2085S. [CrossRef]
54. Manach, C.; Scalbert, A.; Morand, C.; Rémésy, C.; Jiménez, L. Polyphenols: Food Sources and Bioavailability. *Am. J. Clin. Nutr.* **2004**, *79*, 727–747. [CrossRef] [PubMed]
55. Dai, J.; Mumper, R.J. Plant Phenolics: Extraction, Analysis and Their Antioxidant and Anticancer Properties. *Molecules* **2010**, *15*, 7313–7352. [CrossRef]
56. Pietta, P.; Minoggio, M.; Bramati, L. Plant Polyphenols: Structure, Occurrence and Bioactivity. *Stud. Nat. Prod. Chem.* **2003**, *28*, 257–312.
57. Durazzo, A.; Lucarini, M.; Souto, E.B.; Cicala, C.; Caiazzo, E.; Izzo, A.A.; Novellino, E.; Santini, A. Polyphenols: A Concise Overview on the Chemistry, Occurrence, and Human Health. *Phytother. Res.* **2019**, *33*, 2221–2243. [CrossRef] [PubMed]
58. Hichri, I.; Barrieu, F.; Bogs, J.; Kappel, C.; Delrot, S.; Lauvergeat, V. Recent Advances in the Transcriptional Regulation of the Flavonoid Biosynthetic Pathway. *J. Exp. Bot.* **2011**, *62*, 2465–2483. [CrossRef] [PubMed]
59. Scalbert, A.; Morand, C.; Manach, C.; Rémésy, C. Absorption and Metabolism of Polyphenols in the Gut and Impact on Health. *Biomed. Pharmacother.* **2002**, *56*, 276–282. [CrossRef] [PubMed]
60. Lim, C.-G.; AG Koffas, M. Bioavailability and Recent Advances in the Bioactivity of Flavonoid and Stilbene Compounds. *Curr. Org. Chem.* **2010**, *14*, 1727–1751. [CrossRef]

61. Motohashi, N. *Bioactive Heterocycles VI: Flavonoids and Anthocyanins in Plants, and Latest Bioactive Heterocycles I*; Springer: Leipzig, Germany, 2008; Volume 15, ISBN 978-3-540-79217-8. [CrossRef]
62. Aron, P.M.; Kennedy, J.A. Flavan-3-ols: Nature, Occurrence and Biological Activity. *Mol. Nutr. Food Res.* **2008**, *52*, 79–104. [CrossRef] [PubMed]
63. Singla, R.K.; Dubey, A.K.; Garg, A.; Sharma, R.K.; Fiorino, M.; Ameen, S.M.; Haddad, M.A.; Al-Hiary, M. Natural Polyphenols: Chemical Classification, Definition of Classes, Subcategories, and Structures. *J. AOAC Int.* **2019**, *102*, 1397–1400. [CrossRef] [PubMed]
64. Nassiri-Asl, M.; Hosseinzadeh, H. Review of the Pharmacological Effects of *Vitis vinifera* (Grape) and its Bioactive Compounds. *Phytother. Res.* **2009**, *23*, 1197–1204. [CrossRef] [PubMed]
65. Tsao, R. Chemistry and Biochemistry of Dietary Polyphenols. *Nutrients* **2010**, *2*, 1231–1246. [CrossRef]
66. Conde, C.; Silva, P.; Fontes, N.; Dias, A.C.P.; Tavares, R.M.; Sousa, M.J.; Agasse, A.; Delrot, S.; Gerós, H. Biochemical Changes throughout Grape Berry Development and Fruit and Wine Quality. *Food* **2007**, *1*, 1–22.
67. Cho, M.J.; Howard, L.R.; Prior, R.L.; Clark, J.R. Flavonoid Glycosides and Antioxidant Capacity of Various Blackberry, Blueberry and Red Grape Genotypes Determined by High-performance Liquid Chromatography/Mass Spectrometry. *J. Sci. Food Agric.* **2004**, *84*, 1771–1782. [CrossRef]
68. Xia, E.-Q.; Deng, G.-F.; Guo, Y.-J.; Li, H.-B. Biological Activities of Polyphenols from Grapes. *Int. J. Mol. Sci.* **2010**, *11*, 622–646. [CrossRef] [PubMed]
69. Narayana, K.R.; Reddy, M.S.; Chaluvadi, M.; Krishna, D. Bioflavonoids Classification, Pharmacological, Biochemical Effects and Therapeutic Potential. *Indian J. Pharmacol.* **2001**, *33*, 2–16.
70. Heim, K.E.; Tagliaferro, A.R.; Bobilya, D.J. Flavonoid Antioxidants: Chemistry, Metabolism and Structure-Activity Relationships. *J. Nutr. Biochem.* **2002**, *13*, 572–584. [CrossRef] [PubMed]
71. Han, X.; Shen, T.; Lou, H. Dietary Polyphenols and Their Biological Significance. *Int. J. Mol. Sci.* **2007**, *8*, 950–988. [CrossRef]
72. Alzand, K.I.; Mohamed, M.A. Flavonoids: Chemistry, Biochemistry and Antioxidant Activity. *J. Pharm. Res.* **2012**, *5*, 37.
73. Song, P.; Zhang, R.; Wang, X.; He, P.; Tan, L.; Ma, X. Dietary Grape-Seed Procyanidins Decreased Postweaning Diarrhea by Modulating Intestinal Permeability and Suppressing Oxidative Stress in Rats. *J. Agric. Food Chem.* **2011**, *59*, 6227–6232. [CrossRef]
74. Mayer, R.; Stecher, G.; Wuerzner, R.; Silva, R.C.; Sultana, T.; Trojer, L.; Feuerstein, I.; Krieg, C.; Abel, G.; Popp, M. Proanthocyanidins: Target Compounds as Antibacterial Agents. *J. Agric. Food Chem.* **2008**, *56*, 6959–6966. [CrossRef]
75. Goodrich, K.M.; Fundaro, G.; Griffin, L.E.; Hulver, M.W.; Ponder, M.A.; Neilson, A.P. Chronic Administration of Dietary Grape Seed Extract Increases Colonic Expression of Gut Tight Junction Protein Occludin and Reduces Fecal Calprotectin: A Secondary Analysis of Healthy Wistar Furth Rats. *Nutr. Res.* **2012**, *32*, 787–794. [CrossRef]
76. Wang, H.; Xue, Y.; Zhang, H.; Huang, Y.; Yang, G.; Du, M.; Zhu, M. Dietary Grape Seed Extract Ameliorates Symptoms of Inflammatory Bowel Disease in IL 10-deficient Mice. *Mol. Nutr. Food Res.* **2013**, *57*, 2253–2257. [CrossRef]
77. Bagchi, D.; Bagchi, M.; Stohs, S.J.; Das, D.K.; Ray, S.D.; Kuszynski, C.A.; Joshi, S.S.; Pruess, H.G. Free Radicals and Grape Seed Proanthocyanidin Extract: Importance in Human Health and Disease Prevention. *Toxicology* **2000**, *148*, 187–197. [CrossRef]
78. Yamakoshi, J.; Saito, M.; Kataoka, S.; Tokutake, S. Procyanidin-Rich Extract from Grape Seeds Prevents Cataract Formation in Hereditary Cataractous (ICR/f) Rats. *J. Agric. Food Chem.* **2002**, *50*, 4983–4988. [CrossRef]
79. Sato, M.; Bagchi, D.; Tosaki, A.; Das, D.K. Grape Seed Proanthocyanidin Reduces Cardiomyocyte Apoptosis by Inhibiting Ischemia/Reperfusion-Induced Activation of JNK-1 and C-JUN. *Free Radic. Biol. Med.* **2001**, *31*, 729–737. [CrossRef]
80. Harborne, J.B.; Williams, C.A. Advances in Flavonoid Research since 1992. *Phytochemistry* **2000**, *55*, 481–504. [CrossRef]
81. Crozier, A. Classification and Biosynthesis of Secondary Plant Products: An Overview. In *Plants: Diet and Health. The Report of a British Nutrition Foundation Task Force*; Goldberg, G., Ed.; Willey: London, UK, 2003; pp. 27–48. ISBN 9780632059621. [CrossRef]
82. Zhao, J.; Pang, Y.; Dixon, R.A. The Mysteries of Proanthocyanidin Transport and Polymerization. *Plant Physiol.* **2010**, *153*, 437–443. [CrossRef]
83. Santhakumar, A.B.; Battino, M.; Alvarez-Suarez, J.M. Dietary Polyphenols: Structures, Bioavailability and Protective Effects against Atherosclerosis. *Food Chem. Toxicol.* **2018**, *113*, 49–65. [CrossRef] [PubMed]
84. Panche, A.N.; Diwan, A.D.; Chandra, S.R. Flavonoids: An Overview. *J. Nutr. Sci.* **2016**, *5*, e47. [CrossRef] [PubMed]
85. Verries, C.; Guiraud, J.-L.; Souquet, J.-M.; Vialet, S.; Terrier, N.; Olle, D. Validation of an Extraction Method on Whole Pericarp of Grape Berry (*Vitis vinifera* L. Cv. Shiraz) to Study Biochemical and Molecular Aspects of Flavan-3-Ol Synthesis during Berry Development. *J. Agric. Food Chem.* **2008**, *56*, 5896–5904. [CrossRef]
86. Su, C.T.; Singleton, V. Identification of Three Flavan-3-Ols from Grapes. *Phytochemistry* **1969**, *8*, 1553–1558.
87. Souquet, J.-M.; Labarbe, B.; Le Guernevé, C.; Cheynier, V.; Moutounet, M. Phenolic Composition of Grape Stems. *J. Agric. Food Chem.* **2000**, *48*, 1076–1080. [CrossRef]
88. Cantos, E.; Espin, J.C.; Tomás-Barberán, F.A. Varietal Differences among the Polyphenol Profiles of Seven Table Grape Cultivars Studied by LC–DAD–MS–MS. *J. Agric. Food Chem.* **2002**, *50*, 5691–5696. [CrossRef]
89. Gođevac, D.; Tešević, V.; Veličković, M.; Vujisić, L.; Vajs, V.; Milosavljević, S. Polyphenolic Compounds in Seeds from Some Grape Cultivars Grown in Serbia. *J. Serb. Chem. Soc.* **2010**, *75*, 1641–1652. [CrossRef]
90. Freudenberg, K.; Blümmel, F. Hamameli-tannin. III. 17. Mitteilung Über Gerbstoffe Und Ähnliche Verbindungen. *Liebigs Ann. Chem.* **1924**, *440*, 45–59. [CrossRef]

91. Singleton, V.; Draper, D.E.; Rossi, J.A. Paper Chromatography of Phenolic Compounds from Grapes, Particularly Seeds, and Some Variety-Ripeness Relationships. *Am. J. Enol. Vitic.* **1966**, *17*, 206–217. [CrossRef]
92. Czochanska, Z.; Foo, L.Y.; Newman, R.H.; Porter, L.J. Polymeric Proanthocyanidins. Stereochemistry, Structural Units, and Molecular Weight. *J. Chem. Soc. Perkin Trans. 1* **1980**, *12*, 2278–2286. [CrossRef]
93. Romeyer, F.M.; Macheix, J.-J.; Sapis, J.-C. Changes and Importance of Oligomeric Procyanidins during Maturation of Grape Seeds. *Phytochemistry* **1985**, *25*, 219–221. [CrossRef]
94. Meyer, B.J.; Hernandez, R. Seed Tannin Extraction in Cabernet Sauvignon. *Am. J. Enol. Vitic.* **1970**, *21*, 184–188. [CrossRef]
95. Oszmianski, J.; Romeyer, F.M.; Sapis, J.; Macheix, J. Grape Seed Phenolics: Extraction as Affected by Some Conditions Occurring during Wine Processing. *Am. J. Enol. Vitic.* **1986**, *37*, 7–12. [CrossRef]
96. Khanbabaee, K.; Van Ree, T. Tannins: Classification and Definition. *Nat. Prod. Rep.* **2001**, *18*, 641–649. [PubMed]
97. Tsang, C.; Auger, C.; Mullen, W.; Bornet, A.; Rouanet, J.-M.; Crozier, A.; Teissedre, P.-L. The Absorption, Metabolism and Excretion of Flavan-3-Ols and Procyanidins Following the Ingestion of a Grape Seed Extract by Rats. *Br. J. Nutr.* **2005**, *94*, 170–181. [CrossRef]
98. Bogs, J.; Downey, M.O.; Harvey, J.S.; Ashton, A.R.; Tanner, G.J.; Robinson, S.P. Proanthocyanidin Synthesis and Expression of Genes Encoding Leucoanthocyanidin Reductase and Anthocyanidin Reductase in Developing Grape Berries and Grapevine Leaves. *Plant Physiol.* **2005**, *139*, 652–663. [CrossRef]
99. Foo, L.Y.; Porter, L.J. The Phytochemistry of Proanthocyanidin Polymers. *Phytochemistry* **1980**, *19*, 1747–1754. [CrossRef]
100. Foo, L.Y.; Porter, L.J. The Structure of Tannins of Some Edible Fruits. *J. Sci. Food Agric.* **1981**, *32*, 711–716. [CrossRef]
101. Edwin, H. In Vino Veritas: Oligomeric Procyanidins and the Ageing of Red Wines. *Phytochemistry* **1980**, *19*, 2577–2582. [CrossRef]
102. Lin, L.-Z.; Sun, J.; Chen, P.; Monagas, M.J.; Harnly, J.M. UHPLC-PDA-ESI/HRMS n Profiling Method to Identify and Quantify Oligomeric Proanthocyanidins in Plant Products. *J. Agric. Food Chem.* **2014**, *62*, 9387–9400. [CrossRef] [PubMed]
103. Geny, L.; Saucier, C.; Bracco, S.; Daviaud, F.; Glories, Y. Composition and Cellular Localization of Tannins in Grape Seeds during Maturation. *J. Agric. Food Chem.* **2003**, *51*, 8051–8054. [CrossRef] [PubMed]
104. Constabel, P.; Yoshida, K.; Walker, V. *Diverse Ecological Roles of Plant Tannins: Plant Defense and Beyond. Recent Advances in Polyphenol Research*; Willey: London, UK, 2014; pp. 115–142. [CrossRef]
105. Li, M.; Hagerman, A.E. Interactions between Plasma Proteins and Naturally Occurring Polyphenols. *Curr. Drug Met.* **2013**, *14*, 432–445. [CrossRef] [PubMed]
106. Smeriglio, A.; Barreca, D.; Bellocco, E.; Trombetta, D. Proanthocyanidins and Hydrolysable Tannins: Occurrence, Dietary Intake and Pharmacological Effects. *Br. J. Pharmacol.* **2017**, *174*, 1244–1262. [CrossRef]
107. Prior, R.L.; Gu, L. Occurrence and Biological Significance of Proanthocyanidins in the American Diet. *Phytochemistry* **2005**, *66*, 2264–2280. [CrossRef]
108. Terrier, N.; Ollé, D.; Verriès, C.; Cheynier, V. Biochemical & Molecular Aspects of Flavan-3-Ol Synthesis during Berry Development. In *Grapevine Molecular Physiology & Biotechnology*; Roubelakis-Angelakis, K.A., Ed.; Springer: Heidelberg, Germany, 2009; pp. 365–388. ISBN 978-90-481-2304-9. [CrossRef]
109. Hanlin, R.L.; Kelm, M.A.; Wilkinson, K.L.; Downey, M.O. Detailed Characterization of Proanthocyanidins in Skin, Seeds, and Wine of Shiraz and Cabernet Sauvignon Wine Grapes (*Vitis vinifera*). *J. Agric. Food Chem.* **2011**, *59*, 13265–13276. [CrossRef]
110. Pinelo, M.; Arnous, A.; Meyer, A.S. Upgrading of Grape Skins: Significance of Plant Cell-Wall Structural Components and Extraction Techniques for Phenol Release. *Trends Food Sci. Technol.* **2006**, *17*, 579–590. [CrossRef]
111. Creasy, L.; Swain, T. Structure of Condensed Tannins. *Nature* **1965**, *208*, 151–153. [CrossRef] [PubMed]
112. Weinges, K.; Kaltenhäuser, W.; Marx, H.; Nader, E.; Nader, F.; Perner, J.; Seiler, D. Zur Kenntnis Der Proanthocyanidine, X Procyanidine Aus Früchten. *Liebigs Ann. Chem.* **1968**, *711*, 184–204. [CrossRef] [PubMed]
113. Serrano, J.; Puupponen-Pimiä, R.; Dauer, A.; Aura, A.; Saura-Calixto, F. Tannins: Current Knowledge of Food Sources, Intake, Bioavailability and Biological Effects. *Mol. Nutr. Food Res.* **2009**, *53*, S310–S329. [CrossRef] [PubMed]
114. Martinez-Micaelo, N.; González-Abuín, N.; Ardevol, A.; Pinent, M.; Blay, M.T. Procyanidins and Inflammation: Molecular Targets and Health Implications. *Biofactors* **2012**, *38*, 257–265. [CrossRef]
115. Pinent, M.; Castell-Auví, A.; Genovese, M.I.; Serrano, J.; Casanova, A.; Blay, M.; Ardévol, A. Antioxidant Effects of Proanthocyanidin-rich Natural Extracts from Grape Seed and Cupuassu on Gastrointestinal Mucosa. *J. Sci. Food Agric.* **2016**, *96*, 178–182. [CrossRef]
116. Salvadó, M.J.; Casanova, E.; Fernández-Iglesias, A.; Arola, L.; Bladé, C. Roles of Proanthocyanidin Rich Extracts in Obesity. *Food Funct.* **2015**, *6*, 1053–1071. [CrossRef]
117. Jin, Q.; O'Hair, J.; Stewart, A.C.; O'Keefe, S.F.; Neilson, A.P.; Kim, Y.-T.; McGuire, M.; Lee, A.; Wilder, G.; Huang, H. Compositional Characterization of Different Industrial White and Red Grape Pomaces in Virginia and the Potential Valorization of the Major Components. *Foods* **2019**, *8*, 667. [CrossRef]
118. He, F.; Mu, L.; Yan, G.-L.; Liang, N.-N.; Pan, Q.-H.; Wang, J.; Reeves, M.J.; Duan, C.-Q. Biosynthesis of Anthocyanins and Their Regulation in Colored Grapes. *Molecules* **2010**, *15*, 9057–9091. [CrossRef]
119. Montealegre, R.R.; Peces, R.R.; Vozmediano, J.C.; Gascueña, J.M.; Romero, E.G. Phenolic Compounds in Skins and Seeds of Ten Grape *Vitis vinifera* Varieties Grown in a Warm Climate. *J. Food Compos. Anal.* **2006**, *19*, 687–693. [CrossRef]
120. Zhang, Z.-Z.; Che, X.-N.; Pan, Q.-H.; Li, X.-X.; Duan, C.-Q. Transcriptional Activation of Flavan-3-Ols Biosynthesis in Grape Berries by UV Irradiation Depending on Developmental Stage. *Plant Sci.* **2013**, *208*, 64–74. [CrossRef]

121. Zhu, L.; Zhang, Y.; Lu, J. Phenolic Contents and Compositions in Skins of Red Wine Grape Cultivars among Various Genetic Backgrounds and Originations. *Int. J. Mol. Sci.* **2012**, *13*, 3492–3510. [CrossRef] [PubMed]
122. Pasteur, L. *Études Sur Le Vin: Ses Maladies, Causes Qui Les Provoquent, Procédés Nouveaux Pour Le Conserver et Pour Le Vieillir*; Simon Raçou et Comp.: Paris, France, 1873.
123. Willstätter, R.; Everest, A.E. Untersuchungen Über Die Anthocyane. I. Über Den Farbstoff Der Kornblume. *Liebigs Ann. Chem.* **1913**, *401*, 189–232. [CrossRef]
124. Levy, L.F.; Posternack, T.; Robinson, R. Experiments on the Synthesis of the Anthocyanins. Part VIII. A Synthesis of Œnin Chloride. *J. Chem. Soc.* **1931**, 2701–2715. [CrossRef]
125. Ribéreau-Gayon, P. *Recherches Sur Les Anthocyannes Des Végétaux: Application Au Genere Vitis*; Librairie Générale de L'Enseignement: Paris, France, 1959.
126. Rankine, B.C.; Kepner, R.E.; Webb, A.D. Comparison of Anthocyan Pigments of Vinifera Grapes. *Am. J. Enol. Vitic.* **1958**, *9*, 105–110. [CrossRef]
127. Puissant, A.; Léon, H. La Matière Colorante Des Grains de Raisins de Certains Cépages Cultivés En Anjou En 1965. *Ann. Technol. Agric.* **1967**, *3*, 217–226, hal-02731988.
128. Fong, R.A.; Kepner, R.E.; Webb, A.D. Acetic-Acid-Acylated Anthocyanin Pigments in the Grape Skins of a Number of Varieties of *Vitis vinifera*. *Am. J. Enol. Vitic.* **1971**, *22*, 150–155. [CrossRef]
129. Castillo-Muñoz, N.; Gómez-Alonso, S.; García-Romero, E.; Hermosín-Gutiérrez, I. Flavonol Profiles of *Vitis vinifera* Red Grapes and Their Single-Cultivar Wines. *J. Agric. Food Chem.* **2007**, *55*, 992–1002. [CrossRef]
130. Chedea, V.S.; Macovei, Ş.O.; Bocsan, I.C.; Măgureanu, D.C.; Levai, A.M.; Buzoianu, A.D.; Pop, R.M. Grape Pomace Polyphenols as a Source of Compounds for Management of Oxidative Stress and Inflammation—A Possible Alternative for Non-Steroidal Anti-Inflammatory Drugs? *Molecules* **2022**, *27*, 6826. [CrossRef] [PubMed]
131. Mendonça, J.d.S.; Guimarães, R.d.C.A.; Zorgetto-Pinheiro, V.A.; Fernandes, C.D.P.; Marcelino, G.; Bogo, D.; Freitas, K.d.C.; Hiane, P.A.; de Pádua Melo, E.S.; Vilela, M.L.B.; et al. Natural Antioxidant Evaluation: A Review of Detection Methods. *Molecules* **2022**, *27*, 3563. [CrossRef]
132. Kammerer, D.; Claus, A.; Carle, R.; Schieber, A. Polyphenol Screening of Pomace from Red and White Grape Varieties (*Vitis vinifera* L.) by HPLC-DAD-MS/MS. *J. Agric. Food Chem.* **2004**, *52*, 4360–4367. [CrossRef]
133. Baranowski, E.S.; Nagel, C.W. Kinetics of Malvidin-3-glucoside Condensation in Wine Model Systems. *J. Food Sci.* **1983**, *48*, 419–421. [CrossRef]
134. Cheynier, V.F.; Trousdale, E.K.; Singleton, V.L.; Salgues, M.J.; Wylde, R. Characterization of 2-S-Glutathionyl Caftaric Acid and Its Hydrolysis in Relation to Grape Wines. *J. Agric. Food Chem.* **1986**, *34*, 217–221. [CrossRef]
135. Carbonell-Capella, J.M.; Buniowska, M.; Barba, F.J.; Esteve, M.J.; Frígola, A. Analytical Methods for Determining Bioavailability and Bioaccessibility of Bioactive Compounds from Fruits and Vegetables: A Review. *Compr. Rev. Food Sci. Food Sci. Food Saf.* **2014**, *13*, 155–171. [CrossRef] [PubMed]
136. Courraud, J.; Charnay, C.; Cristol, J.-P.; Berger, J.; Avallone, S. In Vitro Lipid Peroxidation of Intestinal Bile Salt-Based Nanoemulsions: Potential Role of Antioxidants. *Free Radic. Res.* **2013**, *47*, 1076–1087. [CrossRef] [PubMed]
137. Palafox-Carlos, H.; Ayala-Zavala, J.F.; González-Aguilar, G.A. The Role of Dietary Fiber in the Bioaccessibility and Bioavailability of Fruit and Vegetable Antioxidants. *J. Food Sci.* **2011**, *76*, R6–R15. [CrossRef] [PubMed]
138. Andreasen, M.F.; Kroon, P.A.; Williamson, G.; Garcia-Conesa, M.-T. Esterase Activity Able to Hydrolyze Dietary Antioxidant Hydroxycinnamates Is Distributed along the Intestine of Mammals. *J. Agric. Food Chem.* **2001**, *49*, 5679–5684. [CrossRef] [PubMed]
139. Bunzel, M.; Ralph, J.; Lu, F.; Hatfield, R.D.; Steinhart, H. Lignins and Ferulate−Coniferyl Alcohol Cross-Coupling Products in Cereal Grains. *J. Agric. Food Chem.* **2004**, *52*, 6496–6502. [CrossRef] [PubMed]
140. Faria, A.; Fernandes, I.; Norberto, S.; Mateus, N.; Calhau, C. Interplay between Anthocyanins and Gut Microbiota. *J. Agric. Food Chem.* **2014**, *62*, 6898–6902. [CrossRef] [PubMed]
141. Corrêa, T.A.F.; Rogero, M.M.; Hassimotto, N.M.A.; Lajolo, F.M. The Two-Way Polyphenols-Microbiota Interactions and Their Effects on Obesity and Related Metabolic Diseases. *Front. Nutr.* **2019**, *6*, 188. [CrossRef]
142. Monagas, M.; Urpi-Sarda, M.; Sánchez-Patán, F.; Llorach, R.; Garrido, I.; Gómez-Cordovés, C.; Andres-Lacueva, C.; Bartolomé, B. Insights into the Metabolism and Microbial Biotransformation of Dietary Flavan-3-Ols and the Bioactivity of Their Metabolites. *Food Funct.* **2010**, *1*, 233–253. [CrossRef]
143. Cermak, R.; Landgraf, S.; Wolffram, S. The Bioavailability of Quercetin in Pigs Depends on the Glycoside Moiety and on Dietary Factors. *J. Nutr.* **2003**, *133*, 2802–2807. [CrossRef] [PubMed]
144. Zubik, L.; Meydani, M. Bioavailability of Soybean Isoflavones from Aglycone and Glucoside Forms in American Women. *The Am. J. Clin. Nutr.* **2003**, *77*, 1459–1465. [CrossRef] [PubMed]
145. Kumar, S.; Pandey, A.K. Chemistry and Biological Activities of Flavonoids: An Overview. *Sci. World J.* **2013**, *2013*, 162750. [CrossRef] [PubMed] [PubMed Central]
146. Eriz, G.; Sanhueza, V.; Roeckel, M.; Fernández, K. Inhibition of the Angiotensin-Converting Enzyme by Grape Seed and Skin Proanthocyanidins Extracted from *Vitis vinífera* L. Cv. País. *LWT-Food Sci. Tech.* **2011**, *44*, 860–865. [CrossRef]
147. Ou, K.; Gu, L. Absorption and Metabolism of Proanthocyanidins. *J. Funct. Foods* **2014**, *7*, 43–53. [CrossRef]

148. Kuhnle, G.; Spencer, J.P.; Schroeter, H.; Shenoy, B.; Debnam, E.S.; Srai, S.K.S.; Rice-Evans, C.; Hahn, U. Epicatechin and Catechin Are O-Methylated and Glucuronidated in the Small Intestine. *Biochem. Biophys. Res. Commun.* **2000**, *277*, 507–512. [CrossRef] [PubMed]
149. Vaidyanathan, J.B.; Walle, T. Glucuronidation and Sulfation of the Tea Flavonoid (−)-Epicatechin by the Human and Rat Enzymes. *Drug Metab. Disp.* **2002**, *30*, 897–903. [CrossRef] [PubMed]
150. Hackman, R.M.; Polagruto, J.A.; Zhu, Q.Y.; Sun, B.; Fujii, H.; Keen, C.L. Flavanols: Digestion, Absorption and Bioactivity. *Phytochem. Rev.* **2008**, *7*, 195–208. [CrossRef]
151. Dangles, O.; Dufour, C.; Manach, C.; Morand, C.; Remesy, C. Binding of Flavonoids to Plasma Proteins. In *Methods in Enzymology*; Academic Press: New York, NY, USA, 2001; Volume 335, pp. 319–333. [CrossRef]
152. Axling, U.; Olsson, C.; Xu, J.; Fernandez, C.; Larsson, S.; Ström, K.; Ahrné, S.; Holm, C.; Molin, G.; Berger, K. Green Tea Powder and Lactobacillus Plantarum Affect Gut Microbiota, Lipid Metabolism and Inflammation in High-Fat Fed C57BL/6J Mice. *Nutr. Metab.* **2012**, *9*, 105. [CrossRef]
153. China, R.; Mukherjee, S.; Sen, S.; Bose, S.; Datta, S.; Koley, H.; Ghosh, S.; Dhar, P. Antimicrobial Activity of Sesbania Grandiflora Flower Polyphenol Extracts on Some Pathogenic Bacteria and Growth Stimulatory Effect on the Probiotic Organism Lactobacillus Acidophilus. *Microbiol. Res.* **2012**, *167*, 500–506. [CrossRef]
154. Queipo-Ortuño, M.I.; Boto-Ordóñez, M.; Murri, M.; Gomez-Zumaquero, J.M.; Clemente-Postigo, M.; Estruch, R.; Diaz, F.C.; Andres-Lacueva, C.; Tinahones, F.J. Influence of Red Wine Polyphenols and Ethanol on the Gut Microbiota Ecology and Biochemical Biomarkers. *Am. J. Clin. Nutr.* **2012**, *95*, 1323–1334. [CrossRef] [PubMed]
155. Etxeberria, U.; Fernández-Quintela, A.; Milagro, F.I.; Aguirre, L.; Martínez, J.A.; Portillo, M.P. Impact of Polyphenols and Polyphenol-Rich Dietary Sources on Gut Microbiota Composition. *J. Agric. Food Chem.* **2013**, *61*, 9517–9533. [CrossRef] [PubMed]
156. Moreno-Indias, I.; Sánchez-Alcoholado, L.; Pérez-Martínez, P.; Andrés-Lacueva, C.; Cardona, F.; Tinahones, F.; Queipo-Ortuño, M.I. Red Wine Polyphenols Modulate Fecal Microbiota and Reduce Markers of the Metabolic Syndrome in Obese Patients. *Food Funct.* **2016**, *7*, 1775–1787. [CrossRef] [PubMed]
157. Parada, J.; Aguilera, J. Food Microstructure Affects the Bioavailability of Several Nutrients. *J. Food Sci.* **2007**, *72*, R21–R32. [CrossRef] [PubMed]
158. Galvano, F.; La Fauci, L.; Vitaglione, P.; Fogliano, V.; Vanella, L.; Felgines, C. Bioavailability, Antioxidant and Biological Properties of the Natural Free-Radical Scavengers Cyanidin and Related Glycosides. *Ann. Ist. Super. Sanita* **2007**, *43*, 382–393. [PubMed]
159. Charron, C.S.; Kurilich, A.C.; Clevidence, B.A.; Simon, P.W.; Harrison, D.J.; Britz, S.J.; Baer, D.J.; Novotny, J.A. Bioavailability of Anthocyanins from Purple Carrot Juice: Effects of Acylation and Plant Matrix. *J. Agric. Food Chem.* **2009**, *57*, 1226–1230. [CrossRef] [PubMed]
160. Mugford, C.A.; Kedderis, G.L. Sex-Dependent Metabolism of Xenobiotics. *Drug Metab. Rev.* **1998**, *30*, 441–498. [CrossRef] [PubMed]
161. Renaud, H.J.; Cui, J.Y.; Khan, M.; Klaassen, C.D. Tissue Distribution and Gender-Divergent Expression of 78 Cytochrome P450 MRNAs in Mice. *Tox. Sci.* **2011**, *124*, 261–277. [CrossRef] [PubMed]
162. Liu, W.; Kulkarni, K.; Hu, M. Gender-Dependent Differences in Uridine 5′-Diphospho-Glucuronosyltransferase Have Implications in Metabolism and Clearance of Xenobiotics. *Expert Opin. Drug Metab. Toxicol.* **2013**, *9*, 1555–1569. [CrossRef]
163. Ruiz, M.L.; Mottino, A.D.; Catania, V.A.; Vore, M. Hormonal Regulation of Hepatic Drug Biotransformation and Transport Systems. *Compr. Physiol.* **2013**, *3–4*, 1721–1740.
164. Dellinger, R.W.; Garcia, A.M.G.; Meyskens Jr, F.L. Differences in the Glucuronidation of Resveratrol and Pterostilbene: Altered Enzyme Specificity and Potential Gender Differences. *Drug Metab. Pharmacok.* **2014**, *29*, 112–119. [CrossRef] [PubMed]
165. Matal, J.; Jancova, P.; Šiller, M.; Masek, V.; Anzenbacherova, E.; Anzenbacher, P. Interspecies Comparison of the Glucuronidation Processes in the Man, Monkey, Pig, Dog and Rat. *Neuro Endocrinol. Lett.* **2008**, *29*, 738.
166. Yamazaki, K.; Suzuki, M.; Itoh, T.; Yamamoto, K.; Kanemitsu, M.; Matsumura, C.; Nakano, T.; Sakaki, T.; Fukami, Y.; Imaishi, H. Structural Basis of Species Differences between Human and Experimental Animal CYP1A1s in Metabolism of 3,3′,4,4′,5-Pentachlorobiphenyl. *J. Biochem.* **2011**, *149*, 487–494. [CrossRef] [PubMed]
167. Helke, K.L.; Swindle, M.M. Animal Models of Toxicology Testing: The Role of Pigs. *Expert Opin. Drug Metab. Toxicol.* **2013**, *9*, 127–139. [CrossRef] [PubMed]
168. Halliwell, B. Biochemistry of Oxidative Stress. *Biochem. Soc. Trans.* **2007**, *35*, 1147–1150. [CrossRef] [PubMed]
169. Frankič, T.; Salobir, J.; Rezar, V. The Effect of Vitamin E Supplementation on Reduction of Lymphocyte DNA Damage Induced by T-2 Toxin and Deoxynivalenol in Weaned Pigs. *Anim. Feed Sci. Tech.* **2008**, *141*, 274–286. [CrossRef]
170. Lin, H.; Decuypere, E.; Buyse, J. Acute Heat Stress Induces Oxidative Stress in Broiler Chickens. *Comp. Biochem. Physiol. Part A Mol. Integr. Physiol.* **2006**, *144*, 11–17. [CrossRef] [PubMed]
171. Georgieva, N.; Koinarski, V.; Gadjeva, V. Antioxidant Status during the Course of *Eimeria tenella* Infection in Broiler Chickens. *Vet. J.* **2006**, *172*, 488–492. [CrossRef] [PubMed]
172. Furukawa, S.; Fujita, T.; Shimabukuro, M.; Iwaki, M.; Yamada, Y.; Nakajima, Y.; Nakayama, O.; Makishima, M.; Matsuda, M.; Shimomura, I. Increased Oxidative Stress in Obesity and Its Impact on Metabolic Syndrome. *J. Clin. Investig.* **2017**, *114*, 1752–1761. [CrossRef]

173. Vuolo, M.M.; Lima, V.S.; Junior, M.R.M. Phenolic Compounds: Structure, Classification, and Antioxidant Power. In *Bioactive Compounds*; Segura-Campos, M.R., Ed.; Health Benefits and Potential Applications; Elsevier: Amsterdam, The Netherlands, 2019; pp. 33–50, ISBN 9780128147740. [CrossRef]
174. Van Hung, P. Phenolic Compounds of Cereals and Their Antioxidant Capacity. *Crit. Rev. Food Sci. Nutr.* **2016**, *56*, 25–35. [CrossRef]
175. Procházková, D.; Boušová, I.; Wilhelmová, N. Antioxidant and Prooxidant Properties of Flavonoids. *Fitoterapia* **2011**, *82*, 513–523. [CrossRef] [PubMed]
176. Khlebnikov, A.I.; Schepetkin, I.A.; Domina, N.G.; Kirpotina, L.N.; Quinn, M.T. Improved Quantitative Structure–Activity Relationship Models to Predict Antioxidant Activity of Flavonoids in Chemical, Enzymatic, and Cellular Systems. *Bioorg. Med. Chem.* **2007**, *15*, 1749–1770. [CrossRef] [PubMed]
177. Pietta, P.-G. Flavonoids as Antioxidants. *J. Nat. Prod.* **2000**, *63*, 1035–1042. [CrossRef] [PubMed]
178. Puiggròs, F.; Llópiz, N.; Ardévol, A.; Bladé, C.; Arola, L.; Salvadó, M.J. Grape Seed Procyanidins Prevent Oxidative Injury by Modulating the Expression of Antioxidant Enzyme Systems. *J. Agric. Food Chem.* **2005**, *53*, 6080–6086. [CrossRef] [PubMed]
179. Balasundram, N.; Sundram, K.; Samman, S. Phenolic Compounds in Plants and Agri-Industrial by-Products: Antioxidant Activity, Occurrence, and Potential Uses. *Food Chem.* **2006**, *99*, 191–203. [CrossRef]
180. Foti, M.; Piattelli, M.; Amico, V.; Ruberto, G. Antioxidant Activity of Phenolic Meroditerpenoids from Marine Algae. *J. Photochem. Photobiol. B Biol.* **1994**, *26*, 159–164. [CrossRef]
181. Figueroa-Espinoza, M.-C.; Villeneuve, P. Phenolic Acids Enzymatic Lipophilization. *J. Agric. Food Chem.* **2005**, *53*, 2779–2787. [CrossRef] [PubMed]
182. Bendary, E.; Francis, R.; Ali, H.; Sarwat, M.; El Hady, S. Antioxidant and Structure–Activity Relationships (SARs) of Some Phenolic and Anilines Compounds. *Ann. Agric. Sci.* **2013**, *58*, 173–181. [CrossRef]
183. Wang, T.; Li, Q.; Bi, K. Bioactive Flavonoids in Medicinal Plants: Structure, Activity and Biological Fate. *Asian J. Pharm. Sci.* **2018**, *13*, 12–23. [CrossRef]
184. Jin, L.-Z.; Dersjant-Li, Y.; Giannenas, I. Application of Aromatic Plants and Their Extracts in Diets of Broiler Chickens. In *Feed Additives*; Academic Press: Cambridge, MA, USA, 2020; pp. 159–185. [CrossRef]
185. Giannenas, I.; Bonos, E.; Christaki, E.; Florou-Paneri, P. Oregano: A Feed Additive with Functional Properties. In *Therapeutic Foods*, 8th ed.; Holban, A.M., Grumezescu, A.M., Eds.; Elsevier: Amsterdam, The Netherlands, 2018; pp. 179–208, ISBN 978-0-12-811517-6. [CrossRef]
186. Valenzuela-Grijalva, N.V.; Pinelli-Saavedra, A.; Muhlia-Almazan, A.; Domínguez-Díaz, D.; González-Ríos, H. Dietary Inclusion Effects of Phytochemicals as Growth Promoters in Animal Production. *J. Anim. Sci. Technol.* **2017**, *59*, 8. [CrossRef]
187. Pantano, C.; Reynaert, N.L.; Vliet, A.V.D.; Janssen–Heininger, Y.M. Redox-Sensitive Kinases of the Nuclear Factor-KB Signaling Pathway. *Antiox. Redox Signal.* **2006**, *8*, 1791–1806. [CrossRef] [PubMed]
188. Tabas, I.; Glass, C.K. Anti-Inflammatory Therapy in Chronic Disease: Challenges and Opportunities. *Science* **2013**, *339*, 166–172. [CrossRef] [PubMed]
189. Hiscott, J.; Marois, J.; Garoufalis, J.; D'Addario, M.; Roulston, A.; Kwan, I.; Pepin, N.; Lacoste, J.; Nguyen, H.; Bensi, G. Characterization of a Functional NF-KB Site in the Human Interleukin 1β Promoter: Evidence for a Positive Autoregulatory Loop. *Moll. Cell. Biol.* **1993**, *13*, 6231–6240. [CrossRef] [PubMed]
190. Barnes, P.J. Nuclear Factor-KB. *Int. J. Biochem. Cell. Biol.* **1997**, *29*, 867–870. [CrossRef] [PubMed]
191. Aggarwal, B.B. Nuclear Factor-KB: The Enemy within. *Cancer Cell* **2004**, *6*, 203–208. [CrossRef] [PubMed]
192. Aggarwal, B.B.; Shishodia, S.; Sandur, S.K.; Pandey, M.K.; Sethi, G. Inflammation and Cancer: How Hot Is the Link? *Biochem. Pharmacol.* **2006**, *72*, 1605–1621. [CrossRef] [PubMed]
193. Dinarello, C. Biologic Basis for Interleukin-1 in Disease. *Blood* **1996**, *87*, 2095–2147. [CrossRef]
194. Dinarello, C.A. Proinflammatory Cytokines. *Chest* **2000**, *118*, 503–508. [CrossRef] [PubMed]
195. Pillarisetti, S.; Saxena, U. Role of Oxidative Stress and Inflammation in the Origin of Type 2 Diabetes—A Paradigm Shift. *Expert Opin. Ther. Targets* **2004**, *8*, 401–408. [CrossRef]
196. Opal, S.M.; DePalo, V.A. Anti-Inflammatory Cytokines. *Chest* **2000**, *117*, 1162–1172. [CrossRef]
197. Ciz, M.; Denev, P.; Kratchanova, M.; Vasicek, O.; Ambrozova, G.; Lojek, A. Flavonoids Inhibit the Respiratory Burst of Neutrophils in Mammals. *Oxid. Med. Cell. Longev.* **2012**, *2012*, 181295. [CrossRef] [PubMed]
198. Dantzer, R.; O'connor, J.C.; Freund, G.G.; Johnson, R.W.; Kelley, K.W. From Inflammation to Sickness and Depression: When the Immune System Subjugates the Brain. *Nat. Rev. Neurosci.* **2008**, *9*, 46–56. [CrossRef]
199. Dinarello, C.A. Anti-Inflammatory Agents: Present and Future. *Cell* **2010**, *140*, 935–950. [CrossRef]
200. Sharma, M. Chemokines and Their Receptors: Orchestrating a Fine Balance between Health and Disease. *Crit. Rev. Biotech.* **2010**, *30*, 1–22. [CrossRef]
201. Rubio-Perez, J.M.; Morillas-Ruiz, J.M. A Review: Inflammatory Process in Alzheimer's Disease, Role of Cytokines. *Sci. World J.* **2012**, 1–15. [CrossRef]
202. Lykkesfeldt, J.; Svendsen, O. Oxidants and Antioxidants in Disease: Oxidative Stress in Farm Animals. *Vet. J.* **2007**, *173*, 502–511. [CrossRef]
203. Lipiński, K.; Mazur, M.; Antoszkiewicz, Z.; Purwin, C. Polyphenols in Monogastric Nutrition—A Review. *Ann. Anim. Sci.* **2017**, *17*, 41–58. [CrossRef]

204. Rubió, L.; Motilva, M.-J.; Romero, M.-P. Recent Advances in Biologically Active Compounds in Herbs and Spices: A Review of the Most Effective Antioxidant and Anti-Inflammatory Active Principles. *Crit. Rev. Food Sci. Nutr.* **2013**, *53*, 943–953. [CrossRef] [PubMed]
205. Sung, B.; Prasad, S.; Gupta, S.C.; Patchva, S.; Aggarwal, B.B. Regulation of Inflammation-Mediated Chronic Diseases by Botanicals. In *Recent Trends in Medicinal Plants Research*; Shyur, L.F., Lau, A.S.Y., Eds.; Elsevier: Amsterdam, The Netherlands, 2012; Volume 62, pp. 57–132, ISBN 978-0-12-394591-4. [CrossRef]
206. Panico, A.; Cardile, V.; Avondo, S.; Garufi, F.; Gentile, B.; Puglia, C.; Bonina, F.; Santagati, N.; Ronsisvalle, G. The In Vitro Effect of a Lyophilized Extract of Wine Obtained from Jacquez Grapes on Human Chondrocytes. *Phytomedicine* **2006**, *13*, 522–526. [CrossRef]
207. Vendrame, S.; Klimis-Zacas, D. Anti-Inflammatory Effect of Anthocyanins via Modulation of Nuclear Factor-KB and Mitogen-Activated Protein Kinase Signaling Cascades. *Nutr. Rev.* **2015**, *73*, 348–358. [CrossRef]
208. Na, H.-K.; Surh, Y.-J. Modulation of Nrf2-Mediated Antioxidant and Detoxifying Enzyme Induction by the Green Tea Polyphenol EGCG. *Food Chem. Toxicol.* **2008**, *46*, 1271–1278. [CrossRef] [PubMed]
209. Chuang, C.-C.; McIntosh, M.K. Potential Mechanisms by Which Polyphenol-Rich Grapes Prevent Obesity-Mediated Inflammation and Metabolic Diseases. *Ann. Rev. Nutr.* **2011**, *31*, 155–176. [CrossRef] [PubMed]
210. Calabrese, V.; Cornelius, C.; Dinkova-Kostova, A.T.; Iavicoli, I.; Di Paola, R.; Koverech, A.; Cuzzocrea, S.; Rizzarelli, E.; Calabrese, E.J. Cellular Stress Responses, Hormetic Phytochemicals and Vitagenes in Aging and Longevity. *BBA-Mol. Basis Dis.* **2012**, *1822*, 753–783. [CrossRef]
211. Son, T.G.; Camandola, S.; Mattson, M.P. Hormetic Dietary Phytochemicals. *Neuromol. Med.* **2008**, *10*, 236–246. [CrossRef]
212. Ishii, T.; Itoh, K.; Yamamoto, M. Roles of Nrf2 in Activation of Antioxidant Enzyme Genes via Antioxidant Responsive Elements. In *Protein Sensors and Reactive Oxygen Species—Part B: Thiol Enzymes and Proteins Series Methods in Enzymology*; Sies, H., Packer, L., Eds.; Academic Press: New York, NY, USA, 2002; Volume 348, pp. 182–190. [CrossRef]
213. Lee, J.; Giordano, S.; Zhang, J. Autophagy, Mitochondria and Oxidative Stress: Cross-Talk and Redox Signalling. *Biochem. J.* **2012**, *441*, 523–540. [CrossRef]
214. Li, W.-G.; Zhang, X.-Y.; Wu, Y.; Tian, X. Anti-Inflammatory Effect and Mechanism of Proanthocyanidins from Grape Seeds. *Acta Pharm. Sin.* **2001**, *22*, 1117–1120.
215. Terra, X.; Valls, J.; Vitrac, X.; Mérrillon, J.-M.; Arola, L.; Ardèvol, A.; Bladé, C.; Fernández-Larrea, J.; Pujadas, G.; Salvadó, J. Grape-Seed Procyanidins Act as Antiinflammatory Agents in Endotoxin-Stimulated RAW 264.7 Macrophages by Inhibiting NFkB Signaling Pathway. *J. Agric. Food Chem.* **2007**, *55*, 4357–4365. [CrossRef]
216. Bitzer, Z.T.; Glisan, S.L.; Dorenkott, M.R.; Goodrich, K.M.; Ye, L.; O'Keefe, S.F.; Lambert, J.D.; Neilson, A.P. Cocoa Procyanidins with Different Degrees of Polymerization Possess Distinct Activities in Models of Colonic Inflammation. *J. Nutr. Biochem.* **2015**, *26*, 827–831. [CrossRef]
217. Williams, A.R.; Klaver, E.J.; Laan, L.C.; Ramsay, A.; Fryganas, C.; Difborg, R.; Kringel, H.; Reed, J.D.; Mueller-Harvey, I.; Skov, S. Co-operative Suppression of Inflammatory Responses in Human Dendritic Cells by Plant Proanthocyanidins and Products from the Parasitic Nematode Trichuris Suis. *Immunology* **2017**, *150*, 312–328. [CrossRef] [PubMed]
218. Midttun, H.L.; Ramsay, A.; Mueller-Harvey, I.; Williams, A.R. Cocoa Procyanidins Modulate Transcriptional Pathways Linked to Inflammation and Metabolism in Human Dendritic Cells. *Food Funct.* **2018**, *9*, 2883–2890. [CrossRef]
219. Skyberg, J.A.; Robison, A.; Golden, S.; Rollins, M.F.; Callis, G.; Huarte, E.; Kochetkova, I.; Jutila, M.A.; Pascual, D.W. Apple Polyphenols Require T Cells to Ameliorate Dextran Sulfate Sodium-Induced Colitis and Dampen Proinflammatory Cytokine Expression. *J. Leukoc. Biol.* **2011**, *90*, 1043–1054. [CrossRef] [PubMed]
220. Yoshioka, Y.; Akiyama, H.; Nakano, M.; Shoji, T.; Kanda, T.; Ohtake, Y.; Takita, T.; Matsuda, R.; Maitani, T. Orally Administered Apple Procyanidins Protect against Experimental Inflammatory Bowel Disease in Mice. *Int. Immunopharmacol.* **2008**, *8*, 1802–1807. [CrossRef]
221. Park, M.-K.; Park, J.-S.; Cho, M.-L.; Oh, H.-J.; Heo, Y.-J.; Woo, Y.-J.; Heo, Y.-M.; Park, M.-J.; Park, H.-S.; Park, S.-H. Grape Seed Proanthocyanidin Extract (GSPE) Differentially Regulates Foxp3+ Regulatory and IL-17+ Pathogenic T Cell in Autoimmune Arthritis. *Immunol. Lett.* **2011**, *135*, 50–58. [CrossRef] [PubMed]
222. Daughenbaugh, K.F.; Holderness, J.; Graff, J.C.; Hedges, J.F.; Freedman, B.; Graff, J.W.; Jutila, M.A. Contribution of Transcript Stability to a Conserved Procyanidin-Induced Cytokine Response in Γδ T Cells. *Genes Immun.* **2011**, *12*, 378–389. [CrossRef] [PubMed]
223. Holderness, J.; Jackiw, L.; Kimmel, E.; Kerns, H.; Radke, M.; Hedges, J.F.; Petrie, C.; McCurley, P.; Glee, P.M.; Palecanda, A. Select Plant Tannins Induce IL-2Rα up-Regulation and Augment Cell Division in Γδ T Cells. *J. Immunol.* **2007**, *179*, 6468–6478. [CrossRef] [PubMed]
224. Witherden, D.A.; Havran, W.L. Cross-Talk between Intraepithelial Γδ T Cells and Epithelial Cells. *J. Leukoc. Biol.* **2013**, *94*, 69–76. [CrossRef]
225. Casanova-Martí, À.; González-Abuín, N.; Serrano, J.; Blay, M.T.; Terra, X.; Frost, G.; Pinent, M.; Ardevol, A. Long Term Exposure to a Grape Seed Proanthocyanidin Extract Enhances L-Cell Differentiation in Intestinal Organoids. *Mol. Nutr. Food Res.* **2020**, *64*, 2000303. [CrossRef] [PubMed]
226. Linden, S.; Sutton, P.; Karlsson, N.; Korolik, V.; McGuckin, M. Mucins in the Mucosal Barrier to Infection. *Mucosal Immunol.* **2008**, *1*, 183–197. [CrossRef]

227. Georgiades, P.; Pudney, P.D.; Rogers, S.; Thornton, D.J.; Waigh, T.A. Tea Derived Galloylated Polyphenols Cross-Link Purified Gastrointestinal Mucins. *PLoS ONE* **2014**, *9*, e105302. [CrossRef]
228. Martins, E.; Ferreira, A.C.F.; Skorupa, A.L.; Afeche, S.C.; Cipolla-Neto, J.; Costa Rosa, L.F.B.P. Tryptophan Consumption and Indoleamines Production by Peritoneal Cavity Macrophages. *J. Leukoc. Biol.* **2004**, *75*, 1116–1121. [CrossRef]
229. Rosillo, M.; Sanchez-Hidalgo, M.; Cárdeno, A.; De La Lastra, C.A. Protective Effect of Ellagic Acid, a Natural Polyphenolic Compound, in a Murine Model of Crohn's Disease. *Biochem. Pharm.* **2011**, *82*, 737–745. [CrossRef]
230. Zhang, G.; Ross, C.; Blecha, F. Porcine Antimicrobial Peptides: New Prospects for Ancient Molecules of Host Defense. *Vet. Res.* **2000**, *31*, 277–296. [CrossRef] [PubMed]
231. Shi, J.; Zhang, G.; Wu, H.; Ross, C.; Blecha, F.; Ganz, T. Porcine Epithelial β-Defensin 1 Is Expressed in the Dorsal Tongue at Antimicrobial Concentrations. *Infect. Immun.* **1999**, *67*, 3121–3127. [CrossRef] [PubMed]
232. Sang, Y.; Patil, A.A.; Zhang, G.; Ross, C.R.; Blecha, F. Bioinformatic and Expression Analysis of Novel Porcine β-Defensins. *Mamm. Genome* **2006**, *17*, 332–339. [CrossRef] [PubMed]
233. Wan, M.L.; Ling, K.; Wang, M.; El-Nezami, H. Green Tea Polyphenol Epigallocatechin-3-gallate Improves Epithelial Barrier Function by Inducing the Production of Antimicrobial Peptide PBD-1 and PBD-2 in Monolayers of Porcine Intestinal Epithelial IPEC-J2 Cells. *Mol. Nutr. Food Res.* **2016**, *60*, 1048–1058. [CrossRef] [PubMed]
234. Gessner, D.K.; Fiesel, A.; Most, E.; Dinges, J.; Wen, G.; Ringseis, R.; Eder, K. Supplementation of a Grape Seed and Grape Marc Meal Extract Decreases Activities of the Oxidative Stress-Responsive Transcription Factors NF-KB and Nrf2 in the Duodenal Mucosa of Pigs. *Acta Vet. Scand.* **2013**, *55*, 18. [CrossRef]
235. Fiesel, A.; Gessner, D.K.; Most, E.; Eder, K. Effects of Dietary Polyphenol-Rich Plant Products from Grape or Hop on pro-Inflammatory Gene Expression in the Intestine, Nutrient Digestibility and Faecal Microbiota of Weaned Pigs. *BMC Vet. Res.* **2014**, *10*, 196. [CrossRef] [PubMed]
236. Li, Q.; Verma, I.M. NF-KB Regulation in the Immune System. *Nat. Rev. Immunol.* **2002**, *2*, 725–734. [CrossRef] [PubMed]
237. Rogler, G.; Brand, K.; Vogl, D.; Page, S.; Hofmeister, R.; Andus, T.; Knuechel, R.; Baeuerle, P.A.; Schölmerich, J.; Gross, V. Nuclear Factor KB Is Activated in Macrophages and Epithelial Cells of Inflamed Intestinal Mucosa. *Gastroenterology* **1998**, *115*, 357–369. [CrossRef]
238. Stadnyk, A.W. Intestinal Epithelial Cells as a Source of Inflammatory Cytokines and Chemokines. *Can. J. Gastroenterol.* **2002**, *16*, 241–246. [CrossRef]
239. Jung, H.C.; Eckmann, L.; Yang, S.; Panja, A.; Fierer, J.; Morzycka-Wroblewska, E.; Kagnoff, M. A Distinct Array of Proinflammatory Cytokines Is Expressed in Human Colon Epithelial Cells in Response to Bacterial Invasion. *J. Clin. Investig.* **1995**, *95*, 55–65. [CrossRef] [PubMed]
240. Son, D.O.; Satsu, H.; Shimizu, M. Histidine Inhibits Oxidative Stress-and TNF-α-Induced Interleukin-8 Secretion in Intestinal Epithelial Cells. *FEBS Lett.* **2005**, *579*, 4671–4677. [CrossRef]
241. Ramiro-Puig, E.; Pérez-Cano, F.; Ramírez-Santana, C.; Castellote, C.; Izquierdo-Pulido, M.; Permanyer, J.; Franch, A.; Castell, M. Spleen Lymphocyte Function Modulated by a Cocoa-Enriched Diet. *Clin. Exp. Immunol.* **2007**, *149*, 535–542. [CrossRef] [PubMed]
242. Taranu, I.; Habeanu, M.; Gras, M.; Pistol, G.; Lefter, N.; Palade, M.; Ropota, M.; Sanda Chedea, V.; Marin, D. Assessment of the Effect of Grape Seed Cake Inclusion in the Diet of Healthy Fattening-finishing Pigs. *J. Anim. Physiol. Anim. Nutr.* **2018**, *102*, e30–e42. [CrossRef]
243. Okazaki, Y.; Kakehi, S.; Xu, Y.; Tsujimoto, K.; Sasaki, M.; Ogawa, H.; Kato, N. Consumption of Sericin Reduces Serum Lipids, Ameliorates Glucose Tolerance and Elevates Serum Adiponectin in Rats Fed a High-Fat Diet. *Biosci. Biotechnol. Biochem.* **2010**, *74*, 1534–1538. [CrossRef] [PubMed]
244. Phalipon, A.; Cardona, A.; Kraehenbuhl, J.-P.; Edelman, L.; Sansonetti, P.J.; Corthésy, B. Secretory Component: A New Role in Secretory IgA-Mediated Immune Exclusion In Vivo. *Immunity* **2002**, *17*, 107–115. [CrossRef]
245. Mantis, N.J.; Rol, N.; Corthésy, B. Secretory IgA's Complex Roles in Immunity and Mucosal Homeostasis in the Gut. *Mucosal Immunol.* **2011**, *4*, 603–611. [CrossRef]
246. Zhang, H.J.; Jiang, X.R.; Mantovani, G.; Lumbreras, A.E.V.; Comi, M.; Alborali, G.; Savoini, G.; Dell'Orto, V.; Bontempo, V. Modulation of Plasma Antioxidant Activity in Weaned Piglets by Plant Polyphenols. *Ital. J. Anim. Sci.* **2014**, *13*, 3242. [CrossRef]
247. Gessner, D.; Ringseis, R.; Eder, K. Potential of Plant Polyphenols to Combat Oxidative Stress and Inflammatory Processes in Farm Animals. *J. Anim. Physiol. Anim. Nutr.* **2017**, *101*, 605–628. [CrossRef]
248. Yan, L.; Kim, I. Effect of Dietary Grape Pomace Fermented by Saccharomyces Boulardii on the Growth Performance, Nutrient Digestibility and Meat Quality in Finishing Pigs. *Asian-Australas. J. Anim. Sci.* **2011**, *24*, 1763–1770. [CrossRef]
249. Han, M.; Song, P.; Huang, C.; Rezaei, A.; Farrar, S.; Brown, M.A.; Ma, X. Dietary Grape Seed Proanthocyanidins (GSPs) Improve Weaned Intestinal Microbiota and Mucosal Barrier Using a Piglet Model. *Oncotarget* **2016**, *7*, 80313. [CrossRef] [PubMed]
250. Kafantaris, I.; Stagos, D.; Kotsampasi, B.; Hatzis, A.; Kypriotakis, A.; Gerasopoulos, K.; Makri, S.; Goutzourelas, N.; Mitsagga, C.; Giavasis, I. Grape Pomace Improves Performance, Antioxidant Status, Fecal Microbiota and Meat Quality of Piglets. *Animal* **2018**, *12*, 246–255. [CrossRef] [PubMed]
251. Ahmed, S.; Hossain, M.; Kim, G.; Hwang, J.; Ji, H.; Yang, C. Effects of Resveratrol and Essential Oils on Growth Performance, Immunity, Digestibility and Fecal Microbial Shedding in Challenged Piglets. *Asian-Australas. J. Anim. Sci.* **2013**, *26*, 683. [CrossRef]
252. Verhelst, R.; Schroyen, M.; Buys, N.; Niewold, T. Dietary Polyphenols Reduce Diarrhea in Enterotoxigenic *Escherichia coli* (ETEC) Infected Post-Weaning Piglets. *Livest. Sci.* **2014**, *160*, 138–140. [CrossRef]

253. Grosu, I.A.; Pistol, G.C.; Taranu, I.; Marin, D.E. The Impact of Dietary Grape Seed Meal on Healthy and Aflatoxin B1 Afflicted Microbiota of Pigs after Weaning. *Toxins* **2019**, *11*, 25. [CrossRef]
254. Choy, Y.Y.; Quifer-Rada, P.; Holstege, D.M.; Frese, S.A.; Calvert, C.C.; Mills, D.A.; Lamuela-Raventos, R.M.; Waterhouse, A.L. Phenolic Metabolites and Substantial Microbiome Changes in Pig Feces by Ingesting Grape Seed Proanthocyanidins. *Food Funct.* **2014**, *5*, 2298–2308. [CrossRef]
255. Hughes, R.; Brooker, J.; Smyl, C. Growth Rate of Broiler Chickens given Condensed Tannins Extracted from Grape Seed. In Proceedings of the 17th Australian Poultry Science Symposium, Sydney, Australia, 7–9 February 2005.
256. Blander, J.M.; Longman, R.S.; Iliev, I.D.; Sonnenberg, G.F.; Artis, D. Regulation of Inflammation by Microbiota Interactions with the Host. *Nat. Immunol.* **2017**, *18*, 851–860. [CrossRef]
257. Peterson, L.W.; Artis, D. Intestinal Epithelial Cells: Regulators of Barrier Function and Immune Homeostasis. *Nat. Rev. Immunol.* **2014**, *14*, 141–153. [CrossRef]
258. Winer, D.A.; Luck, H.; Tsai, S.; Winer, S. The Intestinal Immune System in Obesity and Insulin Resistance. *Cell Metab.* **2016**, *23*, 413–426. [CrossRef]
259. Mosele, J.I.; Macià, A.; Motilva, M.-J. Metabolic and Microbial Modulation of the Large Intestine Ecosystem by Non-Absorbed Diet Phenolic Compounds: A Review. *Molecules* **2015**, *20*, 17429–17468. [CrossRef]
260. Crozier, A.; Jaganath, I.B.; Clifford, M.N. Dietary Phenolics: Chemistry, Bioavailability and Effects on Health. *Nat. Prod. Rep.* **2009**, *26*, 1001–1043. [CrossRef]
261. Bustos, I.; Garcia-Cayuela, T.; Hernandez-Ledesma, B.; Pelaez, C.; Requena, T.; Martínez-Cuesta, M.C. Effect of Flavan-3-Ols on the Adhesion of Potential Probiotic Lactobacilli to Intestinal Cells. *J. Agric. Food Chem.* **2012**, *60*, 9082–9088. [CrossRef]
262. Tzounis, X.; Vulevic, J.; Kuhnle, G.G.; George, T.; Leonczak, J.; Gibson, G.R.; Kwik-Uribe, C.; Spencer, J.P. Flavanol Monomer-Induced Changes to the Human Faecal Microflora. *Br. J. Nutr.* **2008**, *99*, 782–792. [CrossRef] [PubMed]
263. González-Quilen, C.; Rodríguez-Gallego, E.; Beltrán-Debón, R.; Pinent, M.; Ardévol, A.; Blay, M.T.; Terra, X. Health-Promoting Properties of Proanthocyanidins for Intestinal Dysfunction. *Nutrients* **2020**, *12*, 130. [CrossRef]
264. Gonthier, M.-P.; Donovan, J.L.; Texier, O.; Felgines, C.; Remesy, C.; Scalbert, A. Metabolism of Dietary Procyanidins in Rats. *Free Radic. Biol. Med.* **2003**, *35*, 837–844. [CrossRef] [PubMed]
265. Scalbert, A. Antimicrobial Properties of Tannins. *Phytochemistry* **1991**, *30*, 3875–3883. [CrossRef]
266. Anhê, F.F.; Pilon, G.; Roy, D.; Desjardins, Y.; Levy, E.; Marette, A. Triggering Akkermansia with Dietary Polyphenols: A New Weapon to Combat the Metabolic Syndrome? *Gut Microbes* **2016**, *7*, 146–153. [CrossRef]
267. Wu, Y.; Ma, N.; Song, P.; He, T.; Levesque, C.; Bai, Y.; Zhang, A.; Ma, X. Grape Seed Proanthocyanidin Affects Lipid Metabolism via Changing Gut Microflora and Enhancing Propionate Production in Weaned Pigs. *J. Nutr.* **2019**, *149*, 1523–1532. [CrossRef]
268. KruegerChristian, G.; MeudtJennifer, J.; ReedJess, D. Effect of Sweetened Dried Cranberry Consumption on Urinary Proteome and Fecal Microbiome in Healthy Human Subjects. *Omics A J. Integr. Biol.* **2018**, *22*, 145–153. [CrossRef]
269. Liu, Q.; Yu, Z.; Tian, F.; Zhao, J.; Zhang, H.; Zhai, Q.; Chen, W. Surface Components and Metabolites of Probiotics for Regulation of Intestinal Epithelial Barrier. *Microb. Cell Fact.* **2020**, *19*, 23. [CrossRef]
270. Williams, A.R.; Krych, L.; Fauzan Ahmad, H.; Nejsum, P.; Skovgaard, K.; Nielsen, D.S.; Thamsborg, S.M. A Polyphenol-Enriched Diet and *Ascaris suum* Infection Modulate Mucosal Immune Responses and Gut Microbiota Composition in Pigs. *PLoS ONE* **2017**, *12*, e0186546. [CrossRef] [PubMed]
271. Pineiro, M.; Asp, N.-G.; Reid, G.; Macfarlane, S.; Morelli, L.; Brunser, O.; Tuohy, K. FAO Technical Meeting on Prebiotics. *J. Clin. Gastroenterol.* **2008**, *42*, S156–S159. [CrossRef] [PubMed]
272. Christaki, E.; Giannenas, I.; Bonos, E.; Florou-Paneri, P. Chapter 2—Innovative Uses of Aromatic Plants as Natural Supplements in Nutrition. In *Feed Additives*; Florou-Paneri, P., Christaki, E., Giannenas, I., Eds.; Academic Press: New York, NY, USA, 2020; pp. 19–34, ISBN 978-0-12-814700-9. [CrossRef]
273. Mohammadi Gheisar, M.; Kim, I.H. Phytobiotics in Poultry and Swine Nutrition—A Review. *Ital. J. Anim. Sci.* **2018**, *17*, 92–99. [CrossRef]
274. Yang, C.; Chowdhury, M.K.; Hou, Y.; Gong, J. Phytogenic Compounds as Alternatives to In-Feed Antibiotics: Potentials and Challenges in Application. *Pathogens* **2015**, *4*, 137–156. [CrossRef] [PubMed]
275. Park, S.-Y.; Bok, S.-H.; Jeon, S.-M.; Park, Y.B.; Lee, S.-J.; Jeong, T.-S.; Choi, M.-S. Effect of Rutin and Tannic Acid Supplements on Cholesterol Metabolism in Rats. *Nutr. Res.* **2002**, *22*, 283–295. [CrossRef]
276. Guinane, C.M.; Cotter, P.D. Role of the Gut Microbiota in Health and Chronic Gastrointestinal Disease: Understanding a Hidden Metabolic Organ. *Therap. Adv. Gastroenterol.* **2013**, *6*, 295–308. [CrossRef]
277. Kau, A.L.; Ahern, P.P.; Griffin, N.W.; Goodman, A.L.; Gordon, J.I. Human Nutrition, the Gut Microbiome and the Immune System. *Nature* **2011**, *474*, 327–336. [CrossRef]
278. Krajmalnik-Brown, R.; Ilhan, Z.; Kang, D.; DiBaise, J.K. Effects of Gut Microbes on Nutrient Absorption and Energy Regulation. *Nutr. Clin. Pract.* **2012**, *27*, 201–214. [CrossRef]
279. Everard, A.; Cani, P.D. Diabetes, Obesity and Gut Microbiota. *Best Pract. Res. Clin. Gastroenterol.* **2013**, *27*, 73–83. [CrossRef]
280. Ni, J.; Wu, G.D.; Albenberg, L.; Tomov, V.T. Gut Microbiota and IBD: Causation or Correlation? *Nat. Rev. Gastroenterol. Hepatol.* **2017**, *14*, 573–584. [CrossRef]
281. Ridaura, V.K.; Faith, J.J.; Rey, F.E.; Cheng, J.; Duncan, A.E.; Kau, A.L.; Griffin, N.W.; Lombard, V.; Henrissat, B.; Bain, J.R. Gut Microbiota from Twins Discordant for Obesity Modulate Metabolism in Mice. *Science* **2013**, *341*, 1241214. [CrossRef] [PubMed]

282. Zitvogel, L.; Galluzzi, L.; Viaud, S.; Vétizou, M.; Daillère, R.; Merad, M.; Kroemer, G. Cancer and the Gut Microbiota: An Unexpected Link. *Sci. Transl. Med.* **2015**, *7*, 271ps1. [CrossRef] [PubMed]
283. Burdulis, D.; Sarkinas, A.; Jasutiene, I.; Stackevicené, E.; Nikolajevas, L.; Janulis, V. Comparative Study of Anthocyanin Composition, Antimicrobial and Antioxidant Activity in Bilberry (*Vaccinium myrtillus* L.) and Blueberry (*Vaccinium corymbosum* L.) Fruits. *Acta Pol. Pharm.* **2009**, *66*, 399–408.
284. Cushnie, T.T.; Lamb, A.J. Recent Advances in Understanding the Antibacterial Properties of Flavonoids. *Int. J. Antimicrob. Agents* **2011**, *38*, 99–107. [CrossRef] [PubMed]
285. Perumalla, A.; Hettiarachchy, N.S. Green Tea and Grape Seed Extracts—Potential Applications in Food Safety and Quality. *Food Res. Int.* **2011**, *44*, 827–839. [CrossRef]
286. Jayaprakasha, G.; Selvi, T.; Sakariah, K. Antibacterial and Antioxidant Activities of Grape (*Vitis vinifera*) Seed Extracts. *Food Res. Int.* **2003**, *36*, 117–122. [CrossRef]
287. Rhodes, P.; Mitchell, J.; Wilson, M.; Melton, L. Antilisterial Activity of Grape Juice and Grape Extracts Derived from *Vitis vinifera* Variety Ribier. *Int. J. Food Microbiol.* **2006**, *107*, 281–286. [CrossRef] [PubMed]
288. Anastasiadi, M.; Chorianopoulos, N.G.; Nychas, G.-J.E.; Haroutounian, S.A. Antilisterial Activities of Polyphenol-Rich Extracts of Grapes and Vinification Byproducts. *J. Agric. Food Chem.* **2009**, *57*, 457–463. [CrossRef] [PubMed]
289. Gañan, M.; Martínez-Rodríguez, A.J.; Carrascosa, A.V. Antimicrobial Activity of Phenolic Compounds of Wine against Campylobacter Jejuni. *Food Control* **2009**, *20*, 739–742. [CrossRef]
290. Papadopoulou, C.; Soulti, K.; Roussis, I.G. Potential Antimicrobial Activity of Red and White Wine Phenolic Extracts against Strains of Staphylococcus Aureus, Escherichia Coli and Candida Albicans. *Food Technol. Biotechnol.* **2005**, *43*, 41–46.
291. Özkan, G.; Sagdiç, O.; Göktürk Baydar, N.; Kurumahmutoglu, Z. Antibacterial Activities and Total Phenolic Contents of Grape Pomace Extracts. *J. Sci. Food Agric.* **2004**, *84*, 1807–1811. [CrossRef]
292. Silván, J.M.; Mingo, E.; Hidalgo, M.; de Pascual-Teresa, S.; Carrascosa, A.V.; Martinez-Rodriguez, A.J. Antibacterial Activity of a Grape Seed Extract and Its Fractions against *Campylobacter* spp. *Food Control* **2013**, *29*, 25–31. [CrossRef]
293. Daroch, F.; Hoeneisen, M.; González, C.L.; Kawaguchi, F.; Salgado, F.; Solar, H.; García, A. In Vitro Antibacterial Activity of Chilean Red Wines against Helicobacter Pylori. *Microbios* **2001**, *104*, 79–85. [PubMed]
294. Just, J.; Daeschel, M. Antimicrobial Effects of Wine on *Escherichia coli* O157: H7 and *Salmonella typhimurium* in a Model Stomach System. *J. Food Sci.* **2003**, *68*, 285–290. [CrossRef]
295. Zhang, J.; Hou, X.; Ahmad, H.; Zhang, H.; Zhang, L.; Wang, T. Assessment of Free Radicals Scavenging Activity of Seven Natural Pigments and Protective Effects in AAPH-Challenged Chicken Erythrocytes. *Food Chem.* **2014**, *145*, 57–65. [CrossRef] [PubMed]
296. Hervert-Hernández, D.; Pintado, C.; Rotger, R.; Goñi, I. Stimulatory Role of Grape Pomace Polyphenols on Lactobacillus Acidophilus Growth. *Int. J. Food Microbiol.* **2009**, *136*, 119–122. [CrossRef]
297. Morris, G.; Berk, M.; Carvalho, A.; Caso, J.R.; Sanz, Y.; Walder, K.; Maes, M. The Role of the Microbial Metabolites Including Tryptophan Catabolites and Short Chain Fatty Acids in the Pathophysiology of Immune-Inflammatory and Neuroimmune Disease. *Mol. Neurobiol.* **2017**, *54*, 4432–4451. [CrossRef]
298. Leser, T.D.; Amenuvor, J.Z.; Jensen, T.K.; Lindecrona, R.H.; Boye, M.; Møller, K. Culture-Independent Analysis of Gut Bacteria: The Pig Gastrointestinal Tract Microbiota Revisited. *Appl. Environ. Microbiol.* **2002**, *68*, 673–690. [CrossRef]
299. Kim, H.B.; Isaacson, R.E. The Pig Gut Microbial Diversity: Understanding the Pig Gut Microbial Ecology through the next Generation High Throughput Sequencing. *Vet. Microbiol.* **2015**, *177*, 242–251. [CrossRef]
300. Yutin, N.; Galperin, M.Y. A Genomic Update on Clostridial Phylogeny: G Ram-negative Spore Formers and Other Misplaced Clostridia. *Environ. Microbiol.* **2013**, *15*, 2631–2641. [CrossRef]
301. Shah, H.N.; Collins, D.M. Prevotella, a New Genus to Include Bacteroides Melaninogenicus and Related Species Formerly Classified in the Genus Bacteroides. *Int. J. Syst. Evol. Micr.* **1990**, *40*, 205–208. [CrossRef] [PubMed]
302. Hara, H.; Orita, N.; Hatano, S.; Ichikawa, H.; Hara, Y.; Matsumoto, N.; Kimura, Y.; Terada, A.; Mitsuoka, T. Effect of Tea Polyphenols on Fecal Flora and Fecal Metabolic Products of Pigs. *J. Vet. Med. Sci.* **1995**, *57*, 45–49. [CrossRef] [PubMed]
303. Ishihara, N.; Chu, D.-C.; Akachi, S.; Juneja, L. Improvement of Intestinal Microflora Balance and Prevention of Digestive and Respiratory Organ Diseases in Calves by Green Tea Extracts. *Livest. Prod. Sci.* **2001**, *68*, 217–229. [CrossRef]
304. Rastall, R.A.; Gibson, G.R.; Gill, H.S.; Guarner, F.; Klaenhammer, T.R.; Pot, B.; Reid, G.; Rowland, I.R.; Sanders, M.E. Modulation of the Microbial Ecology of the Human Colon by Probiotics, Prebiotics and Synbiotics to Enhance Human Health: An Overview of Enabling Science and Potential Applications. *FEMS Microbiol. Ecol.* **2005**, *52*, 145–152. [CrossRef] [PubMed]
305. Lopetuso, L.R.; Scaldaferri, F.; Petito, V.; Gasbarrini, A. Commensal Clostridia: Leading Players in the Maintenance of Gut Homeostasis. *Gut Pathog.* **2013**, *5*, 23. [CrossRef] [PubMed]
306. Tian, G.; Wu, X.; Chen, D.; Yu, B.; He, J. Adaptation of Gut Microbiome to Different Dietary Nonstarch Polysaccharide Fractions in a Porcine Model. *Mol. Nutr. Food Res.* **2017**, *61*, 1700012. [CrossRef] [PubMed]
307. Zhao, W.; Wang, Y.; Liu, S.; Huang, J.; Zhai, Z.; He, C.; Ding, J.; Wang, J.; Wang, H.; Fan, W. The Dynamic Distribution of Porcine Microbiota across Different Ages and Gastrointestinal Tract Segments. *PLoS ONE* **2015**, *10*, e0117441. [CrossRef]
308. Castrica, M.; Rebucci, R.; Giromini, C.; Tretola, M.; Cattaneo, D.; Baldi, A. Total Phenolic Content and Antioxidant Capacity of Agri-Food Waste and by-Products. *Ital. J. Anim. Sci.* **2019**, *18*, 336–341. [CrossRef]
309. Luehring, M.; Blank, R.; Wolffram, S. Vitamin E-Sparing and Vitamin E-Independent Antioxidative Effects of the Flavonol Quercetin in Growing Pigs. *Anim. Feed Sci. Tech.* **2011**, *169*, 199–207. [CrossRef]

310. Hou, X.; Zhang, J.; Ahmad, H.; Zhang, H.; Xu, Z.; Wang, T. Evaluation of Antioxidant Activities of Ampelopsin and Its Protective Effect in Lipopolysaccharide-Induced Oxidative Stress Piglets. *PLoS ONE* **2014**, *9*, e108314. [CrossRef]
311. Wang, D.; Zhang, N.; Peng, Y.; Qi, D. Interaction of Zearalenone and Soybean Isoflavone in Diets on the Growth Performance, Organ Development and Serum Parameters in Prepubertal Gilts. *J. Anim. Physiol. Anim. Nutr.* **2012**, *96*, 939–946. [CrossRef] [PubMed]
312. Starčević, K.; Krstulović, L.; Brozić, D.; Maurić, M.; Stojević, Z.; Mikulec, Ž.; Bajić, M.; Mašek, T. Production Performance, Meat Composition and Oxidative Susceptibility in Broiler Chicken Fed with Different Phenolic Compounds. *J. Sci. Food Agric.* **2015**, *95*, 1172–1178. [CrossRef] [PubMed]
313. Cho, S.-J.; Jung, U.J.; Park, H.-J.; Kim, H.-J.; Park, Y.B.; Kim, S.R.; Choi, M.-S. Combined Ethanol Extract of Grape Pomace and Omija Fruit Ameliorates Adipogenesis, Hepatic Steatosis, and Inflammation in Diet-Induced Obese Mice. *Evid. Based Complement. Altern. Med.* **2013**, *2013*, 212139. [CrossRef] [PubMed]
314. Han, H.J.; Jung, U.J.; Kim, H.-J.; Moon, B.S.; Cho, S.-J.; Park, Y.B.; Lee, D.G.; Choi, M.-S. Dual Effects of a Mixture of Grape Pomace (Campbell Early) and Omija Fruit Ethanol Extracts on Lipid Metabolism and the Antioxidant Defense System in Diet-Induced Obese Mice. *Nutr. Res. Pract.* **2015**, *9*, 227. [CrossRef] [PubMed]
315. Huang, Z.; Mei, X.; Xu, D.; Xu, S.; Lv, J. Protective effects of polysacchride of *Spirulina platensis* and *Sargassum thunbeergii* on vascular of alloxan induced diabetic rats. *Zhongguo Zhong Yao Za Zhi* **2005**, *30*, 211–215. [PubMed]
316. Evans, M.; Wilson, D.; Guthrie, N. A Randomized, Double-Blind, Placebo-Controlled, Pilot Study to Evaluate the Effect of Whole Grape Extract on Antioxidant Status and Lipid Profile. *J. Funct. Foods* **2014**, *7*, 680–691. [CrossRef]
317. Ngamukote, S.; Mäkynen, K.; Thilawech, T.; Adisakwattana, S. Cholesterol-Lowering Activity of the Major Polyphenols in Grape Seed. *Molecules* **2011**, *16*, 5054–5061. [CrossRef] [PubMed]
318. Baselga-Escudero, L.; Blade, C.; Ribas-Latre, A.; Casanova, E.; Suarez, M.; Torres, J.L.; Salvado, M.J.; Arola, L.; Arola-Arnal, A. Resveratrol and EGCG Bind Directly and Distinctively to MiR-33a and MiR-122 and Modulate Divergently Their Levels in Hepatic Cells. *Nucleic Acids Res.* **2013**, *42*, 882–892. [CrossRef]
319. Catoni, C.; Peters, A.; Schaefer, H.M. Life History Trade-Offs Are Influenced by the Diversity, Availability and Interactions of Dietary Antioxidants. *Anim. Behav.* **2008**, *76*, 1107–1119. [CrossRef]
320. Kannan, G.; Heath, J.; Wabeck, C.; Mench, J. Shackling of Broilers: Effects on Stress Responses and Breast Meat Quality. *Brit. Poult. Sci.* **1997**, *38*, 323–332. [CrossRef]
321. Xing, T.; Gao, F.; Tume, R.K.; Zhou, G.; Xu, X. Stress Effects on Meat Quality: A Mechanistic Perspective. *Compr. Rev. Food Sci. Food Saf.* **2019**, *18*, 380–401. [CrossRef] [PubMed]
322. Lipiński, K.; Antoszkiewicz, Z.; Mazur-Kuśnirek, M.; Korniewicz, D.; Kotlarczyk, S. The Effect of Polyphenols on the Performance and Antioxidant Status of Sows and Piglets. *Ital. J. Anim. Sci.* **2019**, *18*, 174–181. [CrossRef]

MDPI AG
Grosspeteranlage 5
4052 Basel
Switzerland
Tel.: +41 61 683 77 34

*Agriculture* Editorial Office
E-mail: agriculture@mdpi.com
www.mdpi.com/journal/agriculture

www.ingramcontent.com/pod-product-compliance
Lightning Source LLC
Chambersburg PA
CBHW040807250926
51850CB00007BA/1098
*9783725856435*